ADVANCES IN CHEMICAL ENGINEERING

VOLUME FORTY EIGHT

ADVANCES IN CHEMICAL ENGINEERING

Photobioreaction Engineering

Edited by

JACK LEGRAND
GEPEA, Université de Nantes, CNRS, UMR6144, Bd de l'Université, Saint-Nazaire Cedex, France

ELSEVIER

AMSTERDAM • BOSTON • HEIDELBERG • LONDON
NEW YORK • OXFORD • PARIS • SAN DIEGO
SAN FRANCISCO • SINGAPORE • SYDNEY • TOKYO
Academic Press is an imprint of Elsevier

Academic Press is an imprint of Elsevier
50 Hampshire Street, 5th Floor, Cambridge, MA 02139, USA
525 B Street, Suite 1800, San Diego, CA 92101-4495, USA
The Boulevard, Langford Lane, Kidlington, Oxford OX5 1GB, UK
125 London Wall, London, EC2Y 5AS, UK

First edition 2016

ISBN: 978-0-12-803661-7
ISSN: 0065-2377

For information on all Academic Press publications
visit our website at http://store.elsevier.com/

CONTENTS

CONTRIBUTORS

Arnaud Artu
GEPEA, Université de Nantes, CNRS, UMR6144, and AlgoSource Technologies, Bd de l'Université, Saint-Nazaire Cedex, France

Jean-François Cornet
Université Clermont Auvergne, ENSCCF, Clermont-Ferrand, France and CNRS, Institut Pascal, Aubiere, France

Jérémi Dauchet
Université Clermont Auvergne, ENSCCF, Clermont-Ferrand, France and CNRS, Institut Pascal, Aubiere, France

Claude-Gilles Dussap
Université Clermont Auvergne, Université Blaise Pascal, Clermont-Ferrand, France and CNRS, Institut Pascal, Aubiere, France

Fabrice Gros
Université Clermont Auvergne, ENSCCF, Clermont-Ferrand, France and CNRS, Institut Pascal, Aubiere, France

Marcel Janssen
AlgaePARC, Bioprocess Engineering, Wageningen University and Research Centre, Wageningen, The Netherlands

Razmig Kandilian
University of California, Los Angeles, Los Angeles, CA, United States

Francois Le Borgne
AlgoSource Technologies, Bd de l'Université, Saint-Nazaire Cedex, France

Jack Legrand
GEPEA, Université de Nantes, CNRS, UMR6144, Bd de l'Université, Saint-Nazaire Cedex, France

Laurent Pilon
University of California, Los Angeles, Los Angeles, CA, United States

Clemens Posten
Institute of Process Engineering in Life Sciences, Karlsruhe Institute of Technology (KIT), Karlsruhe, Germany

Jeremy Pruvost
GEPEA, Université de Nantes, CNRS, UMR6144, Bd de l'Université, Saint-Nazaire Cedex, France

Matthieu Roudet
Université Clermont Auvergne, ENSCCF, Clermont-Ferrand, France and CNRS, Institut Pascal, Aubiere, France

Matthias Schirmer
Institute of Process Engineering in Life Sciences, Karlsruhe Institute of Technology (KIT), Karlsruhe, Germany

PREFACE

This book provides the main physical and engineering concepts associated with photobioreaction engineering. It aims to apply chemical engineering approach to the design, modeling, and control of photobioreactors. Most of the problems encountered in photobioreactor engineering, such as mixing, medium composition, pH, and temperature control, are common to classical bioprocesses, but light energy supply is highly specific. This is also encountered in any photoreactive process. Chemical nutrients, including carbon dioxide, which could, however, lead to gas–liquid mass transfer challenges, and physical parameters (temperature, pH) can be assumed to be homogeneous in well-mixed conditions. However, irradiance is heterogeneously distributed in the culture due to absorption and scattering by cells, independently of the mixing conditions. This book focuses on autotrophic photobioreaction when light is the only energy source of photosynthesis, which gives microalgae cultivation a real specificity with respect to other classical chemical or biological processes. The different chapters are related to different modeling and experimental approaches to tackle the different aspects of the photobioreactors. The main common point is that the maximal productivity is obtained in light-limited conditions. For that reason, the two first chapters give the basis for the most advanced radiative models applied in photobioreactors with the objective to develop predictive model for the microalgal biomass productivity. Chapter 4 represents a similar approach with more focus on the biological aspects of photosynthesis and a simplified light transfer model. Engineering formulas deduced from these models are used in Chapter 5 for the design and scale-up of photobioreactors. The metabolic fluxes related to the nutrient uptake are modeled in Chapter 3. Another important specificity of photobioreactor engineering is the use of the solar energy (Chapters 1, 4, and 5) for mass production of microalgae. A brief resume of the different chapters is given hereafter.

The first chapter introduces the theoretical framework for constructing predictive knowledge models leading to the calculation of the volumetric and surface rates of biomass production, and the thermodynamic efficiency of the process. Here, the main assumption is that photosynthesis reaction is limited by radiative transfer only. First, the predictive determination of the scattering and absorption properties of photosynthetic microorganisms of various types is addressed. Then, these radiative properties are used to calculate the radiation field within the reaction volume by solving the radiative transfer equation. Finally, the thermokinetic coupling between the radiation field, the photosynthesis reaction rates, and thermodynamic efficiency is investigated. Theoretical calculations of the process performances are shown to be in good agreement with experimental results.

Chapter 2 introduces the physical concepts and gives the experimental and theoretical frameworks to understand and to quantify the interaction between light and photosynthetic microorganisms, able to absorb photons in the photosynthetically active radiation region ranging from 400 to 700 nm thanks to photosynthetic various pigments. The chapter presents state-of-the-art theoretical and experimental methods for determining the scattering phase function and the absorption and scattering cross sections of a wide variety of promising microorganism species with various shapes, sizes, and responses to stresses.

Chapter 3 is devoted to phototrophic processes modeling. The model approach includes the level of metabolic fluxes and of the intracellular control. The appropriate balance equations and kinetics are outlined. The specific features, such as photosynthesis, carbon uptake, and carbon partitioning, are described. Dynamic description of the complex reactions of the cells to environmental changes is also discussed with some examples. The objective of this chapter is to give the basic biological background, to deduce, step by step, the model's governing equations, and to present simulation results with realistic parameter values.

Chapter 4 gives a basis for a model connecting microalgal growth and photobioreactor productivity to light exposure, in the framework of mass production of microalgae. Light exposure is considered as the limiting parameter. The model is connected to photosynthesis models developed by Blackman, Jassby, and Platt. Photosynthesis is then connected to microalgal growth adopting the model of Pirt and distinguishing between maintenance-related respiration and growth-related respiration.

Chapter 5 describes the various parameters that one should consider in designing and operating microalgal cultivation systems, and gives the appropriate engineering design rules. Deduced from rigorous approach developed in Chapter 1, the relevant engineering parameters affecting PBR productivity are the specific illuminated area, nonilluminated volume fraction of the PBR, and light collected. At the end of this chapter, some examples illustrate applications to photobioreactors from lab-scale studies to large solar industrial-scale production systems. The chapter also illustrates how few engineering rules can narrow down the various possibilities of photobioreactor designs (artificial light or natural sunlight, external or internal lighting, high-cell density culture, etc.).

JACK LEGRAND

CHAPTER ONE

Photobioreactor Modeling and Radiative Transfer Analysis for Engineering Purposes

Jérémi Dauchet*,[1], Jean-François Cornet*, Fabrice Gros*, Matthieu Roudet*, Claude-Gilles Dussap[†]

*Université Clermont Auvergne, ENSCCF, Clermont-Ferrand, France and CNRS, Institut Pascal, Aubiere, France

[†]Université Clermont Auvergne, Université Blaise Pascal, Clermont-Ferrand, France and CNRS, Institut Pascal, Aubiere, France

[1]Corresponding author: e-mail address: jeremi.dauchet@univ-bpclermont.fr

Contents

Advances in Chemical Engineering, Volume 48
ISSN 0065-2377
http://dx.doi.org/10.1016/bs.ache.2015.11.003

Abstract

The present chapter introduces the theoretical framework for constructing predictive knowledge-models leading to the calculation of the volumetric rate of biomass production, the surface rate of biomass production and the thermodynamic efficiency of photobioreactors. Here, the main assumption is that photosynthesis reaction is limited by radiative transfer only. First, the predictive determination of the scattering and absorption properties of photosynthetic microorganisms of various types is addressed. Then, these radiative properties are used to calculate the radiation field within the reaction volume by solving the radiative transfer equation. Both the development of approximate solutions appropriated with typical photobioreactor configurations (intermediate scattering optical-thickness) and the rigorous solution of the radiative transfer equation by the Monte Carlo method are addressed, including the treatment of complex geometric structures. Finally, the thermokinetic coupling between the radiation field, the photosynthesis reaction rates and thermodynamic efficiency are investigated. For the special case of the cyanobacterium *Arthrospira platensis*, a complete stoichiometric, kinetic and thermodynamic model is constructed using the linear thermodynamics of irreversible processes to analyze the primary events of photosynthesis (Z-scheme). Comparison between the theoretical calculations presented in this chapter and experimental results confirms the ability of the proposed predictive approach, after parameters reification, to quantify performances of many kinds of photobioreactors (geometry, size) functioning under different operating conditions. An extension of the proposed coupling approach for the more complicated case of eukaryotic (microalgae) micro-organisms is then proposed as further perspective of this work.

1. INTRODUCTION

During the past decades, photobioreactors have found promising applications, in particular for high-value products, for example, in pharmacy, cosmetics, and aquaculture feeds. Nevertheless, the development of industrial photobioreactors still requires optimization efforts, especially for achieving a high volumetric production rate in the context of large-scale implementation. The necessary technological breakthroughs are even more radical for application to energy carrier production for development of algal biorefineries. Here, the main criterion for viability of solar-energy processes is their thermodynamic efficiency. Although it is nowadays believed that such processes should operate at ~10% thermodynamic efficiency to be competitive with other solar-energy technologies, most photobioreactor concepts that are reported in the literature, including state-of-the-art developments, operate between 0.2 and 3%. Photosynthesis of a single microalga is capable of such efficient (>10%) energy conversion (Bolton and Hall, 1991), but here, the obstacle is implementation at the photobioreactor scale, which still requires significant advances in chemical engineering.

We believe that these technological improvements (the optimization of the volumetric production rate and thermodynamic efficiency) can be achieved only by constructing predictive models compatible with the requirements of photobioreactor simulation, design, sizing, scale-up, optimization, and model-based predictive control. On the one hand, such models must be generic to be suitable for testing of a wide range of technological strategies and operating modes; this requirement entails strong theoretical bases and advanced experimental studies. On the other hand, these models must be numerically tractable and compatible with analysis of the process; this requirement means identification of appropriate approximations and numerical methods. This chapter is an overview of our practice of photobioreactor modeling and is aimed at fulfilling the earlier needs. The resulting models are validated at the end of this chapter (in Section 5.6) and used in a separate chapter of this book (chapter "Industrial Photobioreactors and Scale-up Concepts" by Pruvost et al.) for analysis of the process.

Photobioreactor engineering addresses optimization of the volumetric rate of biomass production, the surface rate of biomass production (with respect to the solar-energy collecting surface), and thermodynamic efficiency of the process as well as biomass composition (ie, the biomass quality). Hereafter, we mainly focus on construction of a predictive model for the volumetric rate $< r_x >$ (eg, expressed in kg, or moles of dry biomass per second and per m^3 of the reaction volume). This is the main difficulty with assessing performance of a photobioreactor because most of the other parameters of interest can be deduced only from the value of $< r_x >$, in a rather straightforward manner (see Section 5).

Our model is based on integral formulation of the photobioreactor's volumetric rate; this approach is extremely convenient for analyzing the interaction between the mechanisms involved at different scales of the process:

$$< r_x > = \frac{1}{V} \int_{\mathcal{V}} r_x(\mathbf{x}) \, d\mathbf{x} \tag{1}$$

where $< r_x >$ is the average local volumetric rate $r_x(\mathbf{x})$ at location $\mathbf{x}$, calculated across the geometric domain $\mathcal{V}$ of microorganism culture with volume V. In the text that follows, we focus on perfectly stirred photobioreactors where the microorganism concentration C_x (ie, the dry-biomass concentration) is uniform within $\mathcal{V}$. This assumption may be easily extended to plug-flow photobioreactors, in which intensive variables (such as C_x) are homogeneous within the surface perpendicular to the flow. In this case, $< r_x >$ is obtained as in Eq. (1) from a surface integral, and $< r_x(z) >$ is used in the differential mass balance of the photobioreactor (Pruvost and Cornet, 2012). In these situations, the local volumetric rate is

$$r_x(\mathbf{x}) = C_x J_x(\mathbf{x}) \tag{2}$$

where $J_x(\mathbf{x})$ is the average rate of biomass production by a photosynthetic microorganism within the infinitesimal volume element $d\mathbf{x}$ around the location $\mathbf{x}$ (usually expressed in moles per second per kg of dry biomass[1]). In Eqs. (1) and (2), $< r_x >$ is the production rate at the scale of the process, r_x is the local rate within the reaction volume, and J_x is the production rate of an isolated photosynthetic cell. With this approach, prediction of J_x is the central question.

The construction of a model for the specific rate of biomass production J_x is presented in Section 5. The main assumption of this model is that the photosynthetic reaction is limited by radiative transfer only. Indeed, ensuring proper mixing and maintenance of all physiological needs under their optimal conditions (pH, temperature, dissolved CO_2, and minerals) are quite straightforward at the current state of knowledge in chemical engineering. Under these conditions, it has been clearly demonstrated in the past decades that photobioreactors are mainly governed by light transfer inside the culture volume; this transfer determines the kinetic rates, thermodynamic efficiency, biomass composition, and pigment content (Aiba, 1982; Cornet, 2010; Cornet et al., 1998; Cornet and Dussap, 2009; Cornet et al., 1992b; Csogor et al., 2001; Pilon et al., 2011; Pruvost and Cornet, 2012; Takache et al., 2010). Hereafter, we focus on photobioreactors operating

[1] The specific rate of biomass production by a microorganism is $J_x(\mathbf{x})$, expressed in moles per second per kg of dry biomass, multiplied by the average dry mass of one microbial cell.

under such optimal conditions; therefore, we assume that J_x is a function of radiative quantities only.

The early-primary events during photosynthesis (in photosystems) involve absorption of light within the spectral range of photosynthetically active radiation (PAR); this absorption generates electronic excitation events in the light-harvesting protein and causes water splitting by the oxygen-evolving complex. The characteristic time scale for this complete mechanism (consisting of a five-step cycle) is ~1 ms; therefore, we assume that the earlier processes are a function of local light absorption (cell displacement due to mixing is negligible within this time frame). Moreover, we assume that the contributions of electronic excitation generated by any photon within PAR are identical, regardless of the frequency of the photons.[2] Therefore, our model for photosystems is formulated as a function of the specific rate of photon absorption $\mathcal{A}(\mathbf{x})$, that is, the number of photons within PAR that are absorbed per unit of time and per kg of dry biomass (ie, by the microorganism) at location $\mathbf{x}$. The following mechanisms underlie production of ATP and $NADPH_2$ within the photosynthetic electron transport chain, or the Z-scheme. In Section 5, we propose to use thermodynamics of irreversible processes (Prigogine, 1967), which is appropriate for modeling electron transport chains in a predictive manner, with a limited number of free parameters (Dussap, 1988; Stucki, 1979). In the present case, the validity conditions of the approach require addressing physical quantities averaged across a few minutes (Dussap, 1988; Stucki, 1979); this period is long compared to the typical mixing time within photobioreactors (ie, the time necessary for a microorganism to "explore" the reaction volume $\mathcal{V}$). Therefore, we will construct a model for the Z-scheme that is formulated as a function of averages $<f(\mathcal{A})>$ across the local field of the absorption rate $\mathcal{A}$:

$$<f(\mathcal{A})>=\frac{1}{V}\int_{\mathcal{V}} f(\mathcal{A}(\mathbf{x}))\, d\mathbf{x} \tag{3}$$

where f is a function that is determined by an optimization procedure (see Section 5). The remaining mechanisms correspond to the synthesis of complex organic molecules (the biomass) via metabolic reactions called the dark

[2] This assumption is always true for polychromatic-illumination and for monochromatic-illumination experiments in the case of eukaryotic microalgae, according to the well-known action spectrum of photosynthesis. This is not the case for monochromatic illumination of cyanobacteria (Farges et al., 2009).

reactions because they are independent of radiative transfer (they are driven only by the ATP and $NADPH_2$ generated by the light reactions discussed earlier). Altogether, our model for the specific rate of biomass production J_x is a function of the specific rate of photon absorption $\mathcal{A}$ and averages $<f(\mathcal{A})>$ calculated across the reaction volume:

$$J_x(\mathbf{x}) \equiv J_x(\mathcal{A}(\mathbf{x}), <f(\mathcal{A})>) \tag{4}$$

where J_x depends on the location $\mathbf{x}$ only via the absorption rate $\mathcal{A}(\mathbf{x})$.

The purpose of a photobioreactor is to absorb incident light in order to convert it into biomass via coupling with photosynthesis. On the one hand, efficient light absorption usually corresponds to heterogeneous radiation fields $\mathcal{A}(\mathbf{x})$ within the reaction volume (see Section 3). On the other hand, the coupling law (Eq. (4)) is usually a non-linear function of $\mathcal{A}(\mathbf{x})$ (the law obtained in Section 5 is non-linear, but this is also the case for most of other models reported in the literature). Therefore, the coupling between radiative transfer and photosynthesis must be formulated locally,[3] which implies that determination of the volumetric rate $< r_x >$ requires

1. estimating the radiation field $\mathcal{A}(\mathbf{x})$ within the culture volume (and averages $<f(\mathcal{A})>$),
2. estimating the field of the specific rate of biomass production $J_x(\mathbf{x})$ according to Eq. (4),
3. estimating the field of the local rate $r_x(\mathbf{x})$ according to Eq. (2),
4. solving the integral across the culture volume in Eq. (1).

Then, the surface rate of biomass production $< s_x >$ is obtained as

$$< r_x >= a_{\text{light}} < s_x > \tag{5}$$

where a_{light} is the specific illuminated surface $a_{\text{light}} = S_{\text{light}}/V$, S_{light} is the area of the illuminated surface (eg, the solar-energy collecting surface), and V is the reaction volume (including dark zones). Therefore, the surface rate is calculated from the volumetric rate and a purely geometric characteristic of the process. Finally, the thermodynamic efficiency $< \eta_{\text{th}} >$ of photosynthesis within the reaction volume is obtained from the value of $< r_x >$

[3] If the coupling law is a linear function of $\mathcal{A}(\mathbf{x})$, for example, $J_x(\mathbf{x}) = a\,\mathcal{A}(\mathbf{x}) + b$, where a and b are constants, then determination of r_x requires only the knowledge of $<\mathcal{A}>$ (see Eq. (6)): the coupling does not have to be formulated locally. Indeed, after substitution of the earlier mentioned linear expression for J_x into Eq. (2), Eq. (1) leads to

$$< r_x >= C_x \frac{1}{V} \int_V (a\,\mathcal{A}(\mathbf{x}) + b)\, d\mathbf{x} = C_x(a < \mathcal{A} > + b)$$

and from the average rate of photon absorption $<\mathcal{A}>$ (see Section 5), where

$$<\mathcal{A}>=\frac{1}{V}\int_{\mathcal{V}}\mathcal{A}(\mathbf{x})\,d\mathbf{x} \tag{6}$$

Thus, the construction of predictive models of photobioreactors requires careful formulation of radiative transfer within the reaction volume, in order to obtain the radiation field (cf. step 1 in the earlier procedure). Such analysis is developed this chapter, starting in Section 2 with determination of the light scattering and absorption properties of photosynthetic-microorganism suspensions. Next, these properties are used in Section 3 for analysis of radiative transfer and in Section 4 for rigorous solution of the radiative transfer equation by the Monte Carlo method. Finally, the thermokinetic coupling between radiative transfer and photosynthesis is addressed in Section 5. It should be noted that Sections 2 and 4 mainly summarize works that have been already published elsewhere, whereas Sections 3 and 5 include extensive original work and results.

The main steps in our model and their organization within this chapter are summarized in Fig. 1. Our practice of the Monte Carlo method extends beyond the solution of the radiative transfer equation: in Section 4, we also argue that the Monte Carlo method is well suited for numerical implementation of the entire model, especially in research on photobioreactors with complex geometric structure.

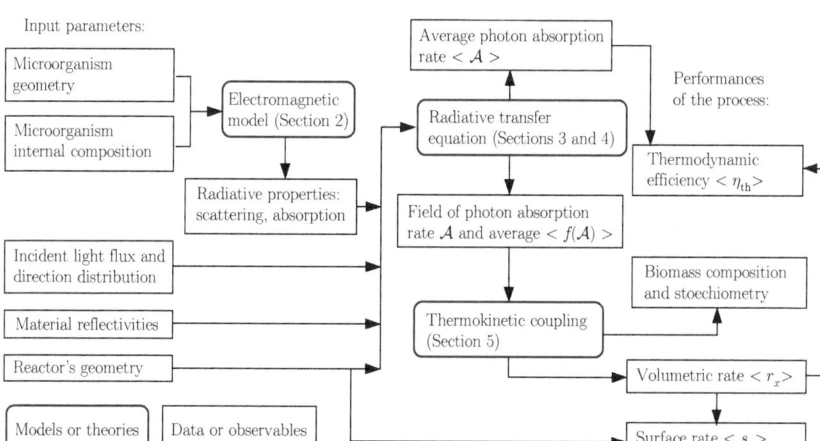

Figure 1 An outline of the predictive model presented in this chapter. Numerical implementation of the entire model by the Monte Carlo method is discussed in Section 4.

2. CALCULATING THE RADIATIVE PROPERTIES OF PHOTOSYNTHETIC MICROORGANISMS

Any radiative analysis of photobioreactors starts with determination of absorption and scattering properties of the photosynthetic microorganism under study. This question is not trivial, and to the best of our knowledge, no available database provides adequate spectral and angular information that is needed, even for the strains of microalgae that are currently widely cultivated. Determination of these properties involves either highly specialized experiments (Berberoglu et al., 2009, 2008; Berg et al., 2014; Chami et al., 2014; Mengüç, 2002; Mishchenko et al., 2009; Muoz and Hovenier, 2011; Pilon et al., 2011; Vaillon et al., 2011) (see also chapter "Interaction Between Light and Photosynthetic Microorganisms" by Pilon and Kandilian) or construction of a model implying solution of Maxwell's equations for particles with the types of heterogeneity, sizes, and shapes for which the usual numerical methods such as Lorenz-Mie, T-Matrix, finite-difference time-domain (FDTD), and discrete dipole approximation (DDA) (Kahnert, 2003; Wriedt, 2009) are still impractical in many cases. The present section addresses the electromagnetic modeling approach, and later, we present the main steps of a methodological chain detailed in Dauchet et al. (2015) for predictive calculation of the radiative properties within PAR. This methodological chain is derived from the expertise of the oceanographic community, and more broadly, of all the research fields that deal with the wave-particle interaction problem in the areas of atmospheric sciences, astrophysics, or engineering. In this broad context, our concern is to take into account not only the specificity of photosynthetic microorganisms but also analysis and optimization requirements of photobioreactor engineering. Therefore, this predictive method is based on limited and "easily" accessible experimental parameters (morphological and structural characteristics as well as photosynthetic-pigment content), allowing us to account for microorganisms' variability from one species to another and as a function of culture conditions (in particular, the illumination conditions) (Cornet et al., 1992a; Farges et al., 2009; Pottier et al., 2005). In Section 2.1 the basic concepts for the construction of such a predictive method are introduced and the choices made in Dauchet et al. (2015) are presented: we construct an electromagnetic model of the light-particle interaction that is consistent with available protocols for determination of input parameters. The results produced by this methodological chain in the case of *Chlamydomonas reinhardtii* and *Rhodospirillum rubrum* under standard subculture conditions are discussed in Section 2.2. Finally, some perspectives for further development of the approach are drawn in Section 2.3.

2.1 The Methodological Chain

The radiative properties are input parameters of the radiative transfer equation (see Section 3): the absorption coefficient $k_{a,\nu}$, the scattering coefficient $k_{s,\nu}$, the extinction coefficient $k_{ext,\nu} = k_{a,\nu} + k_{s,\nu}$ and the single-scattering phase function $p_{\mathbf{\Omega},\nu}(\boldsymbol{\omega}|\boldsymbol{\omega}')$. The coefficients $k_{a,\nu}$, $k_{s,\nu}$, and $k_{ext,\nu}$ (expressed in m^{-1}) characterize attenuation of radiative intensity during passage through a microorganism suspension because of absorption, scattering, and extinction, respectively: they are coefficients from Bouguer's exponential law of attenuation (sometimes called Beer's law). The phase function $p_{\mathbf{\Omega},\nu}(\boldsymbol{\omega}|\boldsymbol{\omega}')$ is the distribution of scattering directions $\boldsymbol{\omega}$ when radiation with incident direction $\boldsymbol{\omega}'$ is scattered. These properties are a function of the frequency ν of radiation; under the assumption of perfect mixing (microorganism locations are statistically distributed uniformly), they are homogeneous within the reaction volume.

Our model implies independent scattering (the assumption that we share with the great majority of photobioreactor researchers). Indeed, typical biomass concentrations within the process are low enough to reasonably assume that each microbial cell interacts with radiation independently. We can therefore define particle cross sections σ that characterize the radiative properties of microbial cells independently of their concentration C_x: $k_{a,\nu} = C_x\, \sigma_{a,\nu}$, $k_{s,\nu} = C_x\, \sigma_{s,\nu}$, and $k_{ext,\nu} = C_x \sigma_{ext,\nu}$, where $\sigma_{ext,\nu} = \sigma_{a,\nu} + \sigma_{s,\nu}$. To be precise, the cross sections and the phase function characterize the interaction between an incident electromagnetic plane wave with frequency ν and a particle with given geometric structure and internal refractive index (as shown in Fig. 2). By solving this electromagnetic problem (ie, by solving Maxwell's equations), one can calculate the radiative properties $\hat{\sigma}_{a,\nu}$, $\hat{\sigma}_{s,\nu}$, $\hat{\sigma}_{ext,\nu}$, and $\hat{p}_{\mathbf{\Omega},\nu}$ of an isolated particle with a specific shape, size, orientation, and refractive index. Then, under the assumption of independent scattering, the radiative properties of a perfectly stirred suspension are the average properties of isolated particles within the suspension. From now on, we will assume that every microbial cell within the photobioreactor has the same shape and refractive index; we use only the average value across orientation and size distributions:

$$\sigma_\nu = \int_{\mathcal{D}_{\mathbf{E}_o}} d\mathbf{e}_o\, p_{\mathbf{E}_o}(\mathbf{e}_o) \int_0^\infty dr_{eq} p_{R_{eq}}(r_{eq})\, \hat{\sigma}_\nu(\mathbf{e}_o, r_{eq}) \tag{7}$$

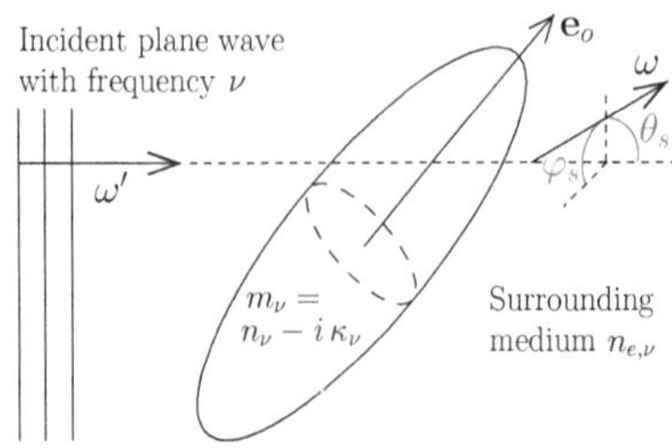

Figure 2 The scattering problem illustrated for a spheroidal particle with orientation $\mathbf{e}_o$ and effective refractive index $m_\nu = n_\nu - i\,\kappa_\nu$. The surrounding medium is nonabsorbing, with the real refractive index $n_{e,\nu}$. The speed of light within the medium is $c = c_0/n_{e,\nu}$, where c_0 is the speed of light in vacuum. The incident plane wave has frequency ν (ie, wavelength $\lambda = c/\nu$) and a wave vector collinear to ω'. A propagation direction of the radiation scattered by the particle is denoted as ω. θ_s is the angle between ω and ω'.

$$p_{\mathbf{\Omega},\nu}(\boldsymbol{\omega}|\boldsymbol{\omega}') = \int_{\mathcal{D}_{\mathbf{E}_o}} d\mathbf{e}_o\, p_{\mathbf{E}_o}(\mathbf{e}_o) \int_0^{\infty} dr_{eq} p_{R_{eq}}(r_{eq}) \frac{\hat{\sigma}_{s,\nu}(\mathbf{e}_o, r_{eq})\, \hat{p}_{\mathbf{\Omega},\nu}(\boldsymbol{\omega}|\boldsymbol{\omega}';\mathbf{e}_o, r_{eq})}{\sigma_{s,\nu}} \tag{8}$$

where

- Eq. (7) is valid for the three cross sections $\sigma_{a,\nu}$, $\sigma_{s,\nu}$, and $\sigma_{ext,\nu}$,
- $\mathbf{e}_o$ is a vector defining orientation of the particle (see Fig. 2), $p_{\mathbf{E_o}}(\mathbf{e_o})$ is the orientation distribution, and $\mathcal{D}_{\mathbf{E}_o}$ is the domain of all possible orientations (for axisymmetric particles, $\mathcal{D}_{\mathbf{E}_o}$ is the total solid angle),
- r_{eq} is the radius of the volume-equivalent sphere (that characterizes the size of the particle), and $p_{R_{eq}}(r_{eq})$ is its distribution (ie, the size distribution),
- the radiative properties $\hat{\sigma}_\nu$ and $\hat{p}_{\mathbf{\Omega},\nu}$ of an isolated particle are a function of its orientation $\mathbf{e}_o$, size r_{eq}, shape, and internal refractive index m_ν as well as the frequency ν of incident radiation and the refractive index of the surrounding medium. Here, the surrounding medium is assumed to be non-absorbing, with the real refractive index $n_{e,\nu}$ equal to that of water (Thormählen et al., 1985).

Actually, the scattering problem in Fig. 2 is not affected by m_ν and $n_{e,\nu}$ but is influenced by the relative refractive index $m_{r,\nu} = m_\nu/n_{e,\nu}$. Similarly, the scattering problem is not affected by r_{eq} and ν but is influenced by the ratio size/wavelength. This value is usually characterized by the size parameter $x = \frac{2\pi\, r_{eq}}{\lambda}$, where $\lambda = c/\nu$ is the wavelength in the surrounding medium (water).

Therefore, determination of the radiative properties implies construction of a model of the microorganism itself (its geometric structure and its internal features in terms of the refractive index) as well as implementation of the standard model of electromagnetism (solution of Maxwell's equations). These two tasks are actually interlocked because according to the literature (Kahnert, 2003; Mishchenko et al., 2002; Wriedt, 2009), internal heterogeneity and shape of the most typical microorganisms correspond to ineluctable numerical difficulties with solution of Maxwell's equations. Consequently, we constructed an approximate electromagnetic model that involves simplification of the particles' description, with the corresponding approximations being chosen in line with the requirements of a photobioreactor study. The first choice is the following: we refuse to accept a compromise on the information about shape because we believe that this characteristic is essential for the interaction with radiation and is the key factor that distinguishes different species of microorganisms. On the other hand, we are willing to define the internal heterogeneity as an approximation of an effective homogeneous medium. Indeed, this approximation has been tested in many situations and appears to distort only the power scattered at large angles (Bernard et al., 2009; Choi et al., 2008; Kolokolova and Gustafsonm, 2001; Wu et al., 2007). We selected this assumption because a very small proportion of the incident power is backscattered by a microorganism (90% of the scattered power is usually confined to a solid angle of roughly 20° aperture around the incident direction, see Section 2.2). As for the targeted radiative configurations, backscattering has a limited influence: the few backscattered photons have a very limited impact on radiative transfer within photobioreactors. Therefore, in order to alleviate the numerical difficulties, we modified our description of the scatterers, making them pseudo-homogeneous particles.

Various methods are available for solving the problem of an electromagnetic wave scattered by a homogeneous particle. Each of these methods is limited by the range of geometric structures and refractive indices that can be tackled (Bohren and Huffman, 1983; Kahnert, 2003; Mishchenko et al., 2000, 2002; van de Hulst, 1981; Wriedt, 2009). Identifying those that are appropriate for the studies of photosynthetic microorganisms (from the pragmatic point of view) involves research on the available formal solutions (including approximations) and numerical approaches enabling their implementation, considering that

1. the size parameter $x = \frac{2\pi r_{eq}}{\lambda}$ of microbial cells cultivated in a photobioreactor ranges from 5 to 200,

2. many microorganisms are strongly elongated particles that can be up to 50 × longer than their width, implying a very small radius of curvature,
3. microorganisms have the relative refractive index $m_{r,\nu} = n_{r,\nu} - i\ \kappa_{r,\nu}$ corresponding to low dielectric contrast with the surrounding medium ($n_{r,\nu} \in [1.02,1.2]$, $\kappa_{r,\nu} \in [10^{-5},10^{-2}]$).

The earlier description corresponds to a defined situation where very little can be done at present (see Nousiainen, 2009 regarding mineral dust, Baran, 2009 regarding ice crystals, and Mishchenko et al., 2000 for an overview). Computation of the rigorous solution of Maxwell's equations is indeed usually achieved by numerical methods that, to our knowledge, fail to address the combination of the first two criteria listed earlier (Kahnert, 2003; Mishchenko et al., 2000, 2002; Wriedt, 2009). The ongoing development of these methods (eg, Bi and Yang, 2013, 2015; Hellmers et al., 2011; Moskalensky et al., 2013; Wriedt, 2009; Yurkin and Hoekstra, 2011) should allow researchers to analyze an increasing number of microorganism species, but currently, it is still necessary to employ approximations for constructing a generic approach: the effective medium approximation is not sufficient. The first strategy is to approximate the scatterer shape in order to rigorously solve Maxwell's equation. This shape usually corresponds to the equivalent sphere approximation, which allows for straightforward resolution of the scattering problem for the whole range of size parameters by means of standard Lorenz-Mie codes for spheres (Bernard et al., 2001, 2009; Bricaud and Morel, 1986; Morel and Bricaud, 1981; Wyatt, 1972). Nevertheless, the use of the equivalent sphere approximation can lead to significant errors, especially in research on elongated particles (see Dauchet et al., 2015; Liu et al., 2008; Nousiainen, 2009; Yang et al., 2004, and Section 2.2). This is why we are keen on preserving the shape description and are thus compelled to simplify the electromagnetic model of the light-particle interaction. It is a common practice in the field of radiative transfer research to use simplifications corresponding to asymptotic approximations: eg, the Rayleigh approximation (when size parameters approach zero) and geometrical optics (when size parameters approach infinity). In the case of photosynthetic microorganisms, size parameters are large and refractive-index contrast is low: we are at the soft-particle limit corresponding to the validity range of the anomalous diffraction approximation (van de Hulst, 1981) and Schiff's approximation (Schiff, 1956). These two approximations are identical, except Schiff's approximation involves formulation of the phase functions (which is not the case for anomalous diffraction). We selected Schiff's approximation because it allows us to analyze a great variety of a microorganism's shapes, within the range of size parameters, with accuracy levels that are suitable for the studies of photobioreactors (see Dauchet et al., 2015 regarding comparisons with spectroscopic measurements and Charon et al., 2015 regarding comparisons with available reference solutions calculated by the T-Matrix method). Our methodological chain uses efficient code for implementation of Schiff's approximation (including simplification for scattering at a large angle θ_s), that is available from Charon et al. (2015). This code has been developed for cylindrical and spheroidal homogeneous particles but is based on the Monte Carlo method, which opens up interesting perspectives on the analysis of particles with more complex shapes. Its extension for the analysis of particles with any shape is currently a work in progress. In the case of spheroids and cylinders, using parallel implementation, we observed accuracy levels and CPU times that are compatible with the production of spectral databases needed for the studies on photobioreactors (less than 2 h for tabulation of radiative properties for 40 wavelengths within PAR, including tabulation of phase functions for 1000 angles) (Charon et al., 2015; Dauchet et al., 2015). Accordingly, the combination of the effective medium approximation and Schiff's approximation makes our model numerically tractable.

The choice of a description in terms of homogeneous-equivalent particles with complex shape (that are analyzed with Schiff's approximation) is pertinent to construction of a generic methodological chain (designed for studies on photobioreactors) that allows for analysis of any photosynthetic microorganism.[4] Nevertheless, in a study on a specific species, rigorous resolution of Maxwell's equation is sometimes tractable, and Schiff's approximation is no longer required. For example, for small microorganisms with modest elongation, most DDA or T-Matrix codes can be used instead of Schiff's approximation. Furthermore, in a study on microorganisms with spherical shape, Lorenz-Mie codes are extremely convenient (see, for example, Bohren and Huffman, 1983; Mishchenko et al., 2002 and Section 2.2).

The rest of our methodological chain is devoted to determination of the shape, the effective refractive index, and the size distribution of the microbial cells (regarding the distribution of orientations, isotropy is always assumed due to the agitation that is needed for mixing). These parameters

[4] Note that the opposite choices are usually made in oceanographic research, during analysis of oceanic albedo. In this case, the backscattered photons have significant effects; therefore, a description of the phytoplankton heterogeneity is required. In order to numerically solve Maxwell's equations for the heterogeneous particles, such models usually simplify the description of the shapes by means of the equivalent sphere approximation (see Bernard et al., 2009 for an example of core-shell model).

must be determined (regardless of the approximation or method that is selected for solving the scattering problem), and we must be able to understand their dependence on the operating mode of the process, by means of either experimentally accessible data (within the scope of the photobioreactor engineering practices) or available databases. Our methodological chain thus implies the following set of characterization procedures (summarized in Table 1) before any implementation of the previously formulated model.

First, the shape and size distributions are determined by optical microscopy and image analysis. Simple rotatory-symmetric parametric shapes are identified and selected in Dauchet et al. (2015) and Charon et al. (2015), and the size distributions are modeled as log-normal ones for the radius r_{eq} of the volume-equivalent sphere. This is not a restriction: analysis of more complex shapes with Schiff's approximation does not correspond to conceptual difficulties, and analysis of other size distributions is straightforward.

The remaining procedures are designed for determination of the effective refractive index, which reflects the microorganism's internal heterogeneity (Bohren and Huffman, 1983; Mishchenko et al., 2000; Sihvola, 1999).

Table 1 The Main Steps of the Characterization Procedure for Determination of Input Parameters of the Model for the Radiative Properties of a Photosynthetic Microorganism

(1)	Determination of the microorganism shape and size distribution by image analysis.
(2)	Determination of concentrations of the photosynthetic pigments (protocols available in Dauchet et al., 2015 that are based on the measurement of the pigments dry-mass fraction and the volume fraction of intracellular water).
(3)	Construction of the imaginary part of the spectrum of the effective refractive index from the results of (2) and from a database containing *in vivo* absorption spectra of pigment molecules (the database is available in StarWest, n.d.).
(4)	Determination of the volume fractions of the anatomic internal structures by image analysis.
(5)	Construction of the real part of the effective refractive index at the anchor point by applying the Bruggeman mixing rule to the results of (4) and the internal-structure real indices obtained from a database (the database is available in Dauchet et al., 2015).
(6)	Construction of the real part of the spectrum of the effective refractive index by applying the singly subtractive Kramers–Krönig approximation to the results of (3) and (5).

First, we determine the imaginary part, which characterizes absorption properties of the continuous medium constituting the microbial cells. For this purpose, we use a model derived from the oceanographic research (Bidigare et al., 1990; Bricaud and Morel, 1986; Hoepffner and Sathyendranath, 1991; Morel and Bricaud, 1981; Wozniak et al., 2000). Within the spectral range of PAR, absorption by photosynthetic cells is assumed to be exclusively due to photosynthetic pigments. Moreover, these pigment molecules are diluted enough to be characterized by an *in vivo* absorption cross section (independently of the microorganism species and the type of study) and by an internal pigment concentration (which is strongly dependent on the species and culture conditions). We therefore modeled the imaginary part κ_ν of the effective refractive index by summing the absorption cross sections $E_{a,pig}(\nu)$ of the pigment molecules (expressed in m^2/kg), with each of these absorption spectra being weighted by the concentration C_{pig} of the corresponding pigment species *pig* within the microorganism in question (C_{pig} is the average local pigment concentration across the cell volume, expressed in kg/m^3; Dauchet et al., 2015):

$$\kappa_\nu = \frac{c_0}{4\pi\,\nu} \sum_{pig} C_{pig}\, E_{a,pig}(\nu) \tag{9}$$

where c_0 is the speed of light in vacuum. Therefore, determination of κ_ν implies:

- measurement of the pigment concentrations C_{pig} for each microorganism species and each culture condition under study: a procedure based on field-tested microbiological protocols is proposed in Dauchet et al. (2015).
- extraction of molecular cross sections $E_{a,pig}$ from a database: a database containing data on 14 of the most important photosynthetic pigments in nature (representative pigments of photosynthetic bacteria, cyanobacteria, and microalgae) was constructed in Dauchet et al. (2015) on the basis of the pioneering work of Bidigare et al. (1990) and is available in StarWest (n.d.).

In Section 2.2, that is dedicated to results, Fig. 4A represents construction of the κ_ν spectrum in the case of *Chlamydomonas reinhardtii*.

The following procedures deal with determination of the real part n_ν of the effective refractive index. Based on the work of Bernard et al. (2001), Naqvi et al. (2004), and Tuminello et al. (1997), our methodological chain uses the singly subtractive Kramers–Krönig approximation (Lucarini, 2005) that yields an expression for n_ν as a function of the spectrum κ_ν of the imaginary part and the value n_{ν_p} of the real part at a particular frequency ν_p (see Fig. 4B in Section 2.2):

$$n_\nu = n_{\nu_p} + \frac{2\left(\nu^2 - \nu_p^2\right)}{\pi} P \int_{\nu_{\min}}^{\nu_{\max}} \frac{\nu_1 \, \kappa_{\nu_1}}{\left(\nu_1^2 - \nu^2\right)\left(\nu_1^2 - \nu_p^2\right)} d\nu_1 \tag{10}$$

where $[\nu_{\min}, \nu_{\max}]$ is PAR, and P means that the Cauchy principal value has to be considered for the singularity ($\nu_1 = \nu$). Therefore, what remains is determination of the anchor point n_{ν_p}. According to the work of Aas (1996) in oceanographic research, we use the Bruggeman mixing rule, which yields the effective refractive index n_ν of a non-absorbing composite particle from the data on the volume fraction and the refractive index of its different structures (Bohren and Huffman, 1983; Mishchenko et al., 2000; Sihvola, 1999):

$$\sum_j f_j \frac{\left(\hat{n}_{j,\nu}\right)^2 - \left(n_\nu\right)^2}{\left(\hat{n}_{j,\nu}\right)^2 + 2\left(n_\nu\right)^2} = 0 \tag{11}$$

where f_j and $\hat{n}_{j,\nu}$ are respectively the volume fraction and the real part of the refractive index for the jth internal structures of the particle. We chose the anchoring frequency ν_p such that the microorganism under study is non-absorbing at ν_p (ie, $\kappa_{\nu_p} = 0$), and n_{ν_p} is determined by solving Eq. (11) for $\nu = \nu_p$, where:

- volume fractions f_j are measured by electron microscopy and image analysis, or far less frequently, are taken from the literature when available,
- refractive indices $\hat{n}_{j,\nu_p}$ of internal anatomic structures are obtained from a small database that is available in Dauchet et al. (2015).

It should be noted that at the current state knowledge, Eq. (11) cannot be used to directly obtain the spectrum n_ν of the refractive index because very little information is available about the spectral properties $\hat{n}_{j,\nu}$ of internal structures. Nonetheless, the choice of non-absorbed anchoring frequency ν_p significantly simplifies the access to these data and allows researchers to estimate the anchor point n_{ν_p}.

This methodological chain is summarized in Fig. 3, and the corresponding characterization procedures are listed in Table 1. Further details and a validation procedure that are based on the analysis of spectroscopic data are presented in Dauchet et al. (2015).

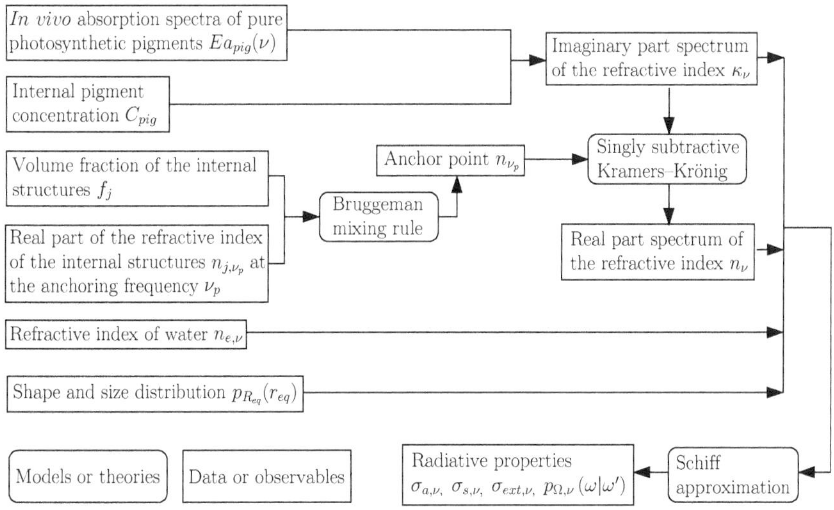

Figure 3 A summary of the methodological chain for determination of radiative properties of a photosynthetic microorganism.

2.2 Results

Figs. 4 and 5 show results obtained by means of the methodological chain presented in Section 2.1. Among the input parameters of the model, the pigment concentrations are extremely sensitive to the culture conditions. They allow researchers to assess dependence of the radiative properties on the operating mode of the process. In the present case, the parameters have been measured for standard subculture conditions: a shaken 250 mL Erlenmeyer flask, 100 rpm, low photon flux density approximately 30 $\mu mol_{h\nu}\ m^{-2}\ s^{-1}$ ($\simeq 7$ W m^{-2}), and optimal pH and temperature. Fig. 4 presents the effective refractive index obtained by implementing the characterization procedure in Table 1 for *C. reinhardtii*. Note that if different culture conditions are considered, then new parameter values have to be determined by measurement or found in the literature. It should also be noted that spectral variations are usually represented as a function of the wavelength λ_0 of radiation in vacuum. This choice can be quite confusing in the present case where the surrounding medium is water, with refractive index $n_{e,\nu} \neq 1$. Indeed, the scattering problem is affected not by λ_0 but by the wavelength λ within the medium (eg, the size parameter x must be calculated with λ): $\lambda = \frac{\lambda_0}{n_{e,\nu}}$. For this reason, we also indicate the frequency ν of radiation, whose value is identical in vacuum and within the medium: $\nu = \frac{c_0}{\lambda_0} = \frac{c}{\lambda}$, where the speed of light is c_0 in vacuum and $c = c_0/n_{e,\nu}$ within the medium.

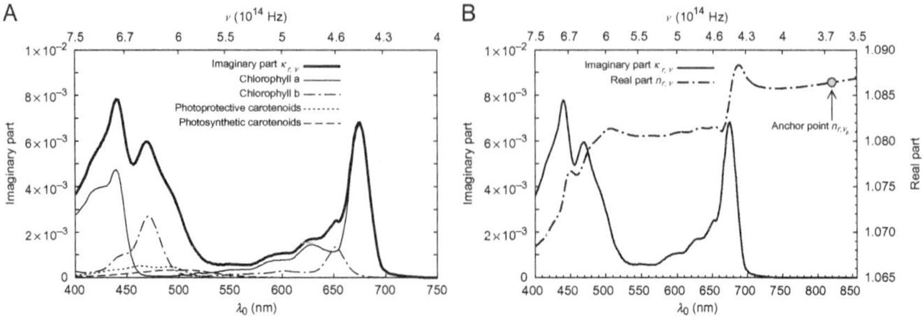

Figure 4 The relative refractive index $m_{r,\nu} = n_{r,\nu} - i\,\kappa_{r,\nu}$ of the homogeneous equivalent medium for *Chlamydomonas reinhardtii* as a function of the wavelength λ_0 in vacuum and the frequency ν of incident radiation. These results were obtained by implementing the characterization procedure summarized in Table 1. The refractive index $m_\nu = n_\nu - i\,\kappa_\nu$ is divided by the real index $n_{e,\nu}$ of water: $m_{r,\nu} = m_\nu / n_{e,\nu}$, where $n_{e,\nu}$ is calculated by means of an empirical relation reported in Thormählen et al. (1985) (assuming that $n_{e,\nu} \simeq 1.33$ leads to significantly different spectral variations for $n_{r,\nu}$ when $\lambda_0 \in [400{,}550 \text{ nm}]$). (A) The imaginary part $\kappa_{r,\nu}$ obtained with Eq. (9) for the molecular cross sections Ea_{pig} obtained from the database available in StarWest (n.d.) and pigment concentrations C_{pig} measured by protocols available in Dauchet et al. (2015): chlorophyll *a* 677.6 kg/m^3, chlorophyll *b* 277.2 kg/m^3, photoprotective carotenoids 184.8 kg/m^3, and photosynthetic carotenoids 30.8 kg/m^3. Contribution of each pigment species is also presented. (B) The real part $n_{r,\nu}$ produced by the singly subtractive Kramers–Krönig approximation (Eq. (10)) for the anchor point $n_{\nu_p} = 1.44$ at wavelength $\lambda_0 = 820$ nm ($\nu_p = 3.656 \times 10^{14}$Hz) calculated in Dauchet et al. (2015) with Bruggeman's mixing rule (Eq. (11)).

Fig. 5 shows the radiative properties obtained for *C. reinhardtii* and *Rhodospirillum rubrum* within PAR. *C. reinhardtii* is a spheroidal unicellular green alga (eukaryote) with PAR ranging from 400 to 700 nm, and *Rs. rubrum* is a rod-shaped purple bacterium with PAR ranging from 400 to 870 nm. Due to the integration over isotropic orientation distribution in Eq. (8), the single scattering phase function is a function of θ_s only: $p_{\mathbf{\Omega},\nu}(\boldsymbol{\omega}|\boldsymbol{\omega}') \equiv p_{\mathbf{\Omega},\nu}(\theta_s)$, where θ_s is defined in Fig. 2. The phase function of each microorganism is represented for the wavelength at the center of PAR, where scattering is predominant over absorption. We verified that the angular distributions are strongly oriented in forward directions (as discussed in Section 2.1): scattering phase functions are presented on a logarithmic scale for the analysis, but the values obtained indicate that 90% of the scattered power is within $\theta_s \in [0,20°]$, ie,

$$2\pi \int_0^{20^\circ} p_{\mathbf{\Omega},\nu}(\theta_s)\,\sin(\theta_s)\,d\theta_s \simeq 0.9$$

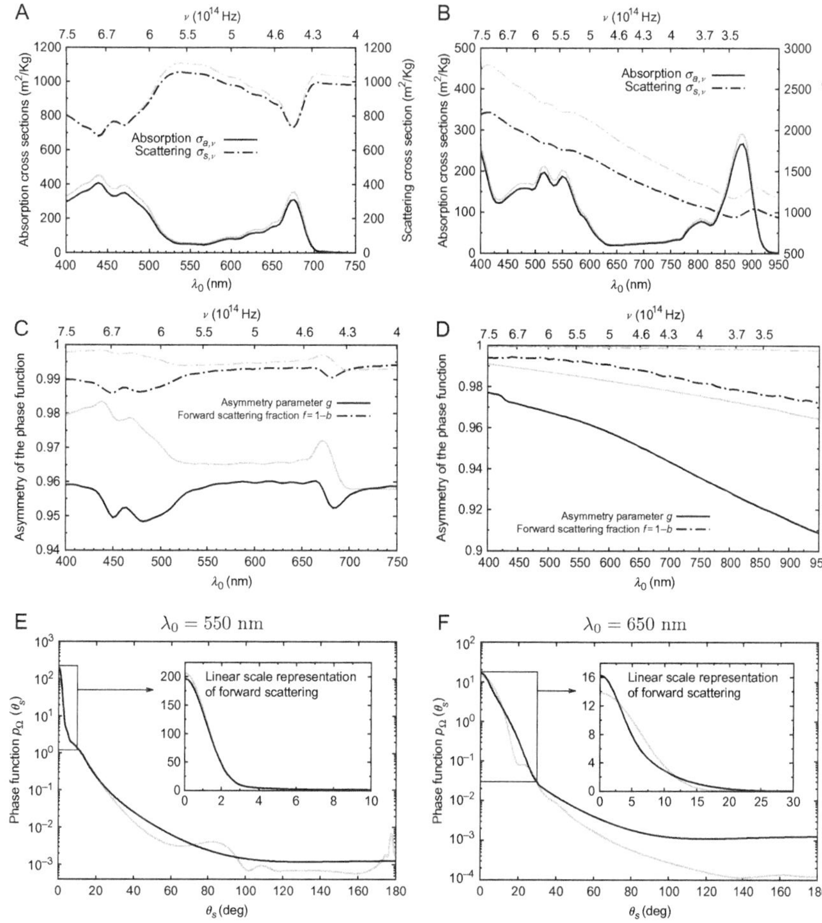

Figure 5 Radiative properties of *Chlamydomonas reinhardtii* (left) and *Rhodospirillum rubrum* (right) obtained with the volume equivalent sphere approximation and a Lorentz-Mie code (gray color) and with a more accurate description of the shape and Schiff's approximation (black color): a prolate spheroid with elongation 1.2 (aspect ratio 0.837) for *C. reinhardtii* and a cylinder with elongation 3.8 (aspect ratio 0.263) for *Rs. rubrum*. These results were obtained by implementing the methodological chain presented in Section 2.1 for a log-normal size distribution $p_{R_{eq}}(r_{eq}) = \frac{1}{\sqrt{2\pi}\, r_{eq} \ln(s)} \exp\left[-\frac{\left(\ln r_{eq} - \ln \bar{r}_{eq}\right)^2}{2\ln^2(s)}\right]$ with $\bar{r}_{eq} = 3.963$ μm and $s = 1.18$ for *C. reinhardtii*, and $\bar{r}_{eq} = 0.983$ μm and $s = 1.1374$ for *Rs. rubrum*. The refractive index is shown in Fig. 4 for *C. reinhardtii* and in Dauchet et al. (2015) for *Rs. rubrum*. The scattering and absorption cross sections are expressed in m^2 per kg of dry biomass by means of division of the particulate cross sections by the effective dry mass M_{eff} of one microbial cell (see Dauchet et al., 2015): $M_{eff} = 9.8615 \times 10^{-14}$ kg for *C. reinhardtii* and $M_{eff} = 1.1354 \times 10^{-15}$ kg for *Rs. rubrum*. In this case, the absorption and scattering coefficients $k_{a,\nu} = C_x \sigma_{a,\nu}$ and $k_{s,\nu} = C_x \sigma_{s,\nu}$ are obtained with the biomass concentration C_x expressed in kg of dry biomass per m^3.

This situation is similar for all frequencies within PAR, as indicated by the spectral variations of the asymmetry parameter g (see Section 3.2) and the forward scattering fraction f:

$$g = 2\pi \int_0^{\pi} p_{\Omega,\nu}(\theta_s) \cos(\theta_s) \sin(\theta_s) d\theta_s$$

and

$$f = 2\pi \int_0^{\pi/2} p_{\Omega,\nu}(\theta_s) \sin(\theta_s) d\theta_s$$

where $2\pi \int_0^{\pi} p_{\Omega,\nu}(\theta_s) \sin(\theta_s) d\theta_s = 1 = f + b$, and b is the backscattering fraction. Therefore, for each scattering event, propagation directions of light are predominantly redistributed within a solid angle with aperture 20°. The influence of this redistribution of propagation directions on radiative transfer within photobioreactors is analyzed in Section 3.

Fig. 5 compares (i) the results obtained with an accurate description of the microorganism's shape, subjected to Schiff's approximation (black color) and (ii) the results obtained with the equivalent sphere approximation and the rigorous solution of Maxwell's equation by means of a Lorenz-Mie code (gray color). For *C. reinhardtii*, which has a near-spherical shape, both approaches lead to extremely similar results. This finding confirms that Schiff's approximation yields accurate results if the samples are compared with available reference solutions. The phase functions at $\lambda_0 = 550$ nm are in good agreement, especially for forward scattering, which has a strong influence on radiative transfer within photobioreactors. The discrepancies observed for large angles θ_s, where the phase function has small values, and for the asymmetry parameter are both due to the effect of the spheroidal shape of *C. reinhardtii* and the error associated with Schiff's approximation (see Charon et al., 2015 for a discussion of scattering at a large angle). These discrepancies are not significant when researchers solve the radiative transfer equation (Dauchet et al., 2015). In contrast, the results obtained for *Rs. rubrum*, which has a cylindrical shape, show significant differences between the scattering properties obtained by the two approaches (eg, relative difference $\simeq 20\%$ for the scattering cross section). On the other hand, the absorption cross section is less sensitive to the shape of *Rs. rubrum*. These results confirm that the equivalent sphere approximation has to be used carefully (or even avoided) when the shape of the microorganism is significantly different from the sphere.

These results are further validated in Dauchet et al. (2015), where the transmittance spectra that were recorded for microorganism suspensions were compared with those predicted by solution of the radiative transfer equation for the radiative properties presented in Fig. 5. In every configuration that has been tested so far, the description of the microorganism's shape increases the accuracy of the results.

2.3 Perspectives

The accuracy of our model can be improved in many ways, but we believe that solution of the scattering problem is the main obstacle for accurate determination of the radiative properties within photobioreactors. Considering the complexity of shapes, the size parameter range, and the refractive indices of photosynthetic microorganisms, it seems evident that Schiff's approximation should receive increasing attention and consideration in the future, even if the existing exact solutions and numerical methods are continuously improved. Accordingly, the capabilities and limitations of Schiff's approximation are actively studied at present, in particular by comparison with experimental measurements in a single-scattering condition (Pilon et al., 2011), including microwave analog measurements (Vaillon et al., 2011).

Another significant challenge for future work is analysis of microorganisms with complex geometric structure. With the Monte Carlo methodology used in Charon et al. (2015) for resolution of Schiff's approximation, the geometric calculations required are closely similar to those used in standard geometric-optics codes (ie, calculation of intersections between rays and surfaces); this situation opens up interesting perspectives on the analysis of particles with complex shape. For example, this approach will enable studies on the effect of the helical shape of *Arthrospira platensis*, whose radiative properties obtained with a straight cylinder model do not lead to satisfactory spectroscopic validation (Dauchet et al., 2015).

Finally, the research into the effect of internal heterogeneity is also an interesting topic for both photobioreactor engineering and natural water/ocean color background analysis (Bernard et al., 2009) (where backscattering is crucial). In order to overcome the difficulty associated with solution of the scattering problem for heterogeneous scatterers, our preliminary studies have been focused on small or spherical microorganisms, but here, the main obstacle is the limited current knowledge about the internal structure of biological cells in terms of the refractive index (this is also a limitation in our method when we calculate the anchor point with a mixing rule).

Despite these areas for improvement, the methodological chain that is presented in the present section already yields radiative properties with a fair level of accuracy for standard culture conditions, when the shape of the microorganism is accurately described (see validation in Charon et al., 2015; Dauchet et al., 2015), including all the spectral and angular data that are needed for formulation of radiative transfer within a photobioreactor.

3. ANALYSIS OF MULTIPLE-SCATTERING RADIATIVE TRANSFER WITHIN PHOTOBIOREACTORS: APPROXIMATE SOLUTIONS FOR THE RADIATION FIELD WITHIN ONE-DIMENSIONAL CARTESIAN PHOTOBIOREACTORS

At the end of the previous section, the radiative properties of photosynthetic microorganisms are already available. Therefore, the aim of the following paragraphs is to analyze multiple-scattering radiative transfer in a simple geometric configuration corresponding to a photobioreactor operating close to its optimum. Today, most photobioreactors under study are flat or cylindrical. Here, we chose to focus on the Cartesian one-dimensional radiative configuration of a flat photobioreactor shown in Fig. 6. This study will allow us to derive analytical approximate solutions to the radiative transfer equation. Cylindrical solar photobioreactors will not be discussed here because obtaining an analytical solution for these devices is extremely difficult (for direct solar illumination, the configuration is not one dimensional). Nonetheless, the stringent analysis of complex geometric structures (including cylindrical solar photobioreactors) will be discussed in Section 4.

Initially, we will focus on the mesoscopic description associated with the radiative transfer equation. Then, we will introduce the single-scattering approximation and two macroscopic approximations: the P1 approximation and two-flux approximation. All of these discussions are based on the configuration shown in Fig. 6. Collimated emission and Lambertian emission will also be considered in the discussion later; they correspond to the direct component and the diffuse component of solar radiation, respectively. Throughout our study, the biomass concentration C_x is homogeneous in the reaction volume $\mathcal{V}$ (assumption of perfect mixing), and the emission phenomena in $\mathcal{V}$ are negligible. The concentration C_x is selected close to the optimum for the operation of the photobioreactor: the local photon absorption rate $\mathcal{A}$ at the rear of the photobioreactor is close to the compensation point $\mathcal{A}_c$ (see Section 5 and chapter "Industrial Photobioreactors and Scale-up Concepts" by Pruvost et al.).

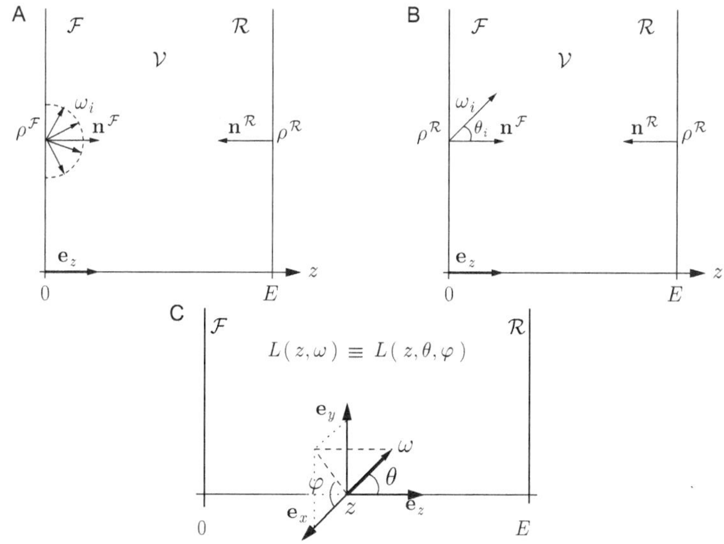

Figure 6 One-dimensional Cartesian radiative configuration that is studied in Section 3. The reaction volume $\mathcal{V}$ is confined by the surfaces $\mathcal{F}$ (front) and $\mathcal{R}$ (rear) located at $z = 0$ and $z = E$, respectively. $\rho^{\mathcal{F}}$ and $\rho^{\mathcal{R}}$ are reflectivity values of the surfaces $\mathcal{F}$ and $\mathcal{R}$; $\mathbf{n}^{\mathcal{F}} = \mathbf{e}_z$ and $\mathbf{n}^{\mathcal{R}} = -\mathbf{e}_z$ are their normals. Emission at $\mathcal{F}$ is either (A) Lambertian (ie, diffuse) or (B) collimated along the direction $\boldsymbol{\omega}_i$, where θ_i is the angle between $\boldsymbol{\omega}_i$ and $\mathbf{n}^{\mathcal{F}}$ (the cosine of θ_i is shown as $\mu_i = \boldsymbol{\omega}_i \cdot \mathbf{n}^{\mathcal{F}}$). In both cases, the surface flux density emitted at $\mathcal{F}$ is denoted as $q_{\cap}$. (C) Light propagates in all directions $\boldsymbol{\omega}$ (three-dimensional scattering): $\boldsymbol{\omega} = \cos(\varphi)\sin(\theta)\mathbf{e}_x + \sin(\varphi)\sin(\theta)\mathbf{e}_y + \sin(\theta)\mathbf{e}_z$. The element of solid angle is $d\omega = d\varphi d\theta \sin(\theta)$. Within this one-dimensional configuration, for Lambertian incident radiation (cf. A) or for collimated normal incidence (ie, $\theta_i = 0$ in B), the intensity is independent of φ and is denoted as $L(z,\theta)$ below. The biomass concentration C_x within $\mathcal{V}$ is homogeneous. $\mathcal{V}$ is nonemitting. Throughout Section 3, the following configuration is studied: $E = 4$ cm, $q_{\cap} = 500\ \mu\text{mol}_{h\nu}\ \text{m}^{-2}\ \text{s}^{-1}$, $C_x = 0.55\ \text{kg}_x\ \text{m}^{-3}$, $\rho^{\mathcal{F}} = 0$, $\rho^{\mathcal{R}} = 0$ or 0.54 (which is close to reflectivity of stainless steel), absorption cross section $\sigma_a = 145\ \text{m}^2\ \text{kg}_x^{-1}$, scattering cross section $\sigma_s = 922\ \text{m}^2\ \text{kg}_x^{-1}$, and a single scattering phase function with asymmetry parameter $g = 0.945$. These radiative properties were obtained by the method presented in Section 2 for *Chlamydomonas reinhardtii*; illumination condition being different than those studied in Section 2.2, these properties are different than those presented in Fig. 5.

3.1 The Radiative Transfer Equation

The objective of this section is to introduce basic concepts of the transport theory in participating media (ie, absorbing and scattering media) and their physical interpretation. These concepts are well established in the radiative transfer research and are detailed in many reference textbooks, such as Siegel and Howell (1981), Case and Zweifel (1967), and Goody and Yung (1964). These concepts are not repeated in detail here; they are simply introduced because of their use below for analysis of typical radiative configurations of a photobioreactor. In these typical configurations, the asymmetry parameter of the phase function is close to 1 and optical thickness is intermediate.

The radiative transfer equation is a simplification of the Boltzmann transport equation (developed by Ludwig Boltzmann in 1872 to describe ideal gas of identical particles) made possible by two characteristics of photons as particles:

1. photons all propagate at a locally identical speed: the speed of light c in the medium,
2. they do not interact with each other but interact only with the medium (here, the microorganisms suspension): we are interested in the *linear transport* phenomenon.

This mesoscopic modeling is a statistical description suitable for complex systems with a large degree of freedom, for example, a set of photons propagating in a scattering medium, fluids, or plasma. This modeling is based on the assumption of repetition of a large number of statistical events within the system. This situation is verified either by the presence of a large number of particles or by replication of a large number of events, for example, scattering events, with a single particle (these two conditions are equivalent in the context of linear transport). The mesoscopic descriptor of the system is the distribution function $f(\mathbf{x}, \boldsymbol{\omega}, t)$, which, up to a normalization factor, is the probability density for a photon to be present at time t and location $\mathbf{x}$ and to have the propagation direction $\boldsymbol{\omega}$. To be precise, $f(\mathbf{x}, \boldsymbol{\omega}, t)d\mathbf{x}d\boldsymbol{\omega}$ is the number of photons within the volume element $d\mathbf{x}$ around the location $\mathbf{x}$, propagating in a direction within the solid-angle element $d\boldsymbol{\omega}$ around $\boldsymbol{\omega}$ (see Fig. 7A). The system is thus described in six-dimensional space: one dimension for time, three for the geometric space $\mathcal{D}_{\mathbf{x}}$ (which is the reaction volume $\mathcal{V}$ in our study), and two for the propagation directions $\mathcal{D}_{\omega}$, which represent the total solid angle (indicated as 4π below). Because all the information about the velocity distribution (or propagation directions) is modeled, Boltzmann-type equations, including the radiative transfer equation, are particularly suitable for description of non-equilibrium situations, even far from equilibrium. We will see that this property is of particular interest in our study because such situations are commonly encountered in photobioreactors. With such mesoscopic description, we can always go back to the usual macroscopic variables, in which only the moments of the velocity distribution are used. For example, the density $\eta(\mathbf{x},t)$ of photons (the number of photons within the volume element $d\,\mathbf{x}$, regardless of their

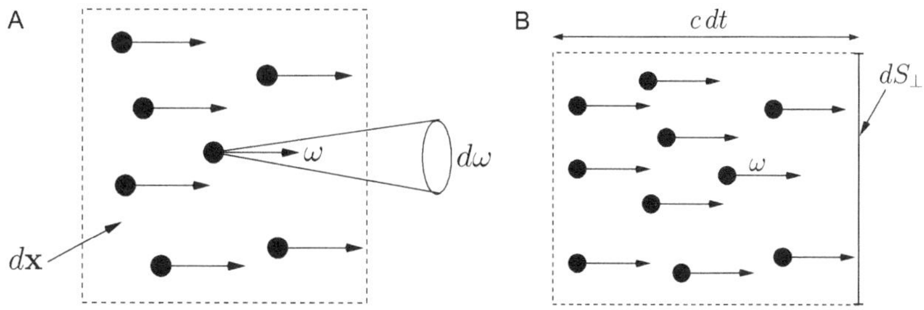

Figure 7 Phase space. (A) The volume element of phase space. (B) The relation between intensity and the distribution function: The amount of radiant energy that crosses the surface $dS_\perp$ during dt is equal to the number of photons propagating in the direction $\boldsymbol{\omega}$ within volume $c\,dtdS_\perp$, multiplied by the energy carried by each photon.

propagation direction) is calculated by integrating the distribution function f over all propagation directions:

$$\eta(\mathbf{x}, t) = \int_{4\pi} f(\mathbf{x}, \boldsymbol{\omega}, t)\, d\boldsymbol{\omega} \tag{12}$$

The radiative transfer equation is the equation of change for the distribution function $f(\mathbf{x},\boldsymbol{\omega},t)$:

$$\partial_t f(\mathbf{x}, \boldsymbol{\omega}, t) + c\,\boldsymbol{\omega} \cdot \mathbf{grad}_\mathbf{x} f(\mathbf{x}, \boldsymbol{\omega}, t) = -c\,k_{ext} f(\mathbf{x}, \boldsymbol{\omega}, t) + c\,k_s \int_{4\pi} f(\mathbf{x}, \boldsymbol{\omega}', t)\, p_\Omega(\boldsymbol{\omega}|\boldsymbol{\omega}')\, d\boldsymbol{\omega}' \tag{13}$$

where c is the speed of light in the medium (c is homogeneous in the context of our study), ∂_t is the partial derivative with respect to time t, $\mathbf{grad}_\mathbf{x}$ is the gradient with respect to $\mathbf{x}$, and the other parameters are the radiative properties obtained in Section 2 (they are also homogeneous in our study): $p_\Omega(\boldsymbol{\omega}|\boldsymbol{\omega}')$ is the phase function, $k_{ext} = k_a + k_s$ is the extinction coefficient, with k_a and k_s the absorption and scattering coefficients, respectively. Temporal variations in the culture conditions that are likely to affect radiative transfer are mainly of two types:

1. variation in incidence and intensity of the solar radiation,
2. changes in the concentration and composition of the biomass, including pigment composition, which strongly influences the radiative properties (see Section 2).

These transitional states are associated with characteristic periods that are much longer than the characteristic duration of establishment of a steady state for radiative transfer. Therefore, throughout this chapter, we will consider steady-state radiative transfer: the distribution function f is independent of time. This approach does not preclude analysis of temporal relations associated with photon propagation (see Fig. 8). Under these conditions, the radiative transfer equation is written as

$$c\,\boldsymbol{\omega} \cdot \mathbf{grad}_{\mathbf{x}} f(\mathbf{x},\boldsymbol{\omega}) = -c\,k_{ext} f(\mathbf{x},\boldsymbol{\omega}) + c\,k_s \int_{4\pi} f(\mathbf{x},\boldsymbol{\omega}')\,p_{\Omega}(\boldsymbol{\omega}|\boldsymbol{\omega}')\,d\boldsymbol{\omega}' \quad (14)$$

This equation formalizes the balance of the photonic phase in the phase space; this balance is found intuitively in each of its terms. For this purpose, we will follow mentally the propagation of the $f(\mathbf{x},\boldsymbol{\omega})d\mathbf{x}d\boldsymbol{\omega}$ photons contained in the phase space volume element $d\mathbf{x}d\boldsymbol{\omega}$ around $(\mathbf{x}, \boldsymbol{\omega})$ during the course of the time interval δt, as shown in Fig. 8:

- **Transport term** $c\,\boldsymbol{\omega} \cdot \mathbf{grad}_{\mathbf{x}} f(\mathbf{x}, \boldsymbol{\omega})$. It indicates variation of f because of free displacement of the photons. The $f(\mathbf{x}, \boldsymbol{\omega})d\mathbf{x}d\boldsymbol{\omega}$ photons located at $\mathbf{x}$ at time point t have the velocity $c\,\boldsymbol{\omega}$. In the absence of absorption or scattering, after the time interval δt, they are located at $\mathbf{x} + c\,\boldsymbol{\omega}\,\delta t$. According

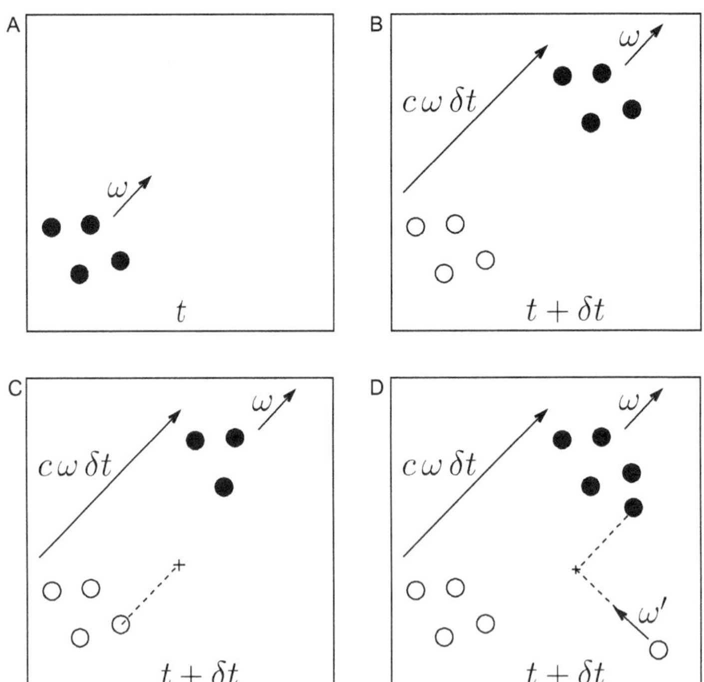

Figure 8 Illustration of a few photons at two time points t and $t + \delta t$, and physical interpretation of the radiative transfer equation. (A and B) The transport term. (C) The extinction term. (D) The collision term.

to Liouville's theorem, the phase space volume $d\,\mathbf{x}\,d\,\boldsymbol{\omega}$ containing the photons is conserved, and we have

$$f(\mathbf{x}+c\,\boldsymbol{\omega}\delta t,\boldsymbol{\omega})=f(\mathbf{x},\boldsymbol{\omega}) \tag{15}$$

If δt is a differential element in time, then $f(\mathbf{x} + c\,\boldsymbol{\omega}\,\delta t,\boldsymbol{\omega})$ can be expressed as its first-order Taylor expansion around $\mathbf{x}$, which gives us

$$f(\mathbf{x},\boldsymbol{\omega})+[c\,\boldsymbol{\omega}\cdot\mathbf{grad}_{\mathbf{x}}f(\mathbf{x},\boldsymbol{\omega})]\delta t=f(\mathbf{x},\boldsymbol{\omega}) \tag{16}$$

hence

$$c\,\boldsymbol{\omega}\cdot\mathbf{grad}_{\mathbf{x}}f(\mathbf{x},\boldsymbol{\omega})=0 \tag{17}$$

- **Extinction term** $-\,c\,k_{ext}f(\mathbf{x},\boldsymbol{\omega})$. It represents the rate at which photons within $d\mathbf{x}d\boldsymbol{\omega}$ are absorbed or scattered in a different direction. They thus leave the phase space volume element under study (see Fig. 8C). This linear formulation assumes that there is always a scale below which the locations of interaction with microorganisms are distributed randomly and uniformly.
- **Collision term** $c\,k_s\int_{4\pi}f(\mathbf{x},\boldsymbol{\omega}')\,p_\Omega(\boldsymbol{\omega}|\boldsymbol{\omega}')d\boldsymbol{\omega}'$. It represents the source of photons in the phase space volume $d\mathbf{x}d\boldsymbol{\omega}$ in relation to photons with propagation direction $\boldsymbol{\omega}'$ that are scattered at $\mathbf{x}$ in the direction $\boldsymbol{\omega}$ (see Fig. 8D). $c\,k_s\,f(\mathbf{x},\boldsymbol{\omega}')d\mathbf{x}d\boldsymbol{\omega}'$ is the rate at which photons within $d\mathbf{x}d\boldsymbol{\omega}'$ are scattered, and $p_\Omega(\boldsymbol{\omega}|\boldsymbol{\omega}')$ is the probability density for their scattering direction to be $\boldsymbol{\omega}$. Integration over the total solid angle accounts for all incoming directions $\boldsymbol{\omega}'$.

This action results in deformation of the distribution function f, which is shown in Fig. 9 for the abscissa $z_0 = 3$ cm of the flat-plate photobioreactor in Fig. 6 with collimated normal incidence. Due to the symmetry of the problem (one-dimensional Cartesian configuration), f is a function of only the abscissa z and the angle θ between the propagation direction $\boldsymbol{\omega}$ and $\mathbf{e}_z$. For collimated illumination, if we ignore scattering, then the photons all propagate in the same direction: the incident direction $\boldsymbol{\omega}_i$. This situation corresponds to Fig. 9A where $f(z_0,\theta)$ is zero in all directions except $\boldsymbol{\omega}_i$ (f is a Dirac distribution centered at $\boldsymbol{\omega}_i$). When we take into account the scattering by microbial cells, as shown on Fig. 9B, the propagation directions of the photons gradually deviate as the photons propagate within the suspension: they arrive at z_0 at different angles of propagation.

As we saw in Section 2, the scattering and absorption properties depend on the frequency ν (or by the same token, on the wavelength $\lambda = c/\nu$). It is therefore necessary to distinguish between photons with different

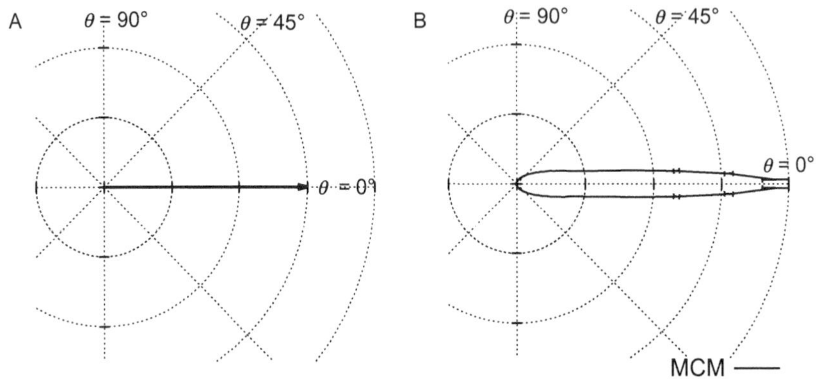

Figure 9 The distribution function $f(z_0,\theta)$ at $z_0 = 3$ cm within the one-dimensional photobioreactor shown in Fig. 6 where $\rho^{\mathcal{F}} = \rho^{\mathcal{R}} = 0$, for collimated normal incidence ($\theta_i = 0$). (A) Without scattering. (B) With scattering. The results were obtained with the Monte Carlo method (MCM, see Section 4).

frequencies: in addition to $\mathbf{x}$ and $\boldsymbol{\omega}$, the distribution function depends on the variable ν. This functional dependence is usually denoted as $f_\nu(\mathbf{x},\boldsymbol{\omega})$ (rather than $f(\mathbf{x},\boldsymbol{\omega},\nu)$) to specify its particular characteristics: for elastic scattering (which is the case in our study), no operator in the radiative transfer equation affects the frequency of radiation (only the radiative properties are a function of ν). The equations of change for the distribution functions f_ν at each frequency are thus independent:

$$c\,\boldsymbol{\omega} \cdot \mathbf{grad}_{\mathbf{x}} f_\nu(\mathbf{x},\boldsymbol{\omega}) = -c\, k_{ext,\nu} f_\nu(\mathbf{x},\boldsymbol{\omega}) + c\, k_{s,\nu} \int_{4\pi} d\boldsymbol{\omega}' f_\nu(\mathbf{x},\boldsymbol{\omega}')\, p_{\Omega,\nu}(\boldsymbol{\omega}|\boldsymbol{\omega}') \tag{18}$$

In other words, photon populations corresponding to different frequencies evolve completely independently from each other. Nonetheless, the frequency of radiation is a dimension of phase space, just as $\mathbf{x}$ and $\boldsymbol{\omega}$ are $f_\nu(\mathbf{x},\boldsymbol{\omega})d\mathbf{x}d\boldsymbol{\omega}\, d\nu$, within the volume element $d\mathbf{x}d\boldsymbol{\omega}$ around $(\mathbf{x},\boldsymbol{\omega})$, is the number of photons that have a frequency within the unit interval $d\nu$ around ν. The distribution function $f(\mathbf{x},\boldsymbol{\omega})$, which describes the photons independently of their frequency, is the integral of f_ν over the spectral range $[\nu_{min},\nu_{max}]$ under study (PAR in this work):

$$f(\mathbf{x},\boldsymbol{\omega}) = \int_{\nu_{min}}^{\nu_{max}} f_\nu(\mathbf{x},\boldsymbol{\omega})\, d\nu \tag{19}$$

The mesoscopic variable describing radiation in engineering sciences is generally intensity rather than the distribution function. The intensity

$L_\nu(\mathbf{x},\boldsymbol{\omega})$ at location $\mathbf{x}$, in the direction $\boldsymbol{\omega}$, and at frequency ν is expressed as W m^{-2} sr^{-1} Hz^{-1}. It is the flux density due to photons with direction $\boldsymbol{\omega}$ crossing the surface normal to $\boldsymbol{\omega}$ at location $\mathbf{x}$ (per unit of the solid angle $d\boldsymbol{\omega}$ and per unit of the frequency interval $d\nu$). To link the intensity and the distribution function, we will consider the radiant energy δQ that crosses the surface $dS_\perp$ (perpendicular to $\boldsymbol{\omega}$) in the direction $\boldsymbol{\omega}$ during dt:

$$\delta Q = \int_{\nu_{\min}}^{\nu_{\max}} L_\nu(\mathbf{x},\boldsymbol{\omega})\, d\nu\, d\boldsymbol{\omega}\, dt\, dS_\perp \tag{20}$$

which is also equal to the number of photons propagating in the direction $\boldsymbol{\omega}$, with speed c, within the volume $dS_\perp c\, dt$ (see Fig. 7B), multiplied by their energy $h\nu$:

$$\delta Q = \int_{\nu_{\min}}^{\nu_{\max}} h\nu \times f_\nu(\mathbf{x},\boldsymbol{\omega})\, d\nu\, d\boldsymbol{\omega}\, dS_\perp c\, dt \tag{21}$$

From Eqs. (20) and (21), we obtain the following relation between f and L:

$$L_\nu(\mathbf{x},\boldsymbol{\omega}) = c\, h\nu f_\nu(\mathbf{x},\boldsymbol{\omega}) \tag{22}$$

Our study of kinetic coupling is based on variables expressed in the number of photons rather than in energy (energetic variables are, for their part, required for formulation of thermodynamic efficiency of the process). Indeed, in a kinetic study, researchers are particularly interested in the flux of photons propagating in the direction $\boldsymbol{\omega}$ at location $\mathbf{x}$; this flux is usually given by $\hat{L}_\nu(\mathbf{x},\boldsymbol{\omega})$ expressed in mol s^{-1} m^{-2} sr^{-1} Hz^{-1}:

$$\hat{L}_\nu(\mathbf{x},\boldsymbol{\omega}) = \frac{L_\nu(\mathbf{x},\boldsymbol{\omega})}{h\nu} = c f_\nu(\mathbf{x},\boldsymbol{\omega}) \tag{23}$$

Despite the different units of measurement, we continue to call $\hat{L}$ intensity. By substituting the earlier definition into the radiative transfer equation (Eq. (18)), we obtain the following equation of change for $\hat{L}$:

$$\boldsymbol{\omega} \cdot \mathrm{grad}_{\mathbf{x}} \hat{L}_\nu(\mathbf{x},\boldsymbol{\omega}) = -k_{ext,\nu} \hat{L}_\nu(\mathbf{x},\boldsymbol{\omega}) + \alpha_{s,\nu}\, k_{ext,\nu} \int_{4\pi} d\boldsymbol{\omega}' \hat{L}_\nu(\mathbf{x},\boldsymbol{\omega}')\, p_{\Omega,\nu}(\boldsymbol{\omega}|\boldsymbol{\omega}') \tag{24}$$

Furthermore, by multiplying Eq. (18) by $h\nu$ and substituting the definition of L (Eq. (22)), we obtain the same equation of change for the intensity L (expressed in energy units). Thus, f, $\hat{L}$, and L obey the same radiative transfer equation.

The boundary conditions associated with the radiative transfer equation usually fix the intensity (or the distribution function) for incoming propagation directions. In the context of solar-energy systems, databases contain the incident surface flux density $q_\cap$ and its spectral distribution $q_{\cap,\nu}$ (rather than the intensity at the boundary):

$$q_\cap = \int_{\nu_{\min}}^{\nu_{\max}} q_{\cap,\nu}\, d\nu \tag{25}$$

In addition to this information, it is necessary to specify angular distribution of the intensity. A Lambertian distribution corresponds to isotropic intensity for the incoming directions (ie, diffuse incidence)

$$L_\nu(\mathbf{x},\boldsymbol{\omega}) = \frac{q_{\cap,\nu}}{\pi} \quad \text{pour} \quad \boldsymbol{\omega}\cdot\mathbf{n} > 0 \tag{26}$$

where $\mathbf{n}$ is the inner normal of the surface. For illumination collimated in the direction $\boldsymbol{\omega}_i$, the intensity is zero for all directions within the inner hemisphere, except for $\boldsymbol{\omega}_i$

$$L_\nu(\mathbf{x},\boldsymbol{\omega}) = \frac{q_{\cap,\nu}}{\mu_i}\,\delta(\boldsymbol{\omega}-\boldsymbol{\omega}_i) \quad \text{pour} \quad \boldsymbol{\omega}\cdot\mathbf{n} > 0 \tag{27}$$

where $\mu_i = \boldsymbol{\omega}_i\cdot\mathbf{n} = cos(\theta_i)$, and $\delta(\boldsymbol{\omega} - \boldsymbol{\omega}_i)$ is the Dirac distribution centered at $\boldsymbol{\omega}_i$ (see Fig. 6). Eqs. (26) and (27) set the intensity for the incoming directions only; the intensity for outgoing directions is a result of the radiative transfer problem. Under reflection boundary conditions, the intensity is not fixed, but there is a relation with the intensity across the inner and outer hemispheres. For specular reflection, we have

$$L_\nu(\mathbf{x},\boldsymbol{\omega}) = \rho_\nu\, L_\nu(\mathbf{x}, -\boldsymbol{\omega}^{spec}) \quad \text{pour} \quad \boldsymbol{\omega}\cdot\mathbf{n} > 0 \tag{28}$$

where ρ_ν is surface reflectivity at the frequency ν, and $-\boldsymbol{\omega}^{spec}$ is the specular direction corresponding to $\boldsymbol{\omega}$ (see Fig. 10). For diffuse reflection, we have

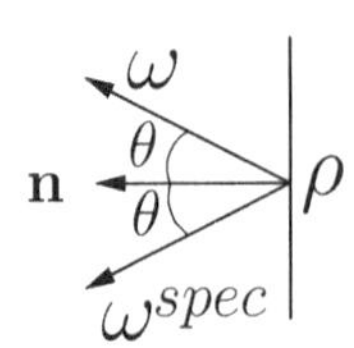

Figure 10 The definition of the specular reflection direction ω^{spec} corresponding to the direction ω for a surface with reflectivity ρ and normal **n**.

$$L_\nu(\mathbf{x},\boldsymbol{\omega}) = \frac{\rho_\nu}{\pi}\int_{\boldsymbol{\omega}'\cdot\mathbf{n}<0} L_\nu(\mathbf{x},\boldsymbol{\omega}')(-\boldsymbol{\omega}'\cdot\mathbf{n})d\boldsymbol{\omega}' \quad \text{pour} \quad \boldsymbol{\omega}\cdot\mathbf{n}>0 \tag{29}$$

where the outer hemisphere is denoted as $\boldsymbol{\omega}'\cdot\mathbf{n} < 0$.

As we have seen, the intensity L expressed in energy units and the intensity $\hat{L}$ expressed in kinetic units obey the same radiative transfer equation. The solutions obtained for these two physical quantities thus have the same formulation and, in case of numerical calculation, either of these variables can be determined (with the same formula) depending on the unit chosen for expressing the incident flux $q_\cap$ or its spectral distribution $q_{\cap,\nu}$. The latter is an input parameter. If we express $q_\cap$ in W m^{-2}, the result determines L, but if we express $q_\cap$ in μmol s^{-1} m^{-2}, then the result determines $\hat{L}$. For this reason, we no longer distinguish L and $\hat{L}$ in the rest of this chapter, except when presenting numerical results.

Let us briefly define, according to the intensity, the usual macroscopic radiative quantities used for photobioreactor analysis. Irradiance $G_\nu(\mathbf{x})$ is the integral of the intensity over propagation directions:

$$G_\nu(\mathbf{x}) = \int_{4\pi} L_\nu(\mathbf{x},\boldsymbol{\omega})d\boldsymbol{\omega} \tag{30}$$

Our study of kinetic-coupling phenomena is based on the specific rate of photon absorption $\mathcal{A}(\mathbf{x})$ (see Sections 1 and 5), that is, the number of photons absorbed by a microbial cell located at $\mathbf{x}$ per unit of the time interval:

$$\mathcal{A}(\mathbf{x}) = \int_{\nu_{\min}}^{\nu_{\max}} \sigma_{a,\nu}\, G_\nu(\mathbf{x})d\nu \tag{31}$$

where G_ν is usually expressed in μmol s^{-1} m^{-2} Hz^{-1} (which is obtained by using $\hat{L}$ rather than L in Eq. (30). Another useful physical quantity is the flux density vector

$$\mathbf{j}_{R,\nu}(\mathbf{x}) = \int_{4\pi} L_\nu(\mathbf{x},\boldsymbol{\omega})\,\boldsymbol{\omega}\, d\boldsymbol{\omega} \tag{32}$$

which is used to define the flux density q_ν through any surface with normal $\mathbf{n}$: $q_\nu(\mathbf{x}) = \mathbf{j}_{R,\nu}(\mathbf{x})\cdot\mathbf{n}$. For example, in our one-dimensional configuration of Fig. 6, the surface flux density q_ν along $\mathbf{e}_z$ at the abscissa z is

$$q_\nu(z) = \int_{4\pi} L_\nu(\mathbf{x},\boldsymbol{\omega})\,\boldsymbol{\omega}\cdot\mathbf{e}_z\, d\boldsymbol{\omega} \tag{33}$$

Note that by substituting the boundary condition Eq. (26) or (27) into this definition, one can verify that $q_\nu(0) = q_\cap + \int_{\boldsymbol{\omega} \cdot \mathbf{e}_z < 0} L_\nu(\mathbf{x}, \boldsymbol{\omega})\, \boldsymbol{\omega} \cdot \mathbf{e}_z \, d\boldsymbol{\omega}$, where $\boldsymbol{\omega} \cdot \mathbf{e}_z < 0$ is the outer hemisphere.

These macroscopic descriptors correspond to integration of the intensity over the space of directions: the irradiance is the 0th moment of the intensity, and the flux density vector is its first moment. This integration results in the loss of a significant portion of information about the propagation directions but reduces the problem to a number of dimensions that is often easier to think about and solve. Most photobioreactor models are based on such variables, as is the case here, with the specific rate of photon absorption $\mathcal{A}$. Rather than solving the mesoscopic model (ie, the radiative transfer equation) and integrating the intensity afterwards, we are also interested in formulating models that directly address the macroscopic descriptors. Integrating the radiative transfer equation (Eq. (24)) over all directions $\boldsymbol{\omega}$, leads to the conservation equation

$$\mathbf{div}\ \mathbf{j}_{R,\nu}(\mathbf{x}) = -k_{a,\nu}\, G_\nu(\mathbf{x}) \tag{34}$$

In addition to Eq. (34) (that is exact), the construction of macroscopic models consisting of a closed set of equations for the moments of the distribution function (or the intensity), usually requires to formulate approximations. In fluid mechanics, this approximation leads, for example, to the Navier–Stokes equation. The most common approximate macroscopic radiative models describe radiative transfer with heat-like equations (eg, see the Rosseland approximation and the P1 approximation). Among them, the P1 approximation leads to Fick's equation of the flux density vector $\mathbf{j}_{R,\nu}$ (see Eq. (73); substituting Fick's equation into Eq. (34) yields the following heat-like equation for the irradiance G (in the absence of a source term):

$$-\tilde{D}\, \nabla^2 G_\nu(\mathbf{x}) = -c\, k_{a,\nu}\, G_\nu(\mathbf{x})$$

where $\tilde{D}$ is the macroscopic diffusion coefficient (expressed in m^2/s), $k_{a,\nu}$ is the absorption coefficient (as defined in Section 2), c is the speed of light in the medium, and ∇^2 is the Laplacian operator with respect to $\mathbf{x}$. Implementation of such models for the purpose of photobioreactor research will be analyzed in Sections 3.4 and 3.5. These macroscopic descriptions are constructed around situations associated with near-equilibrium conditions. In fluid mechanics, the Knudsen numbers are generally small enough for this hypothesis to be tested, but usually, this is not the case in radiative transfer, in particular in photobioreactors. Nonetheless, the macroscopic approximations allow for analysis based on familiar interpretations (such as diffusion), and their implementation generally corresponds to very short calculation time. Thus, we often seek to reduce radiative problems to those descriptions, even in situations that may seem unsuitable (ie, far from equilibrium). This practice requires accurate knowledge of the radiative configuration in question. The analysis of typical photobioreactor configurations that we develop in Section 3.2 will allow us to construct, in Section 3.4, a relevant description of the irradiance field on the basis of the P1 approximation (even in far-from-equilibrium situations).

3.1.1 Note on Analysis of the Spectral Dimension

As we saw in Eq. (18), description of each frequency of the spectrum is completely independent of other frequencies. Thus, all the derivation of equations in the rest of this section, including the approximate solutions developed in Sections 3.3–3.5, are valid regardless of the wavelength of incident radiation. For this reason, we decided to omit the spectral dependencies in our notations. For numerical calculations, however, one should use the value of the radiative properties and the incident flux $q_{\cap,\nu}$ corresponding to the wavelength in question. In this approach, the radiative transfer equation is solved for each frequency, and the spectral solution thus obtained is integrated over PAR in order to calculate the local absorption rate $\mathcal{A}$ according to Eq. (31). This approach can be implemented with approximate solutions from Sections 3.3–3.5.

We made a different choice for the numerical applications presented in this section; namely, we chose to obtain "simple" approximate solutions appropriate for analysis. We will use the approximation of an equivalent gray medium, which defines effective radiative properties that are independent of the frequency. These properties are provided in Fig. 6 and were obtained by averaging the spectral properties from Section 2 weighted by the incident spectrum:

$$\sigma = \frac{\int_{\nu_{\min}}^{\nu_{\max}} q_{\cap,\nu}\, \sigma_\nu \, d\nu}{q_\cap} \tag{35}$$

and

$$p_\Omega(\boldsymbol{\omega}|\boldsymbol{\omega}') = \frac{\int_{\nu_{\min}}^{\nu_{\max}} q_{\cap,\nu}\, \sigma_{s,\nu}\, p_{\Omega,\nu}(\boldsymbol{\omega}|\boldsymbol{\omega}')\, d\nu}{\sigma_s\, q_\cap} \tag{36}$$

where σ is the effective cross section for extinction σ_{ext}, absorption σ_a, or scattering σ_s; $q_\cap$ and $q_{\cap,\nu}$ are defined in Eq. (25). For this approximation, Eq. (31) becomes $\mathcal{A}(\mathbf{x}) = \sigma_a\, G(\mathbf{x})$, where G is the irradiance obtained by direct use of the effective properties in our approximate solutions as well as the flux $q_\cap$ rather than $q_{\cap,\nu}$ (no spectral integration is required).

The spectral dimension will be analyzed precisely in Section 4, which is devoted to obtaining reference solutions by the Monte Carlo method.

3.2 Optical Thickness and Invariance of the Transport Problems

In this section, we introduce a set of dimensionless quantities commonly used to characterize radiative transfer in specific configurations. These quantities are necessary for a detailed understanding of radiative transfer in photobioreactors. Depending on the value of the albedo α_s, the asymmetry parameter g of the phase function, and the optical thickness of the medium, it is possible to conduct first analysis of the scattering regime and to identify the appropriate approximations.

The single-scattering albedo represents the proportion of the interaction events that are scattering events:

$$\alpha_s = \frac{k_s}{k_{ext}} \tag{37}$$

where $k_{ext} = k_s + k_a.\alpha_s$ is equal to 0 in the case of a purely absorbing medium (ie, $k_s = 0$) and equal to 1 in a purely scattering medium (ie, $k_a = 0$). In the radiative configuration of the flat-plate photobioreactor in Fig. 6, $\alpha_s \simeq 0.86$ means that when a photon interacts with a microorganism, the photon is scattered with probability 0.86 and absorbed with probability 0.14. Each scattering event statistically redistributes the propagation direction of the photon according to the single-scattering phase function.[5] $p_\Omega(\boldsymbol{\omega}|\boldsymbol{\omega}')$ For analysis purposes, it is interesting to reduce this angular information to its first moment: the asymmetry parameter

$$g = \int_{4\pi} p_\Omega(\boldsymbol{\omega}|\boldsymbol{\omega}')\; \boldsymbol{\omega}\cdot\boldsymbol{\omega}'\; d\boldsymbol{\omega} \tag{38}$$

[5] $p_\Omega(\boldsymbol{\omega}|\boldsymbol{\omega}')d\boldsymbol{\omega}$ is the probability that when a scattering event occurs, a photon with the propagation direction $\boldsymbol{\omega}'$ is scattered within the element of solid angle $d\boldsymbol{\omega}$ around the direction $\boldsymbol{\omega}$. In the present context, a local thermodynamic equilibrium can be assumed; therefore, $p_\Omega(\boldsymbol{\omega}|\boldsymbol{\omega}') = p_{\Omega'}(\boldsymbol{\omega}'|\boldsymbol{\omega})$.

where $\boldsymbol{\omega} \cdot \boldsymbol{\omega}' = \cos(\theta_s)$, with the notations shown in Fig. 2. g ranges from -1 to 1 and is equal to 0 when the phase function is symmetric for forward and backward directions (ie, for $\boldsymbol{\omega} \cdot \boldsymbol{\omega}' > 0$ and $\boldsymbol{\omega} \cdot \boldsymbol{\omega}' < 0$). This is the case, for example, for an isotropic phase function: all the scattering directions are equiprobable. Furthermore, $g < 0$ in the case of a phase function oriented in backward directions, and $g > 0$ for a phase function oriented in forward directions. In the case of photosynthetic microorganisms, g is close to 1, or more precisely, $g = 0.945$ in the situation under study (see Fig. 6). This means that in a photobioreactor, each scattering event predominantly redistributes the propagation direction within a small solid angle around the incident direction $\boldsymbol{\omega}'$ (aperture $\simeq 20°$, see Section 2.2). Nevertheless, the sum of these successive scattering events may lead to a significant deviation in the propagation directions (see Figs. 9 and 11) and may result in, among other things, a complex residence time distribution (Blanco and Fournier, 2003, 2006). The information about this distribution is commonly reduced to the scattering optical thickness e_s, which is the product of the

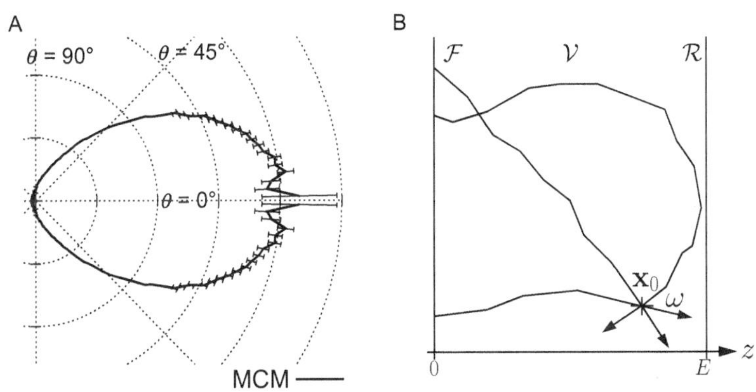

Figure 11 (A) Angular distribution of the intensity $L(z_0, \theta)$ at the location $z_0 = 3$ cm within the one-dimensional photobioreactor from Fig. 6 with Lambertian emission and $\rho^{\mathcal{F}} = \rho^{\mathcal{R}} = 0$. The results were obtained by the Monte Carlo method (see Section 4). (B) Assumption for the corresponding optical paths (z_0 is the abscissa of the location $\mathbf{x}_0$, and θ is the angle between $\boldsymbol{\omega}$ and $\mathbf{e}_z$). The complex angular distribution of the intensity is due to the special shape of the phase function for photosynthetic microorganisms. The multiple-scattering optical path in question is the result of many scattering events corresponding to a small deviation of the propagation direction. Among them, those leading to directions $\boldsymbol{\omega}$ that are significantly different from $\mathbf{e}_z$ are the longest: the probability for absorption to occur along these paths before reaching $\mathbf{x}_0$ is high (regarding attenuation by absorption along optical paths, see Section 4.1).

characteristic dimension of the reaction volume (E in the case of our flat-plate photobioreactor in Fig. 6) and the scattering coefficient:

$$e_s = E\, k_s \tag{39}$$

e_s is the inverse of the Knudsen number. In the radiative configuration under study, $e_s \simeq 20$.

During analysis of radiative transfer, the angular distribution of the intensity is of great significance because its deviation from isotropy defines the validity conditions of various approximations and physical interpretations. Here, it is crucial to distinguish (i) the angular distribution of the phase function, which corresponds to redistribution of propagation directions because of a single scattering event (under the assumption of perfect mixing, this radiative property of the microbial cells is homogeneous within the photobioreactor) and (ii) the angular distribution of the intensity, which corresponds to the distribution of the propagation directions at a given location. The angular distribution of the intensity (which is generally a function of the location within the photobioreactor) results from multiple scattering events: it is formulated by solving the radiative transfer equation. Because the asymmetry parameter is $\simeq 0.95$, it is a significant obstacle for analysis of the angular distribution of intensity within a photobioreactor. Indeed, in this situation, the extent to which the successive forward-scattering events redistribute the propagation directions is difficult to grasp. To simplify the physics involved, it is customary to use a transport problem equivalent to that under study but where the phase function is isotropic. Analysis of this equivalent problem is much easier because information about the initial propagation direction of a photon (ie, the boundary conditions) is lost from the first scattering event: the propagation direction ω is redistributed isotropically, independently of the incoming direction $\boldsymbol{\omega}'$. This equivalent problem is derived by replacing the radiative properties k_{ext}, k_s, k_a, and p_{Ω} obtained in Section 2 by the new radiative properties k^*_{ext}, k^*_s, k^*_a, and p^*_{Ω}, according to the following transformation:

$$\begin{aligned} k^*_{ext} &= k_{ext}(1-\alpha_s g) \\ k^*_s &= k_s(1-g) \\ k^*_a &= k_a \\ p^*_{\Omega} &= \frac{1}{4\pi} \quad \text{(isotropic phase function)} \end{aligned} \tag{40}$$

The dimensionless quantities defined previously become

$$\alpha^*_s = \alpha_s \frac{1-g}{1-\alpha_s g} \tag{41}$$

$$g^* = 0 \tag{42}$$

$$e_s^* = e_s(1-g) \tag{43}$$

The radiative transfer equation is not invariant with this transformation, but we find this invariance in various situations: for example, the diffusion equation obtained with the P1 approximation is invariant with this transformation (see Section 3.4). In addition, we found that solution of this equivalent problem usually provides results that are very close to those obtained by solution of the original problem in the case of a photobioreactor. The approximate solutions that are derived and validated in Sections 3.3 and 3.4 are obtained by addressing this equivalent problem. Note that this transformation is also useful for comparison of very different situations, regardless of the form of the phase function: in the field of transport theory research, when mentioning optical thickness, we are generally referring to e_s^* rather than e_s.

In the radiative configuration shown in Fig. 6, $\alpha_s^* = 0.25$ and $e_s^* = 1.1$; this situation is typical of a photobioreactor operating at its optimum biomass production rate. Such intermediate values of optical thickness mean that scattering plays a significant role but does not systematically ensure that the intensity within the medium is close to isotropy. Accordingly, the angular distribution within the reaction volume depends on both the scattering phenomenon and the boundary conditions. Although analysis of such intermediate situation is not straightforward, the equivalent problem brings us back to situations that are easily manipulated. Instead of reasoning about complex optical paths resulting from multiple forward-scattering events (as in Fig. 11), in the following section, we use the single-scattering approximation, where photons suffer zero or one isotropic scattering event only (see Figs. 12 and 13).

3.3 The Single-Scattering Approximation

It is always possible to formulate the intensity L of the entire photon population as the sum of the intensity values $L^{(j)}$ corresponding to the photons that have undergone exactly j scattering events:

$$L = L^{(0)} + L^{(1)} + L^{(2)} + L^{(3)} + \ \ldots \tag{44}$$

This simply means that the total number of photons within any phase space volume element $d\mathbf{x}d\boldsymbol{\omega}$ is the sum of the photons that have undergone j diffusion events. Each $L^{(j)}$ is governed by an equation of its own, in which the source term corresponds to the lower-order photons $L^{(j-1)}$ that are

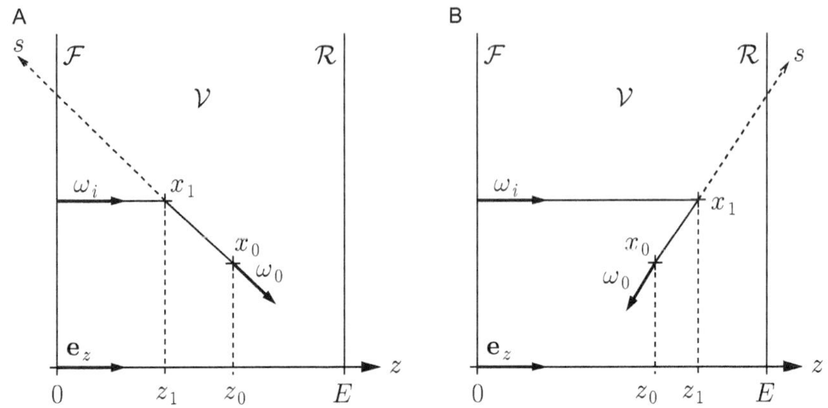

Figure 12 Single-scattering optical paths contributing to $L^{(1)}(z_0, \boldsymbol{\omega}_0)$. (A) $\mu_0 > 0$ and (B) $\mu_0 < 0$, where $\mu_0 = \boldsymbol{\omega}_0 \cdot \mathbf{e}_z$.

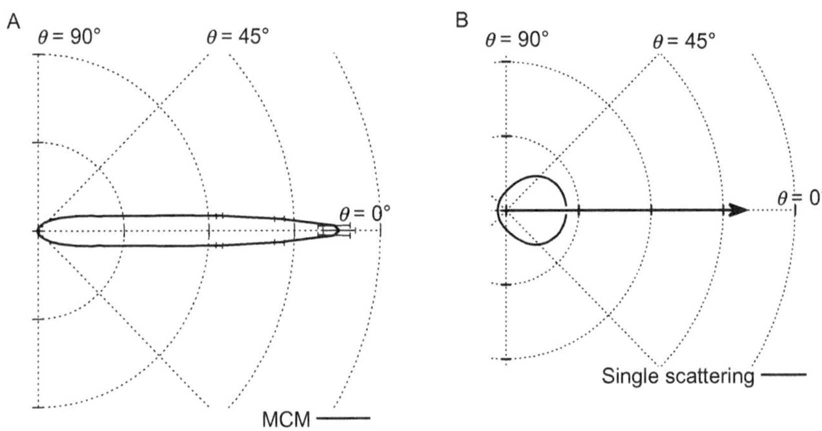

Figure 13 Angular distribution of the intensity $L(z_0, \theta)$ at location $z_0 = 3$ cm within the one-dimensional photobioreactor shown in Fig. 6: $\rho^{\mathcal{F}} = \rho^{\mathcal{R}} = 0$, collimated normal incidence $\mu_i = 1$. (A) Results obtained by the Monte Carlo method (see Section 4). (B) Results obtained for the equivalent transport problem where $\alpha_s^* = 0.25$, $k_{ext}^* = 110$ m^{-1}, and $p_\Omega^* = \frac{1}{4\pi}$, and the single scattering approximation is used. The arrow indicates the part of the distribution that is due to the ballistic photons, ie, the arrow represents a Dirac distribution. This illustration does not allow for analysis of the ratio of ballistic to scattered photons, but we invite the reader to see Fig. 14.

scattered locally and move from population $(j-1)$ to population (j). In this section, we first derive the corresponding infinite system of coupled equations (Eq. (46) and Eq. (54)) and the associated boundary conditions (Eqs. (47) and (48) and Eqs. (56) and (57)) in the special case of the one-dimensional photobioreactor in Fig. 6. Then, we will find a solution to the equivalent transport problem presented in Section 3.2, where scattering optical thickness is $e_s \simeq 1$. In this situation, the single scattering approximation (Ishimaru, 1999) is relevant: only the scattering orders (0) and (1) will be selected, and higher orders will be ignored

$$L \simeq L^{(0)} + L^{(1)} \tag{45}$$

This approach significantly simplifies solution of the mesoscopic problem. Thus, we consider only two subsets in the photon population: the photons that arrive directly from the surface $\mathcal{F}$ (did not undergo any scattering events) will be called "ballistic photons," and the photons that have undergone only one scattering event will be called the "scattered photons."

3.3.1 Expansion of the Radiative Transfer Equation into the Successive Order of Scattering

This section is focused on our one-dimensional photobioreactor with collimated incidence at $z = 0$, in the absence of reflection. Ballistic photons obey an independent radiative transfer equation without a source term; the source is at the boundary $\mathcal{F}$ only:

$$\boldsymbol{\omega} \cdot \mathbf{grad_x} L^{(0)}(\mathbf{x}, \boldsymbol{\omega}) = -k_{ext} L^{(0)}(\mathbf{x}, \boldsymbol{\omega}) \tag{46}$$

with the following boundary conditions (according to Section 3.1):

- At $z = 0$,

$$L^{(0)}(\mathbf{x}, \boldsymbol{\omega}) = \frac{q_\cap}{\mu_i} \delta(\boldsymbol{\omega} - \boldsymbol{\omega}_i) \quad \text{for} \quad \mathbf{x} \in \mathcal{F},\ \boldsymbol{\omega} \cdot \mathbf{n}^{\mathcal{F}} > 0 \tag{47}$$

  where $\mu_i = cos(\theta_i)$, and $\delta(\boldsymbol{\omega} - \boldsymbol{\omega}_i)$ is the Dirac distribution centered at $\boldsymbol{\omega}_i$ (see Fig. 6).

- At $z = E$,

$$L^{(0)}(\mathbf{x}, \boldsymbol{\omega}) = 0 \quad \text{for} \quad \mathbf{x} \in \mathcal{R},\ \boldsymbol{\omega} \cdot \mathbf{n}^{\mathcal{R}} > 0 \tag{48}$$

The solution for $L^{(0)}$ is straightforward: it is the incident intensity $\frac{q_\cap}{\mu_i}$ attenuated by Bouguer's exponential extinction along the ballistic trajectory

$$L^{(0)}(\mathbf{x}, \boldsymbol{\omega}) = \frac{q_\cap}{\mu_i} \exp\left(-k_{ext} \frac{z}{\mu_i}\right) \delta(\boldsymbol{\omega} - \boldsymbol{\omega}_i) \tag{49}$$

where z is the abscissa of the location $\mathbf{x}$.

$L^{(1)}$ obeys the following radiative transfer equation:

$$\boldsymbol{\omega} \cdot \mathbf{grad}_{\mathbf{x}} L^{(1)}(\mathbf{x}, \boldsymbol{\omega}) = -k_{ext} L^{(1)}(\mathbf{x}, \boldsymbol{\omega}) + \mathcal{C}^{(0)}(\mathbf{x}, \boldsymbol{\omega}) \tag{50}$$

where $\mathcal{C}^{(0)}$ is the source term accounting for ballistic photons scattered at $\mathbf{x}$, which then arrive into the population (1) with the direction $\boldsymbol{\omega}$ (according to the collision term of the radiative transfer equation in Section 3.1):

$$\mathcal{C}^{(0)}(\mathbf{x}, \boldsymbol{\omega}) = \alpha_s \, k_{ext} \int_{4\pi} L^{(0)}(\mathbf{x}, \boldsymbol{\omega}') \, p_\Omega(\boldsymbol{\omega}|\boldsymbol{\omega}') d\boldsymbol{\omega}' \tag{51}$$

where $\alpha_s k_{ext} = k_s$. The boundary conditions for $L^{(1)}$ are as follows:

- At $z = 0$,

$$L^{(1)}(\mathbf{x}, \boldsymbol{\omega}) = 0 \quad \text{for} \quad \mathbf{x} \in \mathcal{F}, \ \boldsymbol{\omega} \cdot \mathbf{n}^{\mathcal{F}} > 0 \tag{52}$$

- At $z = E$,

$$L^{(1)}(\mathbf{x}, \boldsymbol{\omega}) = 0 \quad \text{for} \quad \mathbf{x} \in \mathcal{R}, \ \boldsymbol{\omega} \cdot \mathbf{n}^{\mathcal{R}} > 0 \tag{53}$$

The incoming intensity is equal to zero because on the one hand, there is no emission at the boundaries for this population (only ballistic photons are emitted at the boundary $\mathcal{F}$), and on the other hand, reflectivity of $\mathcal{F}$ and $\mathcal{R}$ is zero in the present case.

For the jth order, we have

$$\boldsymbol{\omega} \cdot \mathbf{grad}_{\mathbf{x}} L^{(j)}(\mathbf{x}, \boldsymbol{\omega}) = -k_{ext} L^{(j)}(\mathbf{x}, \boldsymbol{\omega}) + \mathcal{C}^{(j-1)}(\mathbf{x}, \boldsymbol{\omega}) \tag{54}$$

where

$$\mathcal{C}^{(j-1)}(\mathbf{x}, \boldsymbol{\omega}) = \alpha_s \, k_{ext} \int_{4\pi} L^{(j-1)}(\mathbf{x}, \boldsymbol{\omega}') \, p_\Omega(\boldsymbol{\omega}|\boldsymbol{\omega}') d\boldsymbol{\omega}' \tag{55}$$

is the source term corresponding to the transitions from population $(j-1)$ to population (j) because of scattering (with $\mathcal{C}^{(0)} = 0$). The boundary conditions are as follows:

- At $z = 0$,

$$L^{(j)}(\mathbf{x}, \boldsymbol{\omega}) = 0 \quad \text{for} \quad \mathbf{x} \in \mathcal{F}, \ \boldsymbol{\omega} \cdot \mathbf{n}^{\mathcal{F}} > 0 \tag{56}$$

- At $z = E$,

$$L^{(j)}(\mathbf{x}, \boldsymbol{\omega}) = 0 \quad \text{for} \quad \mathbf{x} \in \mathcal{R}, \ \boldsymbol{\omega} \cdot \mathbf{n}^{\mathcal{R}} > 0 \tag{57}$$

The equation for $L^{(0)}$ is independent of the other equations, and each of the higher orders $j > 0$ is coupled only to the order $j - 1$: this system of equations is closed at the 0th order. Therefore, truncation of the expansion Eq. (44) involves simply ignoring the existence of certain photons; this approach will not cause an error in the description of the orders that are selected for analysis.

3.3.2 Implementation of the Single Scattering Approximation for an Equivalent Transport Problem: Application to a Flat-Plate Photobioreactor

In the rest of this section, we address the equivalent transport problem defined by α_s^*, k_{ext}^*, and $p_\Omega(\boldsymbol{\omega}|\boldsymbol{\omega}') = \frac{1}{4\pi}$ (see Section 3.2), and we use only scattering orders 0 and 1 (see Eq. (45)). Under these conditions, Eq. (49) becomes

$$L^{(0)}(z,\boldsymbol{\omega}) = \frac{q_\cap}{\mu_i} \exp\left(-k_{ext}^* \frac{z}{\mu_i}\right) \delta(\boldsymbol{\omega} - \boldsymbol{\omega}_i) \tag{58}$$

and substituting this solution into Eq. (51), we obtain the following source term for population (1):

$$\mathcal{C}^{(0)}(z,\boldsymbol{\omega}) = \frac{\alpha_s^* k_{ext}^* q_\cap}{4\pi \quad \mu_i} \exp\left(-k_{ext}^* \frac{z}{\mu_i}\right) \tag{59}$$

Due to isotropy of the phase function, the source term $\mathcal{C}^{(0)}(z,\boldsymbol{\omega})$ is independent of the direction $\boldsymbol{\omega}$ ($\mathcal{C}^{(0)}$ is isotropic). A solution for $L^{(1)}$ is obtained by solving Eq. (50) under the boundary conditions in Eqs. (52) and (53). This task can be accomplished either by the variation of constants method or by intuitive reasoning: the intensity $L^{(1)}(\mathbf{x}_0,\boldsymbol{\omega}_0)$ is the source term $\mathcal{C}^{(0)}(\mathbf{x}_1,\boldsymbol{\omega}_0)$ attenuated by extinction along the length $\|\mathbf{x}_0 - \mathbf{x}_1\|$, which is integrated over the locations $\mathbf{x}_1$ defined by $\mathbf{x}_1 = \mathbf{x}_0 - s\boldsymbol{\omega}_0$ with $s \in [0, +\infty[$. In Fig. 12, we show that this reasoning indeed involves constructing all the single-scattering optical paths with direction $\boldsymbol{\omega}_0$ at $\mathbf{x}_0$. In the one-dimensional configuration under study, $L^{(1)}(\mathbf{x}_0,\boldsymbol{\omega}_0)$ depends only on the abscissa z_0 at $\mathbf{x}_0$ and on the cosine $\mu_0 = \boldsymbol{\omega}_0 \cdot \mathbf{e}_z$. In addition, $\| \mathbf{x}_0 - \mathbf{x}_1 \| = \left| \frac{z_0 - z_1}{\mu_0} \right|$ (see Fig. 12). For the directions where $\mu_0 > 0$, we obtain

$$L^{(1)}(z_0,\boldsymbol{\omega}_0) = \int_0^{z_0} \mathcal{C}^{(0)}(z_1) \exp\left(-k_{ext}^* \frac{z_0 - z_1}{\mu_0}\right) \frac{dz_1}{\mu_0} \tag{60}$$

and for the directions where $\mu_0 < 0$,

$$L^{(1)}(z_0,\boldsymbol{\omega}_0)=\int_{z_0}^{E} C^{(0)}(z_1)\exp\left(-k^*_{ext}\frac{z_0-z_1}{\mu_0}\right)\frac{dz_1}{-\mu_0} \tag{61}$$

Substituting Eq. (59) into the above equations and solving the integration, we obtain the following:

- for $\mu_0 > 0$,

$$L^{(1)}(z_0,\boldsymbol{\omega}_0)=\frac{\alpha_s^*}{4\pi}\frac{q_\cap}{\mu_0-\mu_i}\left[\exp\left(-k^*_{ext}\frac{z_0}{\mu_0}\right)-\exp\left(-k^*_{ext}\frac{z_0}{\mu_i}\right)\right] \tag{62}$$

- and for $\mu_0 < 0$,

$$L^{(1)}(z_0,\boldsymbol{\omega}_0)=\frac{\alpha_s^*}{4\pi}\frac{q_\cap}{\mu_0-\mu_i}\left[\exp\left(-k^*_{ext}\frac{E}{\mu_i}\right)\exp\left(-k^*_{ext}\frac{z_0-E}{\mu_0}\right)-\exp\left(-k^*_{ext}\frac{z_0}{\mu_i}\right)\right] \tag{63}$$

Finally, the total intensity $L(z_0,\boldsymbol{\omega}_0)$ is estimated as

$$L(z_0,\boldsymbol{\omega}_0)\simeq L^{(0)}(z_0,\boldsymbol{\omega}_0)+L^{(1)}(z_0,\boldsymbol{\omega}_0) \tag{64}$$

where $L^{(0)}$ is given in Eq. (58).

Fig. 13 presents the angular distribution of L resulting from the single-scattering approximation for the equivalent transport problem α_s^*, k^*_{ext} and $p^*_\Omega=\frac{1}{4\pi}$ as well as the reference solution produced by the Monte Carlo method for α_s and k_{ext} and the phase function of *C. reinhardtii*. In the reference situation, at the location in question ($z_0 = 3$ cm), the ballistic beam is completely attenuated: all the photons have undergone at least one scattering event but deviated very little from their incident direction (see Section 3.2). This situation results in a complex angular distribution centered around the incident direction (see Fig. 13A). In our equivalent transport problem, this complex distribution is replaced by the sum of a Dirac distribution (contribution of the ballistic photons, ie, ~ 75% of the photons in the present case, see Fig. 14) and a relatively broad distribution (contribution of the scattered photons) that is simply modeled as Eqs. (62) and (63) under the single-scattering approximation (see Fig. 13B). The angular distribution of the scattered intensity $L^{(1)}$ at different locations is shown in Fig. 15.

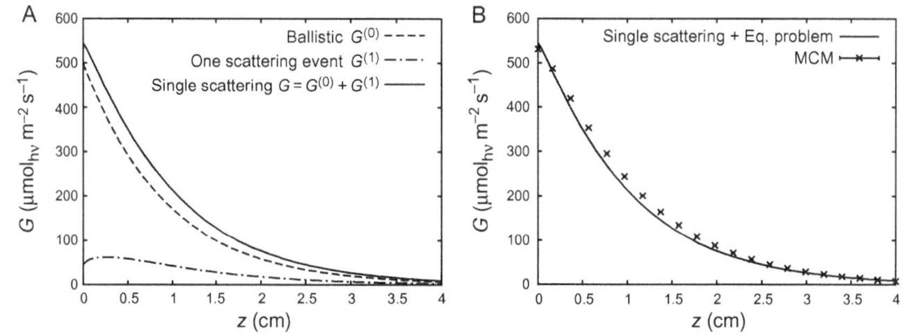

Figure 14 The irradiance field G within the photobioreactor shown in Fig. 6; $\rho^{\mathcal{F}} = \rho^{\mathcal{R}} = 0$ and collimated normal incidence $\mu_i = 1$. The results were obtained for the equivalent transport problem where $\alpha_s^* = 0.25$, $k_{ext}^* = 110\ \text{m}^{-1}$, and $p_\Omega^* = 1/4\pi$: the expression for $G^{(0)}$ is given in Eq. (66), the expression for $G^{(1)}$ is given in Eq. (68) (note that Eq. (67) must be used instead if $\mu_i \neq 1$). The single-scattering approximation is used: $G = G^{(0)} + G^{(1)}$. (A) The proportions of ballistic and scattered photons. (B) Comparison with the reference solution obtained by the Monte Carlo method (MCM) for $\alpha_s = 0.86$, $k_{ext} = 587\ \text{m}^{-1}$, and the phase function of *Chlamydomonas reinhardtii* (see Section 4).

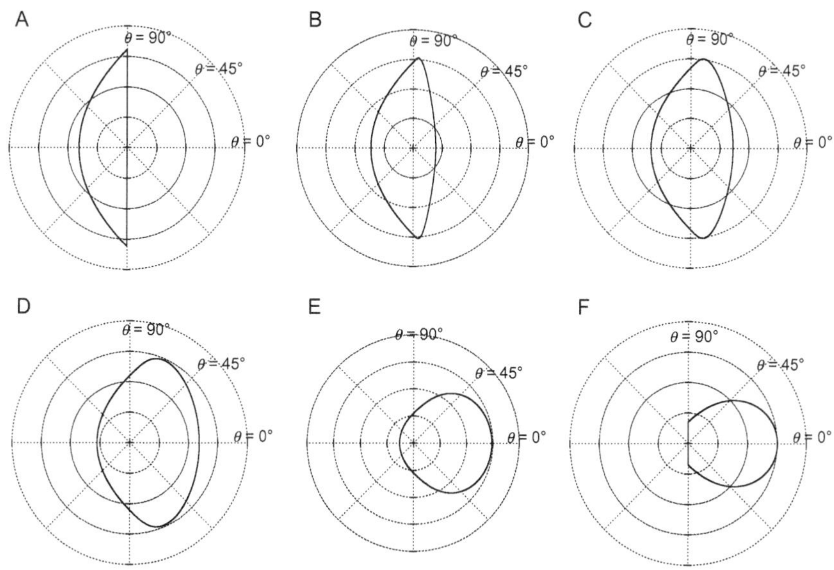

Figure 15 Angular distribution of the scattered intensity $L^{(1)}(z_0, \theta)$ at the abscissa z_0 within the photobioreactor shown in Fig. 6; $\rho^{\mathcal{F}} = \rho^{\mathcal{R}} = 0$; collimated normal incidence $\mu_i = 1$. The results were obtained for the equivalent transport problem where $\alpha_s^* = 0.25$, $k_{ext}^* = 110\ \text{m}^{-1}$, and $p_\Omega^* = 1/4\pi$, according to Eqs. (62) and (63). (A) $z_0 = 0$. (B) $z_0 = 2.5$ mm. (C) $z_0 = 5$ mm. (D) $z_0 = 1$ cm. (E) $z_0 = 2.5$ cm. (F) $z_0 = 4$ cm.

Integration of $L(z,\boldsymbol{\omega})$ over the directions $\boldsymbol{\omega}$ yields the local irradiance $G(z)$:

$$G(z) = \int_{4\pi} L(z,\boldsymbol{\omega})d\boldsymbol{\omega} \simeq \int_{4\pi} L^{(0)}(z,\boldsymbol{\omega})d\boldsymbol{\omega} + \int_{4\pi} L^{(1)}(z,\boldsymbol{\omega})d\boldsymbol{\omega} = G^{(0)}(z) + G^{(1)}(z) \tag{65}$$

where $G^{(0)}$ is the irradiance due to the ballistic photons, and $G^{(1)}$ is the irradiance due to the photons that have undergone only one scattering event. The expression $G^{(0)} = \int_{4\pi} \frac{q_\cap}{\mu_i} \exp\left(-k_{ext}^* \frac{z}{\mu_i}\right) \delta(\boldsymbol{\omega} - \boldsymbol{\omega}_i) d\boldsymbol{\omega}$ is simply

$$G^{(0)}(z) = \frac{q_\cap}{\mu_i} \exp\left(-k_{ext}^* \frac{z}{\mu_i}\right) \tag{66}$$

Obtaining $G^{(1)}$ is usually less straightforward, but the integral in Eq. (65) has a symbolic solution in the present case:

$$\begin{aligned} G^{(1)}(z) = \frac{\alpha_s^*}{2} q_\cap \times \Bigg\{ & \exp\left(-k_{ext}^* \frac{z}{\mu_i}\right) \\ & \left(Ei\left[k_{ext}^* z \frac{1-\mu_i}{\mu_i}\right] - Ei\left[-k_{ext}^* (E-z) \frac{1+\mu_i}{\mu_i}\right] + \ln\left[\frac{1+\mu_i}{1-\mu_i}\right] \right) \\ & - Ei\left[-k_{ext}^* z\right] + \exp\left(-k_{ext}^* \frac{E}{\mu_i}\right) Ei\left[-k_{ext}^* (E-z)\right] \Bigg\} \end{aligned} \tag{67}$$

where Ei is the exponential integral $Ei(x) = -\int_{-x}^{\infty} \frac{e^{-t}}{t} dt$, which is a function available in most scientific computation libraries. In the special case of normal incidence $\mu_i \rightarrow 1$, $G^{(1)}$ becomes

$$\begin{aligned} G^{(1)}(z;\mu_i = 1) = \frac{\alpha_s^*}{2} q_\cap \Big\{ & \exp(-k_{ext}^* z)\left(\gamma + \ln[2] + \ln\left[k_{ext}^* z\right] - Ei\left[-2\, k_{ext}^*(E-z)\right]\right) \\ & - Ei\left[-k_{ext}^* z\right] + \exp(-k_{ext}^* E)\, Ei\left[-k_{ext}^*(E-z)\right] \Big\} \end{aligned} \tag{68}$$

where $\gamma \simeq 0.577$ is the Euler–Mascheroni constant.

The irradiance field obtained with the single-scattering approximation is shown in Fig. 14. Panel (A) shows the proportion of ballistic and scattered photons within the reaction volume. As indicated by the intermediate value of the transport optical thickness $e_s^* = 1.1$, the scattered photons are in the minority but cannot be disregarded. Fig. 14B shows comparison between the reference solution (Monte Carlo method) and the results obtained by combining the equivalent transport problem and the single-scattering approximation. These results indicate that indeed, scattering orders higher that 1 can be ignored during analysis the equivalent problem. Given the agreement observed, we should note that the simple physical interpretations that we developed here are relevant to analysis of the process. In particular, substitution of the complex distribution observed in Fig. 13 by the sum of a Dirac distribution and a wider distribution is extremely convenient. With this approach, description of the ballistic photons is straightforward, and all difficulty of the analysis is reduced to description of the scattered photons. Because the scattered intensity is relatively close to isotropic (see Fig. 15), we can derive the relevant macroscopic description of the scattered photons in the next section.

3.4 The P1 Approximation and Diffusion Equation

In contrast to the single-scattering approximation (previous paragraph), which is mesoscopic, below we adopt a macroscopic point of view. In this section and in Section 3.5, angular distribution of intensity is fixed *a priori*, and the radiative transfer equation is integrated over all propagation directions in order to formulate a closed equation for the irradiance.

3.4.1 The Diffusion Equation in One-Dimensional Cartesian Geometric Configuration

The P1 approximation consists of truncating the spherical-harmonic expansion of the intensity at order 1. In the one-dimensional configuration shown in Fig. 6, for Lambertian or collimated normal incidence, this method is equivalent to fixing the following functional form for the angular dependence:

$$L(\mathbf{x},\boldsymbol{\omega}) = A(z)[1 + C(z)\cos(\theta)] \tag{69}$$

where, given the symmetry of the problem, L is a function of the abscissa z and the angle θ only ($\cos(\theta) = \boldsymbol{\omega} \cdot \mathbf{e}_z$). Substituting this approximation into the radiative transfer equation Eq. (24) (in which we omit the frequency variable) and integrating it over all the propagation directions $\boldsymbol{\omega}$ (ie, across all angles θ), we obtain a diffusion equation for the description of the irradiance field (Ishimaru, 1999):

$$-\widetilde{D}\,\partial_z^2 G(z) = -c\,k_a\,G(z) \tag{70}$$

where $\partial_z^2 G(z)$ is the second derivative of the irradiance with respect to z, and $\widetilde{D}$ is the macroscopic diffusion coefficient: $\widetilde{D}=c/(3\,k_{ext}(1-\alpha_s\,g))=c/(3\,k^*_{ext})$; k^*_{ext} is defined in Section 3.2. Hereafter, we will express the diffusion coefficient in m rather than in m^2/s; this approach is convenient for analysis of steady-state systems. Indeed, in this case, the solution of the radiative transfer equation is independent of the speed of light c; accordingly, it is customary to divide Eq. (70) by c:

$$-D\,\partial_z^2 G(z) = -k_a\,G(z) \tag{71}$$

with the macroscopic diffusion coefficient D expressed in m, defined as $D=\widetilde{D}/c$:

$$D=\frac{1}{3\,k_{ext}(1-\alpha_s\,g)}=\frac{1}{3\,k^*_{ext}} \tag{72}$$

It should be noted that Fick's equation is also obtained by substituting Eq. (69) into the radiative transfer equation, multiplying it by $\cos(\theta)$, and integrating over the propagation directions (Ishimaru, 1999):

$$q(z) = -D\,\partial_z G(z) \tag{73}$$

where $q(z)$ is the surface flux density along $\mathbf{e}_z$ (see Eqs. (33) and (34)).

3.4.1.1 Boundary Conditions

The steady-state diffusion equation (Eq. (71)) is an ordinary differential equation of order 2, whose solution requires two boundary conditions. In radiative transfer, the value of the irradiance or the net flux at the boundary is rarely available. Therefore, a linear relation between G and its derivative, that is, between the irradiance and the flux (see Eq. (73)) is generally used for the boundary conditions: this is what researchers in this field call the Marshak boundary conditions (Marshak, 1947). To our knowledge, in the existing literature, the expression for Marshak's boundary conditions is brought to the following functional form (Case and Zweifel, 1967; Durian, 1994; Ishimaru, 1999):

$$[G \pm \mathcal{L}\,\partial_s\,G] = \mathcal{B} \tag{74}$$

where $\mathcal{L}$ is the extrapolation length, and $\mathcal{B}$ is a constant. Determination of parameters $\mathcal{L}$ and $\mathcal{B}$ that ensure the same order of approximation near the boundaries as within the medium is a topic of research in itself (eg, see Durian, 1994). In the context of photobioreactor analysis, we use a specific expression for $\mathcal{B}$ (derived in Dauchet, 2012), which leads to the following boundary conditions in the case of the flat-plate photobioreactor shown in Fig. 6:

- At $z = 0$,

$$G(0) - \mathcal{L}_0\, \partial_z\, G(0) = 2\,\frac{1+\rho^{\mathcal{F}}}{1-\rho^{\mathcal{F}}}\, q^{(0)}(0) + G^{(0)}(0) \tag{75}$$

- At $z = E$,

$$G(E) + \mathcal{L}_E\, \partial_z\, G(E) = 2\,\frac{1+\rho^{\mathcal{R}}}{1-\rho^{\mathcal{R}}}\, q^{(0)}(E) + G^{(0)}(E) \tag{76}$$

where the exponent $^{(0)}$ deals with the ballistic photons (see Section 3.3), and $\mathcal{L}$ is the extrapolation length that is estimated here as in Durian (1994):

$$\begin{aligned} \mathcal{L}_0 &= \frac{2}{3}\,\frac{1+\rho^{\mathcal{F}}}{1-\rho^{\mathcal{F}}}\,\frac{1}{k^*_{ext}} \\ \mathcal{L}_E &= \frac{2}{3}\,\frac{1+\rho^{\mathcal{R}}}{1-\rho^{\mathcal{R}}}\,\frac{1}{k^*_{ext}} \end{aligned} \tag{77}$$

where ρ is reflectivity of the bounding surface, and k^*_{ext} is defined in Eq. (40). $q^{(0)}$ and $G^{(0)}$ are respectively the surface flux density and the irradiance corresponding to the ballistic photons emitted at the boundary. The analytical solution of the diffusion equation for these boundary conditions in the case of Lambertian illumination is derived in the following section.

3.4.2 The Case of Diffuse Illumination: Direct Solution of the Diffusion Equation

In the text later, we will solve the diffusion equation (Eq. (71)) for the boundary conditions (Eqs. (75) and (76)) for the case of the one-dimensional configuration shown in Fig. 6 with Lambertian emission at $z = 0$ and reflection at $z = E$. The general solution that satisfies Eq. (71) is

$$G(z) = C_0 \exp(\xi\, z) + C_1 \exp(-\xi\, z) \tag{78}$$

where C_0 and C_1 are constants, and $\xi = \sqrt{k_a/D}$ (in the configuration under study $\xi \simeq 161$). The boundary condition at $z = 0$ is (according to Eqs. (75) and (77) where $\rho^{\mathcal{F}} = 0$):

$$G(0) - \mathcal{L}_0\, \partial_z\, G(0) = 2\, q^{(0)}(0) + G^{(0)}(0) \tag{79}$$

with

$$\mathcal{L}_0 = \frac{2/3}{k^*_{ext}} \tag{80}$$

We will now focus on the expression for $G^{(0)}(0)$ and $q^{(0)}(0)$, which are ballistic irradiance and ballistic surface flux density, respectively (see Section 3.3). The mesoscopic definition of Lambertian emission (according to Eq. (26)) results in

$$L^{(0)}(0,\boldsymbol{\omega}) = q_{\cap}/\pi \text{ for } \theta \in [0,\pi/2] \tag{81}$$

where $q_{\cap}$ is the incident surface flux density. Moreover, in the present case, we assume that $L^{(0)}(0,\boldsymbol{\omega}) = 0$ for $\theta \in [\pi/2,\pi]$ because the ballistic optical paths reflected at $z = E$ are completely attenuated when they return at $z = 0$ (the scattering optical thickness is $e_s \simeq 20$). Therefore, the intensity is integrated easily, according to Eqs. (30) and (33):

$$G^{(0)}(0) = \int_{4\pi} L^{(0)}(0,\boldsymbol{\omega})\, d\boldsymbol{\omega} = 2\, q_{\cap} \tag{82}$$

$$q^{(0)}(0) = \int_{4\pi} L^{(0)}(0,\boldsymbol{\omega})\, \boldsymbol{\omega} \cdot \mathbf{e}_z\, d\boldsymbol{\omega} = q_{\cap} \tag{83}$$

Hence

$$G(0) - \mathcal{L}_0\, \partial_z\, G(0) = 4\, q_{\cap} \tag{84}$$

For the boundary condition at $z = E$, ballistic photons can be ignored (they are scattered or absorbed before reaching $z = E$, as in the case of the boundary conditions at $z = 0$ above[6]), ie, $q^{(0)} \simeq 0$ and $G^{(0)} \simeq 0$. Therefore, Eq. (76) becomes

$$G(E) + \mathcal{L}_E\, \partial_z\, G(E) = 0 \tag{85}$$

Constants C_0 and C_1 in the general solution (Eq. (78)) that satisfy the boundary conditions (Eqs. (84) and (85)) are

[6] Contrary to Section 3.3, where we addressed the equivalent transport problem, ballistic photons here are in the minority, except close to $z = 0$, for $\theta \in [0,\pi/2]$. It is possible to take into account all the ballistic photons in our calculations (Eqs. (75) and (76)) because the mesoscopic solution for $L^{(0)}$ is obtained easily, even in the present case, with Lambertian emission and reflection at $z = E$. Nonetheless, except for the term that we used in Eq. (84), their contribution to the boundary conditions is negligible for most photobioreactor configurations during operation close to the optimum biomass growth rate.

$$C_1 = \frac{4\, q_\cap}{\exp(-\xi\, 2E)\dfrac{\mathcal{L}_E\, \xi - 1}{\mathcal{L}_E\, \xi + 1}(1 - \mathcal{L}_0\, \xi) + 1 + \mathcal{L}_0\, \xi} \tag{86}$$

and

$$C_0 = C_1\, \frac{\mathcal{L}_E\, \xi - 1}{\mathcal{L}_E\, \xi + 1} \exp(-\xi\, 2E) \tag{87}$$

Thus, we obtain the following expression for the irradiance field:

$$G(z) = 4\, q_\cap\, C \left(\exp[-\xi\, z] + \frac{\mathcal{L}_E\, \xi - 1}{\mathcal{L}_E\, \xi + 1}\, \exp[-\xi\,(2E - z)] \right) \tag{88}$$

where

$$C = \frac{1}{\exp(-\xi\, 2E)\dfrac{\mathcal{L}_E\, \xi - 1}{\mathcal{L}_E\, \xi + 1}(1 - \mathcal{L}_0\, \xi) + 1 + \mathcal{L}_0\, \xi} \tag{89}$$

From the mesoscopic point of view, the P1 approximation (according to Eq. (69)) gives us

$$L(z,\boldsymbol{\omega}) = \frac{1}{4\pi}[G(z) - D\, \partial_z\, G(z)\, \cos(\theta)] \tag{90}$$

where

$$\partial_z\, G(z) = 4\, q_\cap\, C\, \xi \left(-\exp[-\xi\, z] + \frac{\mathcal{L}_E\, \xi - 1}{\mathcal{L}_E\, \xi + 1}\, \exp[-\xi\,(2E - z)] \right) \tag{91}$$

Figs. 16 and 17 represent respectively the irradiance field and the angular distribution of the intensity[7] obtained with the P1 approximation. Although the situation under study is far from equilibrium, the irradiance field produced by the approximation is in good agreement with the reference solution. This correspondence is surprising because P1 should be unsuitable for such a situation with intermediate optical thickness. The next paragraph is focused on the validity conditions of the P1 approximation.

[7] In the angular distributions presented in Fig. 17, we use the same scale for the P1 approximation and the Monte Carlo method. Note that the area under the curve does not represent the irradiance because the element of solid angle $\sin(\theta)d\theta\, d\varphi$ is not taken into consideration here: $G = 2\pi \int_0^\pi d\theta\, \sin(\theta) L(\theta)$.

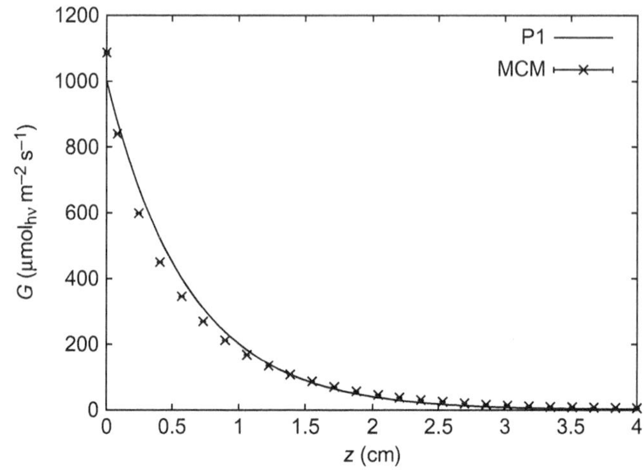

Figure 16 The irradiance field G within the photobioreactor shown in Fig. 6; $\rho^{\mathcal{F}} = 0$, $\rho^{\mathcal{R}} = 0.54$, and Lambertian incidence. Comparison between the P1 approximation (Eq. (88)) and the reference solution (Monte Carlo method, MCM).

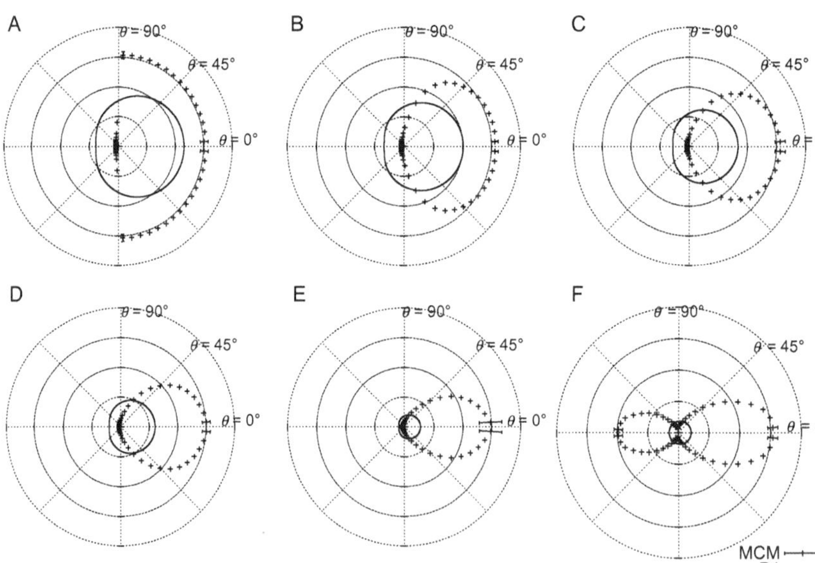

Figure 17 Angular distribution of the intensity $L(z_0,\theta)$ at the abscissa z_0 within the photobioreactor shown in Fig. 6; $\rho^{\mathcal{F}} = 0$, $\rho^{\mathcal{R}} = 0.54$, and Lambertian incidence. Comparison between the P1 approximation (Eq. (90)) and the reference solution (Monte Carlo method, MCM). (A) $z_0 = 0$. (B) $z_0 = 2.5$ mm. (C) $z_0 = 5$ mm. (D) $z_0 = 1$ cm. (E) $z_0 = 2.5$ cm. (F) $z_0 = 4$ cm.

3.4.3 Validity Conditions of the P1 Approximation

These conditions are frequently defined as

$$e_s^* \gg 1$$

where e_s^* is the scattering optical thickness defined in Eq. (43). In the situation studied in Fig. 16, $e_s^* = 1.1$, and the approximation already works well. If we now address the same situation but replace the Lambertian illumination with a collimated source (the situation corresponding to Fig. 18), then the approximation does not work at all. In these two configurations, optical thickness has the same value, and yet the P1 approximation works well in one case but not in the other. In the text later, we explore the validity conditions of the P1 approximation, and the results will lead to a strategy for analysis of collimated illumination.

The P1 approximation postulates the functional form

$$L(\mathbf{x},\boldsymbol{\omega}) = A(z)[1 + C(z)\cos(\theta)]$$

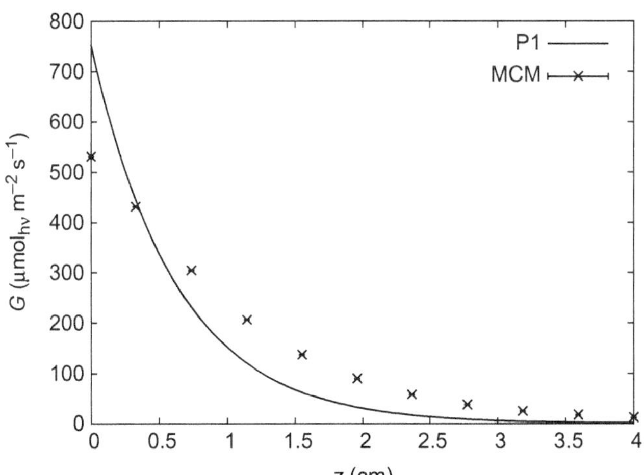

Figure 18 The irradiance field G within the photobioreactor shown in Fig. 6, with the same parameters as in Fig. 16, but the Lambertian emission is replaced by collimated emission at $\theta_i = 0$. Comparison between the P1 approximation (Eq. (88)) and the reference solution (Monte Carlo method, MCM). For collimated incidence, only the boundary condition at $z = 0$ is modified, in comparison with the solution used in Fig. 16. We still have $q^{(0)}(z=0) = q_\cap$, but the ballistic irradiance becomes $G^{(0)}(z=0) = q_\cap/\mu_i$. Therefore, the same solution as in Fig. 16 can be used, but with replacement of $4q_\cap$ with $(2 + 1/\mu_i)q_\cap$ in Eq. (88).

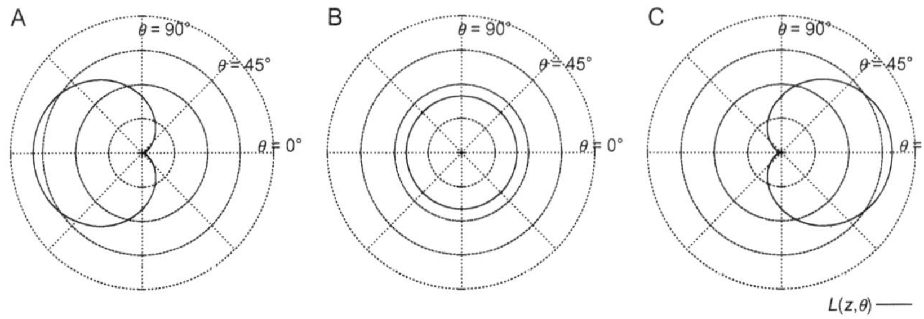

Figure 19 Angular distribution of the intensity for the P1 approximation: $L(z,\theta) = A(z)[1 + C(z)\cos(\theta)]$, where $C(z) \in [-1,1]$. (A) $C(z) = -1$, (B) isotropic distribution $C(z) = 0$, and (C) $C(z) = 1$.

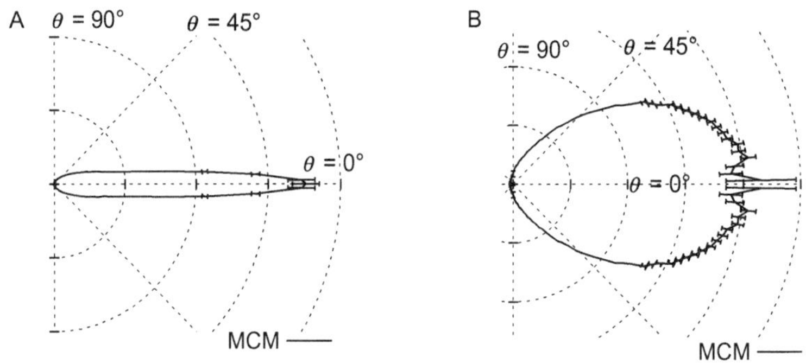

Figure 20 Angular distribution of the intensity $L(z,\theta)$ within the photobioreactor shown in Fig. 6. The results were obtained by the Monte Carlo method (A) for collimated normal incidence and (B) for Lambertian incidence (diffuse illumination).

for the intensity L (ie, Eq. (69)). It is therefore valid if the intensity can be represented by this functional form: this is the only strict definition that can be formulated for validity of the P1 approximation. This form corresponds to situations where the intensity is close to isotropy (ie, near-equilibrium situations.[8]) In fact, in the above equation, $C(z) \in [-1,1]$ because otherwise the intensity may be negative. Thus, we see in Fig. 19 that the range of angular distributions resulting from the P1 approximation is not compatible with the description of a photobioreactor during collimated illumination (see Fig. 20A). The figure shows the angular distribution of the intensity as a function of the boundary conditions: for Lambertian emission, light enters the medium from all directions, resulting in intensity that is much closer to

[8] We will remind readers that here, the angular distribution of the intensity in question is completely different from the angular distribution of the phase function most of the time.

isotropy, in comparison with collimated incidence. We thus understand that the P1 approximation allows us to analyze Lambertian emission (Section 3.4.2), even in a situation far from equilibrium. We also see that this approximation fails in the case of collimated incidence in Fig. 18. The condition $e_s^* \gg 1$ (mentioned earlier) usually ensures that the intensity is close to isotropy (because of a large number of isotropic scattering events), but the P1 approximation is actually less restrictive: it is sufficient that the intensity is compatible with Eq. (69), regardless of the scattering phenomenon. This situation will allow us to develop an approach to analysis of collimated illumination phenomena in the paragraphs that follow.

3.4.4 The Case of Collimated Illumination: Separation of Ballistic and Diffusive Contributions

3.4.4.1 Separation Between Ballistic and Scattered Photons

In Section 3.3, Fig. 13, we saw that the equivalent transport problem allows us to separate our radiative study into two simple systems: the ballistic photons, for which the exact solution is analytical, and the scattered photons, which correspond to intensity close to isotropy. This relative isotropy of the scattered intensity in the equivalent transport problem suggests that the P1 approximation is relevant. Therefore, to formulate the collimated incidence phenomena, we will address the equivalent transport problem and separate the analysis of ballistic photons from that of scattered photons: only scattered photons will be subjected to the P1 approximation. In the rest of the chapter, the ballistic population is denoted as (0), whereas the scattered photons will be called "the diffuse population" and denoted as (*d*).

It is always possible to formulate the irradiance of the entire photon population as the sum of ballistic irradiance and diffuse irradiance (as is the case for the intensity in Eq. (44):

$$G(z) = G^{(0)}(z) + G^{(d)}(z) \tag{92}$$

where $G^{(d)}$ is defined as the sum of the irradiance values for all scattering orders $j \geq 1$, $G^{(d)}(z) = \sum_{j=1}^{\infty} G^{(j)}(z)$ (see Section 3.3.1). The rigorous solution for the ballistic irradiance $G^{(0)}$ is easy to obtain: we already did so in the context of the single-scattering approximation, in Eq. (66). The diffuse irradiance $G^{(d)}$ is the solution to a radiative transfer problem in which the source $S_{G^{(d)}}$ is isotropic and distributed throughout all the reaction volume: this source represents ballistic photons that are scattered for the first time in the medium,[9] exactly as in Section 3.3. Therefore, the diffusion equation for $G^{(d)}$ is the same as in Eq. (71) but with addition of the source term $S_{G^{(d)}}$:

$$-D\,\partial_z^2 G^{(d)}(z) = -k_a\, G^{(d)}(z) + S_{G^{(d)}}(z) \tag{93}$$

where the macroscopic diffusion coefficient D and the absorption coefficient k_a are identical to those used in Section 3.4.2 for the analysis of Lambertian emission. This is because they are invariant with the transformation Eqs. (72) and (40).

In the text later, we address the equivalent transport problem defined by α_s^*, k_{ext}^*, and $p_\Omega^*(\omega|\omega') = \frac{1}{4\pi}$ (see Section 3.2), and we analyze the one-dimensional configuration shown in Fig. 6 with collimated incidence at $z = 0$ and a non-reflecting surface at $z = E$. We use the solution obtained in Section 3.3 for ballistic irradiance (according to Eq. (66)):

$$G^{(0)}(z) = \frac{q_\cap}{\mu_i} \exp\left(-k_{ext}^* \frac{z}{\mu_i}\right) \tag{94}$$

where $\mu_i = \cos(\theta_i)$ is the cosine of the angle of incidence. Then, we address the diffuse irradiance $G^{(d)}$ by solving the diffusion equation (Eq. (93)) where the source term is

$$S_{G^{(d)}}(z) = \int_{4\pi} d\omega\, \mathcal{C}^{(0)}(z) = 4\pi\, \mathcal{C}^{(0)}(z) = \alpha_s^*\, k_{ext}^* \frac{q_\cap}{\mu_i} \exp\left(-k_{ext}^* \frac{z}{\mu_i}\right) \tag{95}$$

where $\mathcal{C}^{(0)}(z)$ is the mesoscopic source term discussed in Section 3.3 (see Eq. (59)). Compared to Section 3.4.2, here, we replaced the sources at the boundaries (responsible for the strong anisotropy of the intensity) by an isotropic source distributed throughout the entire volume. The general solution that satisfies the diffusion equation (Eq. (93)) is

$$G^{(d)}(z) = \frac{q_\cap}{\mu_i}\left[C_0 \exp(\xi z) + C_1 \exp(-\xi z) - \frac{\xi^2 - k_{ext}^*/D}{\xi^2 - (k_{ext}^*/\mu_i)^2} \left(\exp\left[-k_{ext}^* \frac{z}{\mu_i}\right] - \exp[-\xi z] \right) \right] \tag{96}$$

where C_0 and C_1 are constants, and $\xi = \sqrt{k_a/D}$ (in the configuration under study $\xi \simeq 161$).

[9] These scattering events are isotropic because the phase function is isotropic in the equivalent transport problem as defined in Section 3.2.

3.4.4.2 Boundary Conditions

Here, the ballistic photons are analyzed separately; therefore, $G^{(0)} = 0$ and $q^{(0)} = 0$ in Eqs. (75) and (76) (there is no source at the boundary for the diffuse population). In addition, we ignore reflectivity (ie, $\rho = 0$); thus, we have the boundary conditions as follows:

- At $z = 0$,

$$G^{(d)}(0) - \mathcal{L}_0\, \partial_z\, G^{(d)}(0) = 0 \tag{97}$$

- At $z = E$,

$$G^{(d)}(E) + \mathcal{L}_E\, \partial_z\, G^{(d)}(E) = 0 \tag{98}$$

where

$$\mathcal{L}_0 = \mathcal{L}_E = \frac{2/3}{k^*_{ext}} \tag{99}$$

Constants C_0 and C_1 in Eq. (96) that satisfy the boundary conditions (Eqs. (97) and (98)) are

$$C_1 = -\frac{\xi^2 - k^*_{ext}/D}{\xi^2 - (k^*_{ext}/\mu_i)^2} \times \frac{\mathcal{L}_0(k^*_{ext}/\mu_i - \xi)\dfrac{1+\mathcal{L}_E\,\xi}{1-\mathcal{L}_0\,\xi} + \exp[-\xi\, 2E]\left\{1 - \mathcal{L}_E\,\xi - \left(1 - \dfrac{\mathcal{L}_E\, k^*_{ext}}{\mu_i}\right)\exp\left[(\xi - k^*_{ext}/\mu_i)E\right]\right\}}{\exp[-\xi\, 2E](1-\mathcal{L}_E\,\xi) - (1+\mathcal{L}_0\,\xi)\dfrac{1+\mathcal{L}_E\,\xi}{1-\mathcal{L}_0\,\xi}} \tag{100}$$

and

$$C_0 = -C_1\,\frac{1+\mathcal{L}_0\,\xi}{1-\mathcal{L}_0\,\xi} - \mathcal{L}_0\,\frac{\xi^2 - k^*_{ext}/D}{(\xi + k^*_{ext}/\mu_i)(1-\mathcal{L}_0\,\xi)} \tag{101}$$

Finally, the irradiance $G(z)$ is obtained by adding up the ballistic irradiance (Eq. (94)) and the diffuse irradiance (Eq. (96)):

$$G(z) = \frac{q_\cap}{\mu_i}\left[\exp\left(-k^*_{ext}\frac{z}{\mu_i}\right) + C_0\exp(\xi\, z) + C_1\exp(-\xi\, z) - \frac{\xi^2 - k^*_{ext}/D}{\xi^2 - (k^*_{ext}/\mu_i)^2}\left(\exp\left[-k^*_{ext}\frac{z}{\mu_i}\right] - \exp[-\xi\, z]\right)\right] \tag{102}$$

From the mesoscopic point of view, the P1 approximation yields the following diffuse intensity:

$$L^{(d)}(z,\boldsymbol{\omega}) = \frac{1}{4\pi}\left[G^{(d)}(z) - D\,\partial_z\,G^{(d)}(z)\,\cos(\theta)\right] \tag{103}$$

The total intensity L is the sum of $L^{(d)}$ and of the contribution of ballistic photons, that is, a Dirac distribution centered at the incident direction.

Figs. 21 and 22 show respectively the irradiance field and the angular distribution of the diffuse intensity obtained with the P1 approximation of the equivalent transport problem. Fig. 21B shows comparison with the reference solution. As we expected in the previous paragraph, the agreement here is significantly improved in comparison with Fig. 18. On the one hand, the solution for the ballistic irradiance is exact; on the other hand, the description of the diffuse population is now compatible with restrictions of the P1 approximation. The angular distributions of the diffuse intensity are compared with the results obtained for the single-scattering approximation in Fig. 15. We show this comparison because these are the two solutions that we obtained for the equivalent transport problem, but these formulae cannot serve as a reference solution. We see, however, that the single-scattering approximation is more likely to describe the phenomena at the boundary because it takes into account the discontinuity phenomena of the intensity angular distribution, whereas the P1 approximation requires

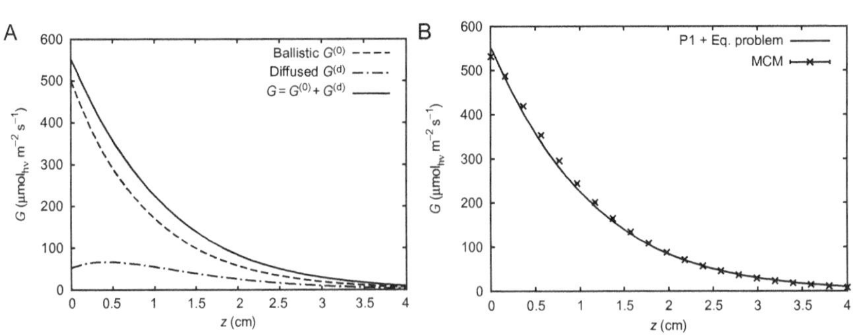

Figure 21 The irradiance field G within the photobioreactor shown in Fig. 6; $\rho^{\mathcal{F}} = \rho^{\mathcal{R}} = 0$ and collimated normal incidence $\mu_i = 1$. The results were obtained by means of the P1 approximation of the equivalent transport problem where $\alpha_s^* = 0.25$, $k_{ext}^* = 110\ \mathrm{m}^{-1}$, and $p_\Omega^* = 1/4\pi$: the expression for $G^{(0)}$ is provided in Eq. (94), the expression for $G^{(d)}$ is shown in Eq. (96), and the total irradiance $G = G^{(0)} + G^{(d)}$ is shown in Eq. (102). (A) The proportions of ballistic and diffused photons. (B) Comparison with the reference solution obtained by the Monte Carlo method (MCM) for $\alpha_s = 0.86$, $k_{ext} = 587\ \mathrm{m}^{-1}$, and the phase function of *Chlamydomonas reinhardtii* (see Section 4).

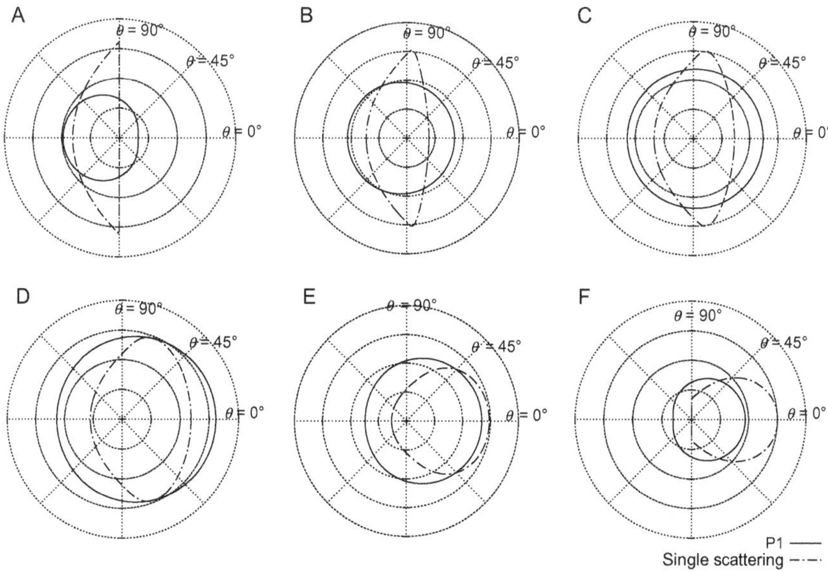

Figure 22 Angular distribution of the diffuse intensity $L^{(d)}(z_0,\theta)$ at the abscissa z_0 within the photobioreactor shown in Fig. 6; $\rho^{\mathcal{F}} = \rho^{\mathcal{R}} = 0$; collimated normal incidence. The results were obtained for the equivalent transport problem where $\alpha_s^* = 0.25$, $k_{ext}^* = 110\ \mathrm{m}^{-1}$, and $p_{\Omega}^* = 1/4\pi$. Comparison between the P1 approximation and the single-scattering approximation (for which $L^{(d)} \simeq L^{(1)}$, see Section 3.3). (A) $z_0 = 0$. (B) $z_0 = 2.5$ mm. (C) $z_0 = 5$ mm. (D) $z_0 = 1$ cm. (E) $z_0 = 2.5$ cm. (F) $z_0 = 4$ cm.

spherical-harmonic expansion. Inside the reaction volume, the two approximations yield the diffuse intensity close to isotropy.

To sum up, the P1 approximation was shown to efficiently model the irradiance fields in typical flat-plate photobioreactor configurations with intermediate optical-thickness values. Yet the validity condition of the P1 approximation is often associated with high scattering optical thickness (and with low absorption) because this criterion ensures situations near equilibrium. Nonetheless, P1 requires only that the angular dependence of the intensity can be formulated as the cosine of the propagation angle. Using a well-known invariance property of transport (in order to construct an equivalent transport problem), we proposed an approach enabling the use of the P1 approximation in a relevant manner in studies on photobioreactors, even for collimated incidence.

3.5 Two-Flux Approximation

This approximation is widely used for analytical purposes in spectroscopy (where it is called the Kubelka–Munk theory), astrophysics, and photobioreactor engineering. Its implementation and capabilities in the context of photobioreactor research are detailed in Pruvost and Cornet (2012), Cornet et al. (1992b, 1994, 1995), Cornet and Dussap (2009), Cornet (2010), Takache et al. (2010, 2012), Farges et al. (2009), and Pottier et al. (2005). The two-flux approximation is a macroscopic approximation of radiative transfer in the sense that it formulates a set of equations for description of the irradiance and flux density. Its advantage over other macroscopic approximations is that the assumption of near-equilibrium is not required. As in the case of the P1 approximation, the angular distribution of the intensity L is fixed *a priori*, but here, L does not have to be close to isotropy. Indeed, the functional form is fixed independently for forward and backward hemispheres, and this situation yields discontinuity at $\theta = \pi/2$ (see Fig. 23). The approximation was originally developed under the assumption of isotropic intensity across both hemispheres (Schuster, 1905) (see Fig. 23A) and was then formulated for collimated intensity (Hottel and Sarofim, 1967) (Dirac distributions in the forward and backward directions) and later extended to the intermediate anisotropic situation (eg, see Meng and Viskanta, 1983). In the text later, we focus on a recent generalization (Cornet, 2010; Cornet and Dussap, 2009; Pruvost and Cornet, 2012) (developed in the context of photobioreactor engineering) that consists of the following functional form for the intensity, when applied to the configuration shown in Fig. 6:

$$\begin{aligned} L(\mathbf{x},\boldsymbol{\omega}) &= A^+(z)\left|\cos^{n^+}(\theta)\right| \text{for } \theta \in [0,\pi/2] \\ L(\mathbf{x},\boldsymbol{\omega}) &= A^-(z)\left|\cos^{n^-}(\theta)\right| \text{for } \theta \in [\pi/2,\pi] \end{aligned} \qquad (104)$$

where $A^+(z)$ and $A^-(z)$ are functions of z, and the value of the parameters n^+ and n^- determines the form of the angular distribution ($n = 0$ for isotropic intensity and $n \to \infty$ for collimated intensity). These parameters have to be fixed *a priori* and allow us to assume a wide range of the angular distribution with the same formula, as shown in Fig. 23B. For example, different values can be chosen for n^+ and n^-, leading to different assumptions for the angular distributions in the forward and backward hemispheres: Fig. 23C represents the case of an isotropic distribution for one hemisphere ($n^- = 0$), and collimated for the other ($n^+ \to \infty$). Nevertheless, we observed very low sensitivity of the results to the value of n^- for typical photobioreactor configurations. For this reason, we restrict the rest of our study to situations with $n^+ = n^- = n$. Unlike with P1, the angular distribution here is identical

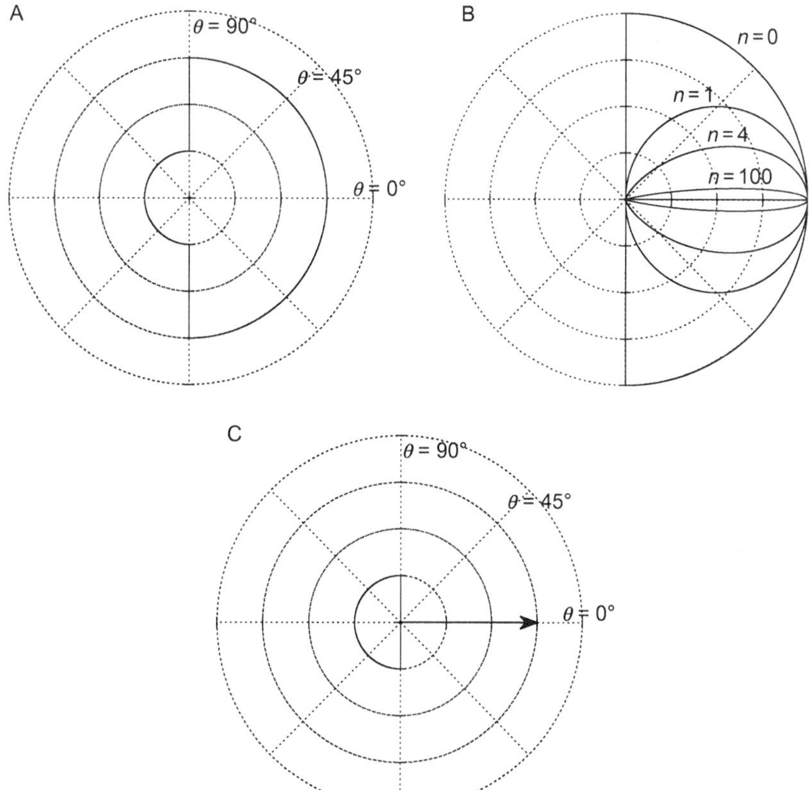

Figure 23 The two-flux approximation: different angular distributions of the intensity that can be postulated. (A) Isotropic distribution for each hemisphere. (B) Distribution (Eq. (104)) for $\theta \in [0,\pi/2]$ at different values of n. (C) Collimated distribution for one hemisphere and isotropic distribution for the other.

within the whole reaction volume (n^+ and n^- are independent of the location) with discontinuity at the junction between the two hemispheres. On the basis of Fig. 17, we should note that this discontinuity may be justified at the boundary of the system, but within the volume, this assumption is plausible only for optically thin media (ie, $e_s^* \ll 1$), or equivalently, for locations close to the boundaries.

Substituting Eq. (104) into the radiative transfer equation Eq. (24) (in which we omit the frequency variable) and integrating over propagation directions $\boldsymbol{\omega}$ (ie, over the propagation angles θ), we obtain the following equation for the irradiance within the photobioreactor shown in Fig. 6 with $\rho^{\mathcal{F}} = 0$ (Pruvost and Cornet, 2012):

$$G(z)=\frac{q_\cap}{\mu_i}2\,\frac{n+2}{n+1}\times$$
$$\frac{\rho^{\mathcal{R}}(1+\alpha)\,\exp(-\delta E)-(1-\alpha)\exp(-\delta(E-z))+\left[(1+\alpha)\exp(\delta E)-\rho^{\mathcal{R}}(1-\alpha)\exp(\delta E)\right]\exp(-\delta z)}{(1+\alpha)^2\,\exp(\delta E)-(1-\alpha)^2\,\exp(-\delta L)+\rho^{\mathcal{R}}(1-\alpha)^2[\exp(-\delta E)-\exp(\delta E)]} \tag{105}$$

where

$$\alpha=\sqrt{\frac{\sigma_a}{\sigma_a+2b\,\sigma_s}} \tag{106}$$

$$\delta=\frac{\alpha}{\mu_i}\,C_x\,(\sigma_a+2b\,\sigma_s) \tag{107}$$

where μ_i is the cosine of the incidence angle θ_i, b is the backscattering coefficient

$$b=2\pi\int_{\pi/2}^{\pi}p(\theta_s)\sin(\theta_s)d\theta_s \tag{108}$$

ie, the integral of the phase function p over backscattering directions (see Section 2 and Fig. 2), $b=0.008$ for the phase function in question, and all other notations are defined in Fig. 6.

In Fig. 24, the irradiance field obtained with the two-flux approximation for $n\rightarrow\infty$ is compared with the Monte Carlo reference solution in the case of collimated solar-light incidence. The two-flux approximation will be used in Section 5.6 to analyze the coupling between radiative transfer and photosynthesis thermokinetics in photobioreactors with simple geometric structure.

3.6 Implementation of the Analytical Approximate Solutions Developed in this Section for the Field of Specific Absorption Rate $\mathcal{A}$

During the previous radiative-transfer analysis, the following approximate solutions were obtained for the irradiance field within the typical photobioreactor configuration in Fig. 6:

1. The single-scattering approximation applied to the equivalent transport problem for analysis of collimated illumination in the case of nonreflecting surfaces: Eqs. (65) to (68).
2. The P1 approximation for analysis of diffuse illumination: Eq. (88).

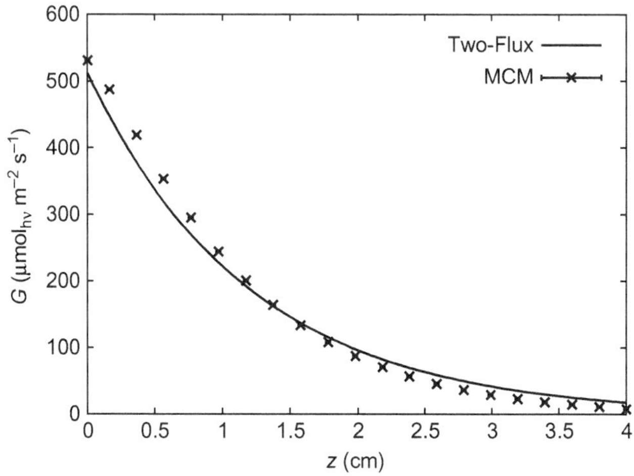

Figure 24 The irradiance field G in the photobioreactor shown in Fig. 6; $\rho^{\mathcal{F}} = \rho^{\mathcal{R}} = 0$ and collimated normal incidence $\mu_i = 1$. Comparison between the two-flux approximation (Eq. (105)) at $b = 0.008$ and the reference solution obtained by the Monte Carlo method (MCM; see Section 4).

3. The P1 approximation applied to the equivalent transport problem for analysis of collimated illumination in the case of non-reflecting surfaces: Eq. (102).
4. The two-flux approximation: Eq. (105).

These solutions can be used in various ways to calculate the field of specific absorption rate $\mathcal{A}$:

- In the numerical calculations presented in the figures of this section, we used the approximation of an equivalent gray medium (see Eqs. (35) and (36)). In this case, the gray radiative properties are used directly in the expressions, and the variable $q_\cap$ is the value of the incident surface flux density (integrated over PAR). Finally, the solution for the irradiance is simply multiplied by the gray absorption cross section to obtain the specific rate of photon absorption $\mathcal{A} = \sigma_a \, G$ (see the discussion at the end of Section 3.1). This approach allows us to obtain simple analytical solutions appropriate for such analysis.
- On the other hand, the spectral integration can also be analyzed. In this case, the spectral radiative properties are used in the expressions, and the variable $q_\cap$ assumes the value of the spectral distribution $q_{\cap,\nu}$ of the incident flux density (see Eq. (25)). This situation leads to an expression for the spectral distribution G_ν of the irradiance, which is multiplied by the

spectral absorption cross section and integrated over PAR in order to obtain the specific rate of photon absorption: $\mathcal{A} = \int_{\nu_{\min}}^{\nu_{\max}} \sigma_{a,\nu} G_{\nu} \, d\nu$ (see Eq. (31)).

In both of the earlier cases, $\mathcal{A}$ and G (according to Section 3.1) can be expressed

- either in moles of photons per second if the variable $q_{\cap}$ is expressed in moles of photons per second,
- or in Watts if the variable $q_{\cap}$ is expressed in Watts.

It should be noted that the approximations explored in this section can also be used to obtain analytical solutions for one-dimensional cylindrical configurations (eg, see Pruvost and Cornet, 2012 regarding the case of the two-flux approximation). Moreover, in the case of solutions 1 and 3 in the earlier list, we chose to focus on non-reflecting surfaces in order to simplify the mathematical expressions. These approximations, however, are not restricted to non-reflecting surfaces. For example, when extended to reflecting surfaces on both sides, solution 3 is still analytical.

It is also important to note that due to the linearity of the radiative transfer equation, the solutions for configurations illuminated on both sides (or for mixtures of collimated and diffuse illumination) are obtained simply by adding up the solutions obtained in this section. For example, for incident solar radiation with direct and diffuse components, the radiation field can be obtained by adding up solutions 2 and 3. For a photobioreactor illuminated on both sides, the radiation field is $G(z) = \widetilde{G}(z) + \widetilde{G}(E - z)$, where $\widetilde{G}$ is any solution obtained for an emitting surface at $z = 0$. Again, these solutions can be linearly combined in order to analyze a photobioreactor illuminated on both sides by a mixture of collimated and diffuse radiation.

4. NUMERICAL IMPLEMENTATION OF PHOTOBIOREACTOR MODELS BY THE MONTE CARLO METHOD, INCLUDING RIGOROUS SOLUTION OF THE RADIATIVE TRANSFER EQUATION FOR COMPLEX GEOMETRIC STRUCTURE

Since Metropolis' original work in 1949 (Metropolis and Ulam, 1949), numerous monographs and review articles have been devoted to the Monte Carlo method. In the present study, we are concerned both with simulation of a linear transport phenomenon (namely radiative transfer) and with a solution to our integral model for a photobioreactor (see Section 1). Here, we arbitrarily chose to point out Hammersley and Handscomb's book

(Hammersley and Handscomb, 1964) because of the everlasting influence of this short synthesis on this area of research as well as Howell's review (Howell, 1998) because of its proximity to our more specific engineering-application concerns in the field of radiative transfer. These texts provide a sufficient theoretical framework for most of the algorithms encountered in photobioreactor research and may serve as a meaningful starting point for any further bibliographic research.

Among more recent methodological advances, simulation of nonlinear processes, sensitivity estimation, and the zero-variance concept are discussed in Section 4.3, Section 4.4, and Delatorre et al. (2014) and Dauchet et al. (2013), respectively, because they can be at least partially translated into simple systematic procedures for simulation and analysis of photobioreactor models. They rely on explicit definition of the strict relation between a linear Monte Carlo algorithm and an integral formulation. Indeed, the Monte Carlo method is, above all, a numerical approach to solving integrals. In our context, this approach implies that the method is not only pertinent to simulation of radiative transfer (it is generally thought to be the reference method for solution of the radiative transfer equation) but also very well suited for solving our photobioreactor model, which is based on integral formulations. Let us briefly illustrate this relation between the Monte Carlo algorithms and integral formulations with a simple example that does not imply radiative transfer: estimation of the average rate of biomass production $< r_x >$ in a photobioreactor, as formulated in Eq. (1) (Section 1):

$$< r_x >= \int_{\mathcal{V}} \frac{1}{V} r_x(\mathbf{x}_0)\, d\mathbf{x}_0$$

where the local rate r_x is assumed to be known at any location $\mathbf{x}_0$ (for illustration purposes). This integral formula can be interpreted statistically as the expectation of the random variable $W = \hat{w}(\mathbf{X}_0) = r_x(\mathbf{X}_0)$, where $\mathbf{X}_0$ is a random location within $\mathcal{V}$, with uniform probability density function $p_{\mathbf{X}_0} = \frac{1}{V}$:

$$< r_x >= \int_{\mathcal{V}} p_{\mathbf{X}_0}\, \hat{w}(\mathbf{x}_0)\, d\mathbf{x}_0 \tag{109}$$

The corresponding Monte Carlo algorithm consists of sampling N independent realizations w_1 and $w_2 \cdots w_N$ of the random variable W by repeating N times the following sampling procedure (where $i = 1, 2, \ldots N$):

Step (1) A location, $\mathbf{x}_0$, is sampled within $\mathcal{V}$ according to the probability density function $p_{\mathbf{X}_0} = \frac{1}{V}$ (ie, uniform sampling).

Step (2) The weight w_i is calculated according to $w_i = r_x(\mathbf{x}_0)$.

Then, $< r_x >$ is estimated as

$$< r_x > \simeq \tilde{b}_N = \frac{1}{N}\sum_{i=1}^{N} w_i \tag{110}$$

Because realizations w_i are independent, meaningful statistical uncertainty (evaluation of the standard deviation of the estimator) is systematically available as

$$\tilde{\sigma}_N = \frac{1}{\sqrt{N-1}}\sqrt{\left(\frac{1}{N}\sum_{i=1}^{N} w_i^2\right) - \tilde{b}_N^2} \tag{111}$$

that is, directly related to the numerical error. In general terms, during analysis of the physical quantity B (in our example $B = < r_x >$), any approximation $\tilde{b}_N$ of B corresponding to a linear Monte Carlo algorithm involving N sampled events is constructed as Eq. (110), with the statistical uncertainty Eq. (111). The events can be simple, as in our example. In contrast, as illustrated in Section 4.1, the events rapidly become quite complex as soon as radiative transfer in multiple-reflection and multiple-scattering configurations is simulated. In all cases, however, the reason why the Monte Carlo method is so popular is its intuitive nature: in the earlier example, the average production rate $< r_x >$ is estimated simply as the average of N local production rates evaluated at uniformly sampled locations. The method is nonetheless mathematically rigorous: the meaning of the integral formulation Eq. (109) is that when $N \rightarrow +\infty$, the estimator $\tilde{b}_N$ evaluates $< r_x >$ as the expectation of the random variable W. When we simulate radiative transfer (ie, when B is a radiative quantity), the events are more complex, but these advantages (mathematical rigor and the ease of understanding) are preserved: the integral solution of the radiative transfer equation is estimated by "tracing photon trajectories" in the photobioreactor.

In the earlier example, the local production rate r_x was assumed to be known for the purposes of illustration. Nevertheless, as stated in Section 1 (and detailed in Section 5), r_x is a function of the specific rate of photon absorption $\mathcal{A}$. This is why photobioreactor studies require solution of the radiative transfer equation prior to estimation of the production rate $< r_x >$. A radiative-transfer Monte Carlo algorithm for rigorous estimation of $\mathcal{A}$ is presented in Section 4.1, and its implementation for complex geometric structure is discussed in Section 4.2. Based on this algorithm, the estimation of the production rate $< r_x >$ of photobioreactors is addressed in Section 4.3. Finally, in Section 4.4, we briefly explore the expected benefits of sensitivity estimation, in relation to the analysis and optimization of the process.

Monte Carlo integral-formulations such as Eq. (109) lie at the root of the work that is presented later. Nevertheless, we chose to avoid the details of these kinds of formulations in the text that follows because we believe that they are beyond the scope of the present book: integral formulae will be used for illustrative purposes only. Readers wishing to explore the formal basis of our work more deeply are invited to read Delatorre et al. (2014) and Dauchet et al. (2013).

4.1 An Algorithm for Evaluating the Specific Rate of Photon Absorption

In this section, we present a Monte Carlo algorithm for estimation of the specific photon absorption rate $\mathcal{A}(\mathbf{x}_0)$ at any location $\mathbf{x}_0$ within any photobioreactor's reaction volume confined by two diffuse-reflective surfaces ($\mathcal{R}$ and $\mathcal{F}$) with uniform reflectivity $\rho^{\mathcal{R}}$ and $\rho^{\mathcal{F}}$, respectively, where $\mathcal{F}$ is Lambertian emitting with uniform surface flux density $q_{\cap,\nu}$ and $\mathcal{R}$ is non-emitting. Let us recall the definition of $\mathcal{A}(\mathbf{x}_0)$ from Section 3.1:

$$\mathcal{A}(\mathbf{x}_0) = \int_{\nu_{\min}}^{\nu_{\max}} d\nu \int_{4\pi} d\boldsymbol{\omega}_0 \; \sigma_{a,\nu} \, L_\nu(\mathbf{x}_0, -\boldsymbol{\omega}_0) \tag{112}$$

where $L_\nu(\mathbf{x}_0, -\boldsymbol{\omega}_0)$ is the intensity at $\mathbf{x}_0$ in the direction $-\boldsymbol{\omega}_0$ at frequency ν. You may recall that the integral over the reaction volume was translated into a location-sampling procedure in our introductory example. In exactly the same way, the Monte Carlo algorithm for estimation of Eq. (112) starts with the sampling of a frequency, that is, with translation of the integral over PAR $[\nu_{\min}, \nu_{\max}]$, followed by sampling of a direction ($\boldsymbol{\omega}_0$), that is, translation of the integral over the total solid angle 4π. Then, in order to estimate $L(\mathbf{x}_0, -\omega_0)$, we design a *reverse* Monte Carlo procedure consisting of sampling of a multiple scattering and reflection optical path starting from $\mathbf{x}_0$ with direction ω_0 until it is "absorbed" at the emitting surface $\mathcal{F}$ (ie, the optical paths are sampled backward). The sampling procedure is detailed next and illustrated in Fig. 25 with two examples: the one-dimensional

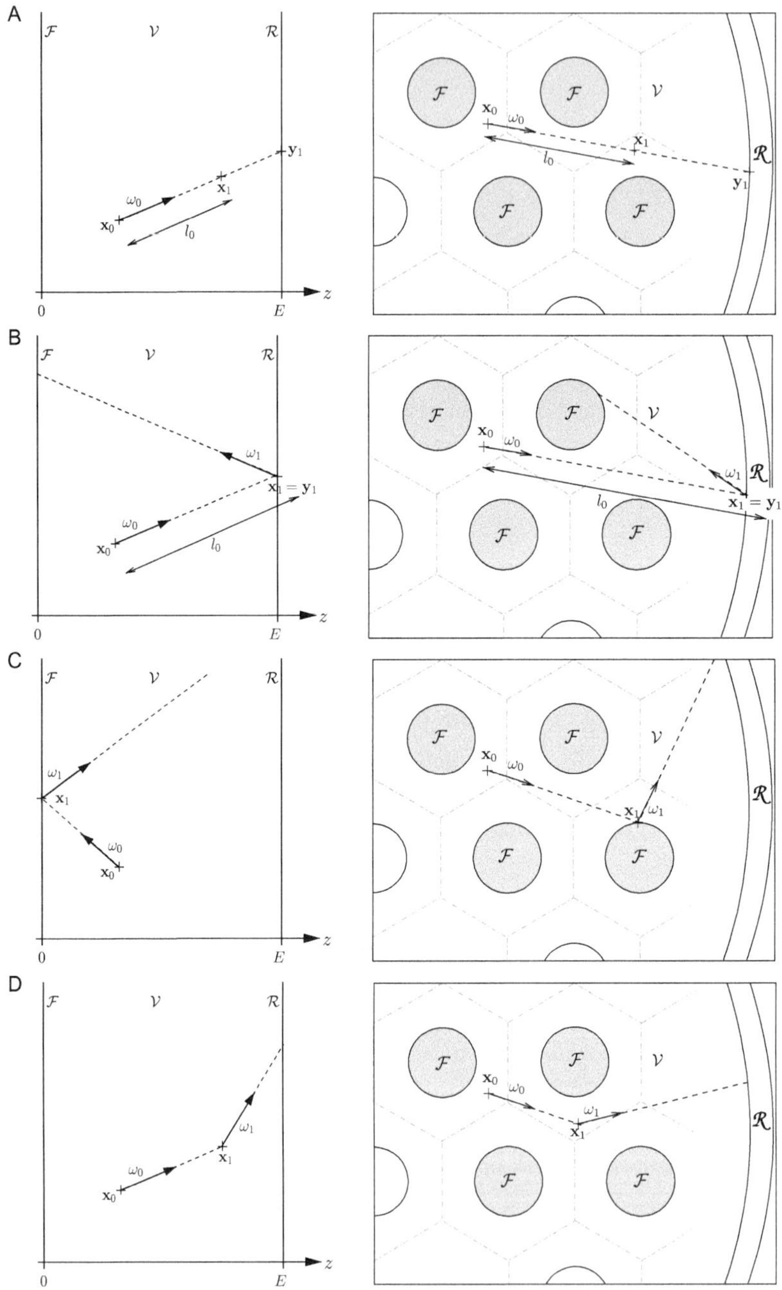

Figure 25 Illustration of the Monte Carlo algorithm presented in Section 4.1 for evaluation of the local rate of photon absorption $\mathcal{A}(\mathbf{x}_0)$ at any location $\mathbf{x}_0$ within the culture volume: (left panel) the case of the one-dimensional Cartesian configuration shown in Fig. 6 and (right panel) the case of the DiCoFluV photobioreactor presented in Fig. 26.

photobioreactor from Fig. 6 and a prototype of the volumetrically illuminated photobioreactor from Fig. 26. The corresponding integral formulation is reported in Delatorre et al. (2014) and Dauchet et al. (2013).

The sampling procedure:

Step (1) A frequency is sampled across $[\nu_{\text{min}}, \nu_{\text{max}}]$ (PAR) according to the uniform probability density function $p_\nu(\nu) = \frac{1}{\nu_{\text{max}} - \nu_{\text{min}}}$. This frequency determines all the spectral properties for the current optical path: scattering and absorption properties of the reaction volume (ie, the radiative properties calculated in Section 2), reflectivity of surfaces, and surface flux density $q_{\cap,\nu}$ emitted at the surface $\mathcal{F}$.

Step (2) Starting from the location $\mathbf{x}_0$, the first propagation direction ω_0 is sampled across the total solid angle according to the isotropic probability density function $p_{\Omega_0}^{\mathcal{V}}(\omega_0) = \frac{1}{4\pi}$, and the first scattering length l_0 is sampled across $[0, +\infty]$ according to the Bouguer extinction law $p_{L_0,\nu}(l_0; k_{s,\nu}) = k_{s,\nu} \exp(-k_{s,\nu}\, l_0)$, where $k_{s,\nu}$ is the scattering coefficient calculated in Section 2.

Step (3) Now that $\{\mathbf{x}_0, \omega_0, l_0\}$ has been sampled, the first interaction location $\mathbf{x}_1 \equiv \mathbf{x}_1(\mathbf{x}_0, \omega_0, l_0)$ is determined. As discussed in Section 4.2, purely geometric considerations are easily translated into scientific

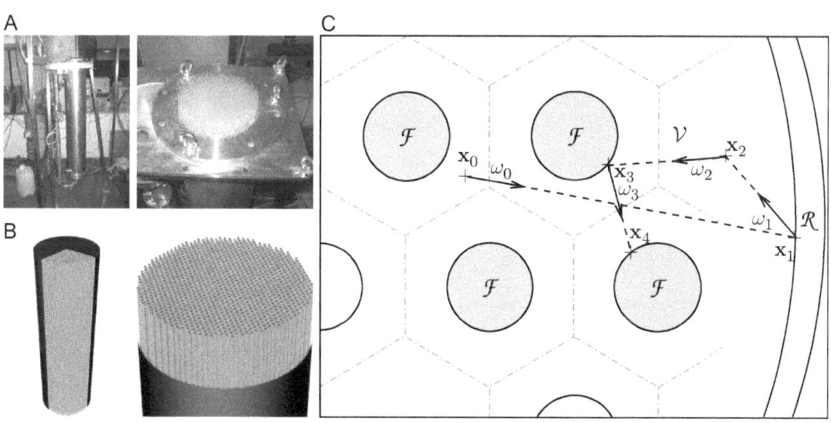

Figure 26 (A) A 25 L prototype of the solar volumetrically illuminated photobioreactor DiCoFluV (Cornet, 2010). (B) EDStar geometric structure: both the reactor ($\mathcal{R}$) and the 979 light-diffusing optical fibers ($\mathcal{F}$) are cylinders 1 m high; the reactor's diameter is 16.5 cm, the distance between two fiber axes is $d_{\mathcal{F}} = 4.8$ mm, and the fiber radius is $r_{\mathcal{F}} = 1.2$ mm. $\mathcal{R}$ and $\mathcal{F}$ are diffuse-reflective with uniform reflectivity $\rho^{\mathcal{R}}$ and $\rho^{\mathcal{F}}$, respectively. $\mathcal{F}$ is Lambertian emitting with the uniform surface flux density $q_{\cap,\nu}$. (C) Two-dimensional hexagonal lattice fiber arrangement; an optical-path example in the culture medium $\mathcal{V}$.

computation libraries. For a given couple, $\{\mathbf{x}_0,\omega_0\}$, such libraries provide us with the location $\mathbf{y}_1 \equiv \mathbf{y}_1(\mathbf{x}_0,\omega_0)$ of the first time the half-line (starting at $\mathbf{x}_0$ in the direction ω_0) intersects the total bounding surface $\mathcal{R} \cup \mathcal{F}$ (see Fig. 25A). If the distance $\|\mathbf{y}_1 - \mathbf{x}_0\|$ to the bounding surface is smaller than the scattering length l_0, then the optical path interacts with the surface (see Fig. 25B); otherwise, scattering occurs inside the volume of culture (Fig. 25A):

$$\mathbf{x}_1 = \begin{cases} \mathbf{y}_1 & \text{if } \| \mathbf{y}_1 - \mathbf{x}_0 \| < l_0 \\ \mathbf{x}_0 + l_0\omega_0 & \text{otherwise} \end{cases}$$

Step (4) A branching test is performed depending on the nature of the interaction:

- In case of an interaction with the non-emitting surface $\mathcal{R}$ of the photobioreactor, the Bernoulli test is performed: a random number, r_1, is uniformly sampled across the unit of interval and
  - if r_1 is less than the surface reflectivity $\rho^{\mathcal{R}}$, then the optical path is reflected (see Fig. 25B): the reflection direction ω_1 is sampled according to the diffuse angular distribution $p^{\mathcal{R}}_{\Omega_1}(\omega_1) = \frac{\omega_1 . \mathbf{n}_1}{\pi}$ ($\mathbf{n}_1$ being the normal at the location $\mathbf{x}_1$), and a new scattering length l_1 is sampled according to the same extinction law as for l_0 ($p_{L_1,\nu} \equiv p_{L_0,\nu}$);
  - otherwise, the optical path sampling procedure is terminated and the weight $\hat{w}_1$ is calculated according to Eq. (113) (in this case, the photon is dissipated at the reflecting surface and therefore $\hat{w}_1 = 0$).
- In case of an interaction with the emitting surface $\mathcal{F}$, a Bernoulli test is performed: a random number, r_1, is uniformly sampled across the unit of the interval and
  - if r_1 is less than the surface reflectivity $\rho^{\mathcal{F}}$, then the optical path is reflected (see Fig. 25C): the reflection direction ω_1 and the scattering length l_1 are sampled as described earlier ($p^{\mathcal{R}}_{\Omega_1} \equiv p^{\mathcal{F}}_{\Omega_1}$);
  - otherwise, the optical-path-sampling procedure is terminated, and the weight $\hat{w}_1$ is calculated according to Eq. (113) (in this case, the optical path contributes to local absorption, and therefore $\hat{w}_1$ is not zero).
- Finally, if $\mathbf{x}_1$ is within the volume of culture $\mathcal{V}$, a scattering direction (ω_1) is sampled according to the single-scattering phase function $p^{\mathcal{V}}_{\Omega_1,\nu}(\omega_1|\omega_0)$ calculated in Section 2, and l_1 is sampled as described earlier (see Fig. 25D).

Step (5) At this stage, if the optical-path-sampling procedure is not terminated, the algorithm loops to (3) and evaluates the next interaction position (the index 1 being incremented to 2, and the index 0 being incremented to 1) and so on until absorption occurs at the surface $\mathcal{R}$ or $\mathcal{F}$.

An example of an optical path sampled by this procedure is shown in Fig. 26C. Altogether, each sampled optical path leads to evaluation of the weight according to the weight function $\hat{w}_j$ (see below), and $A(\mathbf{x}_0)$ is estimated as the average of all weights.

$$\begin{aligned} \hat{w}_j &= 0 && \text{if } \mathbf{x}_j \in \mathcal{R} \\ \hat{w}_j &= (\nu_{\max} - \nu_{\min}) 4\pi\, \sigma_{a,\nu} \frac{q_{\cap,\nu}}{\pi\left(1-\rho^{\mathcal{F}}\right)} e^{-k_{a,\nu}\, d_j} && \text{if } \mathbf{x}_j \in \mathcal{F} \end{aligned} \tag{113}$$

where $[\nu_{\min}, \nu_{\max}]$ is PAR, $\sigma_{a,\nu}$ is the absorption cross section[10] determined in Section 2, $q_{\cap,\nu}$ is the spectral distribution of the surface flux density emitted at $\mathcal{F}$ (see Section 3.1), $\rho^{\mathcal{F}}$ is reflectivity of $\mathcal{F}$, $q_{\cap,\nu}/\pi\left(1-\rho^{\mathcal{F}}\right)$ is the equivalent blackbody intensity for emission at $\mathcal{F}$, and $e^{-k_{a,\nu}\, d_j}$ is the transmission along the optical path (accounting for attenuation due to absorption), with $d_j = \sum_{q=0}^{j-1} \| \mathbf{x}_{q+1} - \mathbf{x}_q \|$ being the total length of the sampled optical path.

The Monte Carlo method is usually preferred to other numerical simulation approaches because of its flexibility (in terms of inclusion of new physical phenomena) and its ability to deal with geometrically complex realistic systems. Next, we briefly illustrate the flexibility of the earlier algorithm; analysis of complex geometric structures is specifically addressed in Section 4.2. Monte Carlo codes for standard radiative transfer can be depicted as close translations of well-established physical situations of photon emission, scattering, reflection, refraction and absorption. Therefore, such codes are indeed easy to design and easy to upgrade toward representation of additional (or more accurate) physical phenomena. For example, the earlier algorithm can be easily modified for analysis of specular-reflective surfaces: in this case, at Step (4), the reflection direction $\boldsymbol{\omega}_1$ is set to the specular direction corresponding to the incident direction $\boldsymbol{\omega}_0$ (see Fig. 10) instead of sampling $\boldsymbol{\omega}_1$ according to the diffuse angular distribution. Extension of our algorithm to heterogeneous illumination is also straightforward (eg, simulation of a tubular photobioreactor): the very same algorithm is used except in

[10] In Section 3, we estimate the local irradiance $G(\mathbf{x}_0)$ by the same algorithm, only replacing $\hat{w}_j$ with the new weight function $\hat{w}_j^G = \hat{w}_j / \sigma_{a,\nu}$.

the weight function (Eq. (113)), the homogeneous surface flux density $q_{\cap,\nu}$ is replaced by the surface flux density $q_{\cap,\nu}(\mathbf{x}_j)$ at location $\mathbf{x}_j$. Among all the possible refinements of our algorithm, the addition of solar-light collection (accounting, for example, for shading and blocking effects or collector orientation) is certainly an interesting perspective. In this case, the optical-path-sampling procedure does not stop at the surface $\mathcal{F}$ but continues (eg, within the atmosphere) according to standard Monte Carlo algorithms for concentrated solar applications, such as those presented in Delatorre et al. (2014).

4.2 Practical Implementation for Complex Geometric Structure

In this section, we address the practical question of numerically implementing the earlier algorithm within photobioreactors with complex geometric structure. Fig. 25 shows an example of a DiCoFluV photobioreactor (Cornet, 2010), in which the incident solar-light flux density is diluted within the volume of culture because of a thousand light-diffusing optical fibers emitting a quasi-homogeneous density flux on the totality of their surface (see Fig. 26).

With the Monte Carlo method, the difficulty with the geometric complexity is reduced to calculation of the intersections between the straight rays and surfaces. In Section 4.1, this situation corresponds to calculation of the intersection between half-lines starting at $\mathbf{x}_j$ in the direction ω_j and the surface $\mathcal{R} \cup \mathcal{F}$ [Step (3) of our algorithm]. Indeed, the calculation of these intersections for the complex surface $\mathcal{R} \cup \mathcal{F}$ of a DiCoFluV is the only additional difficulty, in contrast to the implementation for a one-dimensional Cartesian photobioreactor (see Fig. 25).

In Delatorre et al. (2014) and Dauchet et al. (2013), this practical difficulty is alleviated by implementation in the EDStar development environment (StarWest, n.d.), which makes available to radiation physicists a set of computational tools produced by the computer graphics research community during the last 20 years, in particular within the framework of the Physically Based Rendering Techniques (PBRT) project (Pharr and Humphreys, 2010). These tools are helpful in the process of geometric definition of complex *scenes* and for accelerating photon tracking in such scenes. The implementation in EDStar completely separates the geometry and physics: the geometric structure is defined (or imported) in a specific file, whereas the Monte Carlo algorithm describing the physical phenomena (such as absorption, scattering, and reflection) is written in a separate file where researchers have access to abstractions such as *ray*, *intersection*, or *shape* that are used to code the optical-path-sampling procedure regardless of the geometric structure specified in the first file. This separation perfectly meets the needs of the modern engineering studies: first, the algorithm is developed without worries about the technical characteristics that have no direct relation to physical reasoning (EDStars scientific computation libraries handle statistical methods, parallelization, and pure geometric questions), then the algorithm is validated in simple geometric structure, and finally it can be directly implemented for any geometric structure of a photobioreactor (without modification of the sampling procedure).

Both the file for the optical-path-sampling algorithm corresponding to the procedure in Section 4.1 and the scene description file corresponding to the DiCoFluV photobioreactor in Fig. 26 are provided on the EDStar website (StarWest, n.d.). This code was used in Delatorre et al. (2014) and Dauchet et al. (2013) to estimate the specific absorption rate at any location within the DiCoFluV. For estimation with 1% accuracy, we observed calculation time of ~5 s per location with the processor Quad-Core Intel Xeon 2.66 GHz (the time decreases linearly with the number of processors in a parallel implementation). Additionally, Fig. 27 shows the influence of the number of optical fibers in the photobioreactor. These results indicate that because of the computer graphics tools for acceleration of ray tracing,

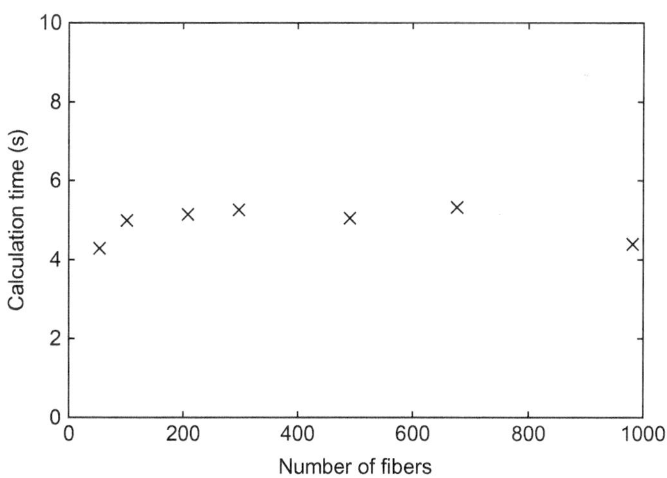

Figure 27 Calculation time for estimating the specific rate of photon absorption $\mathcal{A}$ within the reaction volume of the DiCoFluV photobioreactor, as a function of the number of light-diffusing optical fibers. The results were obtained in EDStar (Delatorre et al., 2014) by implementation of the algorithm presented in Section 4.1 for different versions of the geometric structure in Fig. 26B (containing different numbers of fibers).

the calculation time is independent of the geometric complexity. These results were obtained for the configuration presented in Fig. 26, in which both the photobioreactor $\mathcal{R}$ and fibers $\mathcal{F}$ are modeled as cylinders. In Rochatte et al. (2015), we tested our ability to extend this practice to simulation of performance of a photobioreactor on the basis of its computer-aided design (CAD), which is a tool used by engineers in the process of designing an innovative reactor, from initial sketching to final realization. We imported the CAD of the DiCoFluV prototype into EDStar and estimated the radiation field within this geometric structure, which is composed of 73,000 triangles describing the surfaces of fibers and a stainless-steel vessel. Common bugs encountered with such complex geometric structure were tracked, in particular, photon losses at the triangles' edges. No major difficulties were encountered in the present case, and the stability of calculation time was preserved.

Therefore, available computer graphics tools allow us to simulate any geometric structure of a photobioreactor with the same Monte Carlo algorithm and with acceptable calculation time. This recent advance in the field of the Monte Carlo method opens up interesting perspectives on photobioreactor design, for example, in relation to the optimization algorithm.

4.3 Coupling of Radiative Transfer with Photosynthesis

The following paragraphs address estimation of the production rate $< r_x >$ of a photobioreactor. To be precise, we discuss how a recent methodological advance in the simulation of nonlinear processes by the Monte Carlo method allows us to estimate $< r_x >$ without constructing the field of the absorption rate $\mathcal{A}$ within the reaction volume. On the one hand, this method obviates the cumbersome task of constructing an appropriate volume mesh for estimation of the $\mathcal{A}$ field; this approach is especially convenient in the case of internally lightened technologies such as DiCoFluV. On the other hand, this approach allows researchers to rigorously evaluate $< r_x >$ with acceptable calculation time (a few minutes in the case of DiCoFluV).

Among the numerical methods for evaluation of integrals, the Monte Carlo method is famous because its convergence is independent of the dimension of the integration domain (Hammersley and Handscomb, 1964). In the present context, this means, for example, that the calculation time for evaluating the average absorption rate $<\mathcal{A}>= \int_V d\mathbf{x}_0 \frac{1}{V} \mathcal{A}(\mathbf{x}_0)$ is almost the same as that for evaluating $\mathcal{A}(\mathbf{x}_0)$ at one location $\mathbf{x}_0$. This is because the integration domains are linearly combined in the integral formula for $<\mathcal{A}>$:

$$<\mathcal{A}>=\int_{\mathcal{V}} d\mathbf{x}_0 \frac{1}{V}\int_{\nu_{\min}}^{\nu_{\max}} d\nu \int_{4\pi} d\boldsymbol{\omega}_0\ \sigma_{a,\nu}\ L_\nu(\mathbf{x}_0,-\boldsymbol{\omega}_0) \tag{114}$$

where we substituted Eq. (112) into the definition of $<\mathcal{A}>$. Therefore, evaluating $<\mathcal{A}>$ requires only addition of a new step to the sampling procedure from Section 4.1 before Step (1): first, the location $\mathbf{x}_0$ is uniformly sampled across the reaction domain $\mathcal{V}$ (as in our introductory example); then, the following steps and the weight function are strictly identical to those for evaluation of $\mathcal{A}(\mathbf{x}_0)$ in Section 4.1. Each optical path that is sampled in this way contributes to the specific absorption at a different location $\mathbf{x}_0$. With this Monte Carlo algorithm, $<\mathcal{A}>$ is estimated easily, with calculation time $t_{<\mathcal{A}>} \simeq t_{\mathcal{A}} \simeq 5$ s, without calculating $\mathcal{A}(\mathbf{x}_0)$ for a set of M locations $\mathbf{x}_0$ (leading to calculation time $t_{<\mathcal{A}>} = Mt_{\mathcal{A}}$), where $t_{\mathcal{A}}$ is the calculation time for estimating $\mathcal{A}$ at one location (we estimate that $M \simeq 10^5$ in the case of DiCoFluV). Actually, this algorithm estimates $<\mathcal{A}>$ without calculating $\mathcal{A}(\mathbf{x}_0)$ at any location $\mathbf{x}_0$: the integration over the reaction volume and integration over the optical paths are simultaneously statistically sampled. This property makes the Monte Carlo method well suited for numerical implementation of our photobioreactor model, which is based on integral formulae with many dimensions (see Section 1).

In contrast, during evaluation of the average production rate $<r_x>=\int_{\mathcal{V}} d\mathbf{x}_0 \frac{1}{V} r_x(\mathcal{A}(\mathbf{x}_0))$, integration domains are no longer combined linearly:

$$<r_x>=\int_{\mathcal{V}} d\mathbf{x}_0 \frac{1}{V} r_x\left(\int_{\nu_{\min}}^{\nu_{\max}} d\nu \int_{4\pi} d\boldsymbol{\omega}_0\ \sigma_{a,\nu}\ L_\nu(\mathbf{x}_0,-\boldsymbol{\omega}_0)\right) \tag{115}$$

where the local production rate $r_x(\mathcal{A}(\mathbf{x}_0))$ is a nonlinear function of $\mathcal{A}(\mathbf{x}_0)$. Because the coupling law $r_x(\mathcal{A})$ is nonlinear, it leads to well-known difficulties (Dimov, 2008; Kalos and Whitlock, 2008). Practically, this means that construction of a Monte Carlo algorithm for evaluating $<r_x>$ is much more subtle than in the case of $<\mathcal{A}>$. The same difficulty is encountered during solution of Schiff's approximation in Section 2 for calculation of the radiative properties of photosynthetic microbial cells. The Monte Carlo code that is used in Section 2 is based on a method presented in Charon et al. (2015) that allows for analysis of quadratic functions. In Dauchet (2012), this method is extended to formulation of any analytic[11] nonlinear function and is successfully implemented for solution of Eq. (115). The resulting Monte Carlo algorithm is used in Section 5.7 (see Fig. 29) for estimation of the production rate $< r_x >$ of a DiCoFluV photobioreactor (cultivating *C. reinhardtii*), including rigorous solution of the radiative transfer equation for the spectral radiative properties obtained in Section 2. The sampling procedure of this algorithm consists of, first, uniform sampling of a location ($\mathbf{x}_0$) within the reaction volume and then, sampling of a few optical paths starting from this location, according to the optical-path-sampling procedure presented in Section 4.1. Altogether, for estimation with 1% accuracy, we observed calculation time ranging from 1 to 6 min with the processor Intel Core i7-2720QM 2.20 GHz, depending on the biomass concentration under study (see Fig. 29). This algorithm retains all the features of the Monte Carlo method: the numerical error is systematically evaluated as in Eq. (111); addressing integrals of $< r_x >$ is conceptually straightforward (in terms of, for example, annual averages); and the theoretical framework of recent methodological advances such as the zero-variance concept (see Dauchet et al., 2013; Delatorre et al., 2014) and sensitivity estimation (see Section 4.4) remains accessible. The present study on the capabilities of this novel methodology yielded promising results, in particular for solar-energy applications; however, formulation of coupling laws with discontinuity phenomena, such as the law derived for cyanobacteria in Section 5, remains an open question.

4.4 Sensitivity Analysis

When a Monte Carlo algorithm is used for estimation of any physical quantity (B), a simple and fast additional procedure can be implemented that simultaneously estimates sensitivity of B to any parameter (Delatorre et al., 2014). This practically means that when Monte Carlo code is available that computes B, only a few additional lines of code are needed so that partial derivatives of B are also computed with respect to all the parameters of interest. We are interested either in physical analysis (how does B evolve when a parameter is modified ?) or in optimal design (what is the optimal value of the parameter for a target value of B ?). A general overview of sensitivity estimation is available in Delatorre et al. (2014). This methodology was implemented in Dauchet et al. (2013) and Delatorre et al. (2014) to evaluate sensitivity of the radiation field within a DiCoFluV photobioreactor (see

[11] Here we mean continuous and infinitely differentiable functions.

Fig. 26) to radiative properties of microorganisms, to biomass concentration, and to the reflectivity of the optical fibers. In Section 5.7, the same methodology was applied in a straightforward manner to the algorithm from Section 4.3 in order to evaluate sensitivity of the production rate $< r_x >$ of a DiCoFluV photobioreactor to the absorption cross section, to biomass concentration, and to reflectivity of the optical fibers.

Sensitivity estimation is a methodological advance that we consider mature enough for immediate use in a photobioreactor study. In particular, this approach can be used to accelerate optimization procedures by providing simultaneously $< r_x >$ and its gradient in the parameter space, in relation, for example, to the method of steepest descent. Nevertheless, the specific case of sensitivity to parameters defining the geometric structure of the system leads to well-known difficulties that were characterized in Roger et al. (2005). To date, practical implementation of such sensitivity to geometric parameters has been restricted to academic configurations.

5. STOICHIOMETRIC, THERMOKINETIC, AND ENERGETIC COUPLING WITH A RADIATION FIELD: CALCULATION OF THE MAIN AVERAGED RATES AND EFFICIENCY FOR THE PHOTOBIOREACTOR

As explained in Introduction, the mean volumetric biomass growth rate $< r_x >$, which is linked to the stoichiometric equation for the biomass synthesis, is the key process variable appearing in all the other observable variables representative of photobioreactor performance (for instance, the surface growth rate and thermodynamic efficiency). The model then requires formulation of the thermokinetic coupling between the local rate of photon absorption and the local rate of biomass growth, as it is done in the engineering community for any other type of photoreactions (Aiba, 1982; Cassano et al., 1995; Irazoqui et al., 1976; Spadoni et al., 1978). Here, the word "thermokinetic" means that it is impossible to obtain purely kinetic coupling for a light-matter interaction process and for conversion of radiant light energy into biomass through definition of quantum or energetic yields. As already discussed (see Section 1), because photosynthesis shows non-linear behavior ($< J(\mathbf{x}) >$ is a non-linear function of $\mathcal{A}(\mathbf{x})$ and $< f(\mathcal{A}(\mathbf{x})) >$), this coupling must be formulated at the local scale (the medium is considered a continuum) before averaging the resulting rates across the total volume of the photobioreactor. As an important consequence, any attempt to formulate direct coupling of a mean spatial growth rate using a given averaged radiative quantity will lead only to representative model formulation. Application of the latter is strongly limited to geometric structures, the range of process variables, and experimental conditions used to identify the model's parameters.

Obtaining a predictive and generic knowledge model of the growth rate in a photobioreactor requires then (i) first to clearly define the main controlling steps involved (and their corresponding yields) in the coupling with the radiation field and (ii) to calculate *ab initio* all the resulting parameters appearing in this formulation (the reification procedure reduces the parametric space of the model). In this section, we intend to demonstrate that it is possible to formulate such a predictive model of thermokinetic coupling in the limited domain of photosynthesis modeling only, ie, ignoring the effect of respiration on the global metabolism of the photosynthetic microorganisms. This assumption is not restrictive in the case of prokaryotic cyanobacteria, in which respiration is indeed inhibited by light (De La Vara and Gomez-Lojero, 1986) as soon as it has irradiance levels in the photobioreactor that are at least fivefold higher than the compensation point. This assumption, however, must be considered only as a chloroplast level coupling model for eukaryotic microalgae (see the last part of this section for perspectives on the coupling radiation field and rates for microalgae). In the text later, it will become clear to the reader that such predictive and knowledge coupling formulation relies on a sound and accurate description of the radiation field, thus explaining the special attention paid to this subject matter in the previous sections of this chapter. Finally, the link between rates and stoichiometry will be explicitly determined via the crucial role played by the well-known $P/2e^-$ ratio (see Section 5.2). At this stage only, the analysis requiring to deal with a given microorganism will be restricted to the case of *A. platensis*, for which the authors have accumulated a considerable amount of experimental data.

5.1 Specific Rates and Thermokinetic Coupling with Radiation Field Formulation

As already explained (see Section 1), local thermokinetic coupling with the radiation field must be formulated at the microorganism scale requiring us to work with molar specific rates J_i (in $\text{mol}_i\ \text{kg}_x^{-1}\ \text{s}^{-1}$), which are defined (as a generalization of Eq. (2) for any compound i from the local molar volumetric rate r_i by

$$J_i = \frac{r_i}{C_x} \tag{116}$$

In particular, if the specific rate for biomass growth J_x is specified in the model, then any molar volumetric rate r_i must be deduced from the data on the dry-mass concentration C_x and the stoichiometric coefficients involved in the associated stoichiometric equation for biomass synthesis (requiring nevertheless predictive formulation giving a C-molar formula of the produced biomass). The mass volumetric rates are then easily deduced using the molar mass of the compound i being considered. Particularly, the volumetric biomass growth rate R_x (in kilograms of dry mass per unit of volume and per unit of time) is given by the following relation:

$$R_x = J_x C_x M_x \tag{117}$$

where M_x is C-molar mass of the biomass produced in the photobioreactor (which will be shown later to depend on the P/2e$^-$ ratio, ie, on the radiation field inside the photobioreactor).

On the other hand, the specific local photon absorption rate (in $\mu mol_{h\nu}\, kg_x^{-1}\, s^{-1}$) $\mathcal{A}(\mathbf{x}) = \int_{PAR} \sigma_{a,\nu}\, G_\nu(\mathbf{x})\, d\nu$ is obtained from the radiative approaches described in Sections 3 and 4 above by an integral over all the frequencies under study (in PAR). Finally, the thermokinetic coupling between kinetic rates and radiant energy absorption rates can be easily formulated from the definition, as for any photo-reactive process (Cassano et al., 1995; Cornet and Dussap, 2009; Cornet et al., 2003; Pruvost and Cornet, 2012), of the overall quantum yield $\mathbf{\Phi}$ as follows:

$$J_x(\mathbf{x}) = \mathbf{\Phi}(\mathbf{x})\mathcal{A}(\mathbf{x}) = \rho(\mathbf{x})\, \overline{\phi}_x\, \mathcal{A}(\mathbf{x}) \tag{118}$$

First, this equation clearly establishes, as explained in Introduction, that the coupling law inside the photobioreactor is a non-linear local law (depending on the location $\mathbf{x}$), which will be shown to be in the form of Eq. (4). The earlier equation also shows that proper analysis of coupling with radiant energy absorption rates is not compatible with the use of the classical "local growth rate" μ as a time constant. Second, one should always keep in mind that the overall quantum yield $\mathbf{\Phi}$ has been split in two kinds of yield. The purely energetic yield ρ, which is a local parameter, takes into consideration all the dissipative phenomena in the light-to-chemical energy conversion processes by the primary mechanisms of the photosynthesis (oxygen evolving complex OEC, reaction centers in photosystems). In contrast, the "stoichiometric" quantum yield $\overline{\phi}_x$ is associated with the conservative photon-to-electron mechanisms involved in the Z-scheme for synthesis of ATP and $NADPH_2$. We will see later that this is a time-averaged parameter, ie, in the linear domain assumption, $\overline{\phi}_x$ depends on a spatially averaged function of the radiation field (see Eq. (3)).

The thermokinetic coupling (Eq. (118)) must be applied directly in the case of eukaryotic microalgae in the limited situation of photosynthesis modeling in chloroplasts (respiration in mitochondria requires an additional part for the coupling model). In case of prokaryotic cyanobacteria, which have common electron carrier chains (De La Vara and Gomez-Lojero, 1986), respiration is inhibited by light, and this law is applicable to the whole-cell metabolism above the compensation point of photosynthesis (corresponding to the specific absorption rate $\mathcal{A}_c$), ie, in the form (Cornet and Dussap, 2009; Pruvost and Cornet, 2012):

$$J_x(\mathbf{x}) = \rho(\mathbf{x})\,\overline{\phi}_x\,\mathcal{A}(\mathbf{x})\,H(\mathcal{A} - \mathcal{A}_c) \tag{119}$$

where the Heaviside function $H(\mathcal{A} - \mathcal{A}_c)$ is introduced ($J_x = 0$ if $\mathcal{A} < \mathcal{A}_c$).

5.2 Structured Stoichiometry, Biomass Composition, and the P/2e⁻ Ratio

As discussed earlier, the kinetic rate for biomass must be linked to a stoichiometric equation of biomass growth to enable calculation of the rates of all the abiotic compounds involved in the metabolism of the microorganism. In photosynthesis, this stoichiometric equation can be formulated without any degree of freedom as soon as the mean C-molar formula for the microorganism is known. It was established a long time ago (Cornet et al., 1998) that the biomass composition of a given photosynthetic microorganism depends on the radiation field inside the photobioreactor, regardless of any mineral or carbon source limitations. We will demonstrate in the next part of this section that indeed, the biomass composition is fixed by the well-known $P/2e^-$ ratio (the ratio of the mean specific rate of photophosphorylation to the mean specific rate of cofactor reduction, see Eq. (120)) in the Z-scheme of photosynthesis (see Fig. 28). This ratio is itself a function of the radiation field controlling subsequently quality of the biomass produced in the photobioreactor:

$$P/2e^- = P_{2e} = \frac{\overline{J}_{ATP}}{\overline{J}_{COF}} = \; < f(\mathcal{A}) > \tag{120}$$

In the case of *A. platensis* considered here as a model organism of cyanobacteria, it is well known (Cornet et al., 1998; Mouhim et al., 1993) that

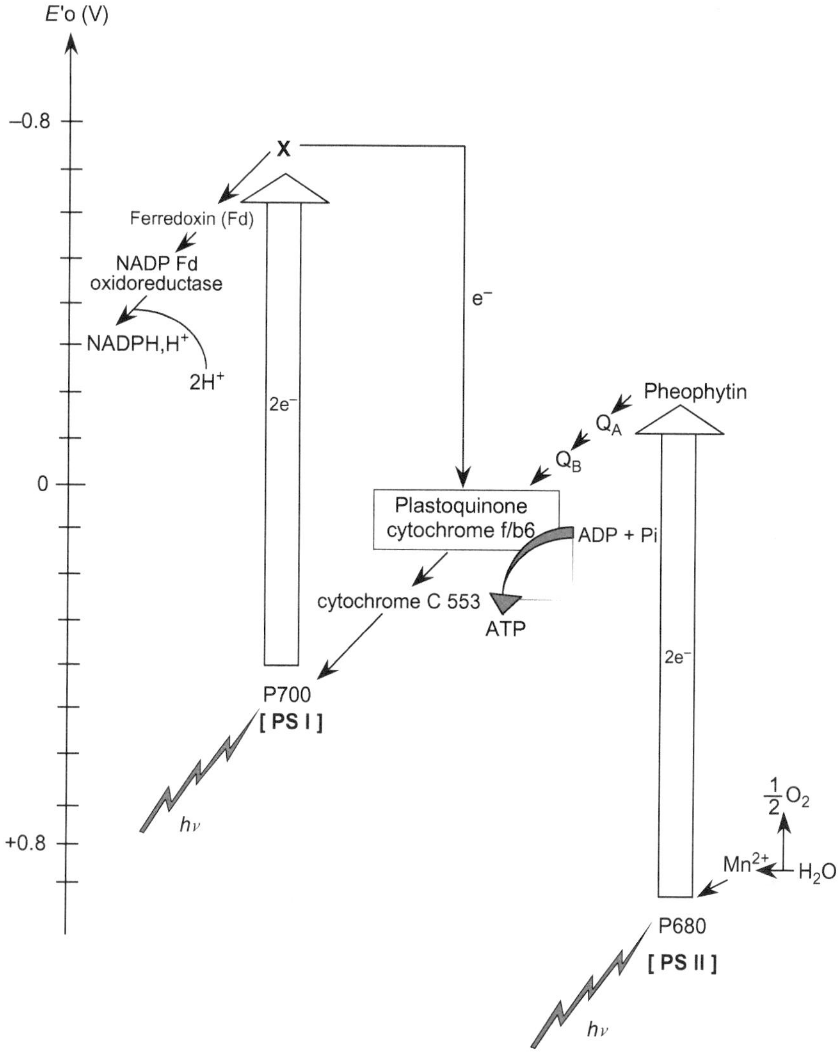

Figure 28 The Z-scheme of photosynthesis for cyanobacteria (from Cornet et al., 1998). The cyclic photophosphorylation pathway enabling values of P/2e$^-$ greater than 1.0 is indicated.

the P/2e$^-$ ratio deviations are balanced by the synthesis of an exopolysaccharide (EPS). Thus, the only degree of freedom for the micro-organism's metabolism (corresponding to the value of the P/2e$^-$ ratio imposed by the radiation field on the Z-scheme of photosynthesis) is filled by the corresponding ratio of a specific rate of synthesis (of "active biomass" with constant composition) to the specific EPS synthesis rate (Cornet et al., 1998). The C-molar formulas of active biomass (as the averaged sum of carbohydrates, proteins, lipids, and nucleic acids [classes of macromolecules]) and EPS for *A. platensis* together with their structured stoichiometry (Roels, 1983) have been reported elsewhere (Cornet et al., 1998). In this section, we provide the resulting structured stoichiometric equation for the total biomass synthesis (active biomass plus EPS) averaged by their respective molar fractions appearing then as a function of the ratio $P/2e^-$:

$$\begin{aligned} &CO_2 + (1.806P_{2e} - 0.885)H_2O + (0.507 - 0.256P_{2e})HNO_3 \\ &\quad + (0.013P_{2e} - 0.011)H_2SO_4 + (4.475 - 1.291P_{2e})NADPH,H^+ \\ &\quad + (3.146 + 0.330P_{2e})ATP \\ &\qquad\qquad \xrightarrow{J_X} \\ &CH_{(1.428+0.112P_{2e})}O_{(0.727P_{2e}-0.489)}N_{(0.507-0.256P_{2e})}S_{(0.013P_{2e}-0.011)}P_{(0.016-0.008P_{2e})} \\ &\quad + (3.130 + 0.338P_{2e})Pi + (4.475 - 1.291P_{2e})NADP^+ \\ &\quad + (3.146 + 0.330P_{2e})ADP \end{aligned} \tag{121}$$

The associated couple of structured stoichiometric equations for photosynthesis (Z-scheme) is then expressed as

$$\begin{aligned} &(4.475 - 1.291P_{2e})NADP^+ + (4.475 - 1.291P_{2e})H_2O \\ &\qquad \xrightarrow{J_{COF}} \\ &(4.475 - 1.291P_{2e})NADPH,H^+ + (2.238 - 0.645P_{2e})O_2 \\ &\qquad *** \\ &(3.146 + 0.330P_{2e})[ADP + Pi] \\ &\qquad \xrightarrow{J_{ATP}} \\ &(3.146 + 0.330P_{2e})ATP + (3.146 + 0.330P_{2e})H_2O \end{aligned} \tag{122}$$

Of course, summing Eq. (121) and (122) enables us to obtain the following unstructured equation of biomass synthesis:

$$\begin{aligned} &CO_2 + (0.185P_{2e} + 0.445)H_2O + (0.507 - 0.256P_{2e})HNO_3 \\ &\quad + (0.013P_{2e} - 0.011)H_2SO_4 + (0.016 - 0.008P_{2e})Pi \\ &\qquad \xrightarrow{J_X} \\ &CH_{(1.428+0.112P_{2e})}O_{(0.727P_{2e}-0.489)}N_{(0.507-0.256P_{2e})}S_{(0.013P_{2e}-0.011)}P_{(0.016-0.008P_{2e})} \\ &\quad + (2.238 - 0.645P_{2e})O_2 \end{aligned} \tag{123}$$

It must be noted that these stoichiometric equations have been established with the preferred nitrate NO_3^- ion as an N source, leading to the value of the stoichiometric coefficient:

$$\upsilon_{NADPH,H^+-X} = 4.475 - 1.291\mathrm{P}_{2e} \tag{124}$$

If ammonia NH_4^+ is used as the N source, the same analysis leads to the following stoichiometry:

$$\begin{aligned} &CO_2 + (0.439P/2e - 0.059)H_2O + (0.507 - 0.256P/2e)NH_3 \\ &\quad + (0.013P_{2e} - 0.011)H_2SO_4 + (0.016 - 0.008P_{2e})Pi \\ &\quad \xrightarrow{J_x} \\ &CH_{(1.428+0.112P_{2e})}O_{(0.727P_{2e}-0.489)}N_{(0.507-0.256P_{2e})}S_{(0.013P_{2e}-0.011)}P_{(0.016-0.008P_{2e})} \\ &\quad + (1.225 - 0.134P_{2e})O_2 \end{aligned} \tag{125}$$

in which the stoichiometric coefficient $\upsilon_{NADPH,H^+-X} = 2\,\upsilon_{O_2-X}$ is sharply different:

$$\upsilon_{NADPH,H^+-X} = 2.450 - 0.268\,\mathrm{P}_{2e} \tag{126}$$

At this stage, we must make the following important observation regarding the Z-scheme for photosynthesis (summarizing all the primary biochemical reactions of the metabolism, ie, the light reactions of photosynthesis). If we consider only the "stoichiometric" photons involved in this scheme, then we can simply define the mean quantum yield $\overline{\phi}_x$ using the data on the stoichiometric coefficient $\upsilon_{h\upsilon-X}$, directly linked by the structured equations (eg, Eq. (121)) to the stoichiometric coefficient υ_{NADPH,H^+-X} and to the value of the P/2e$^-$ ratio from

$$\overline{\phi}_x = \frac{10^{-6}}{\upsilon_{h\upsilon-X}} = \frac{10^{-6}}{2\,\upsilon_{NADPH,H^+-X}(1+P_{2e})} \quad (\mathrm{C-mol}_x\ \mu\mathrm{mol}_{h\nu}^{-1}) \tag{127}$$

This result is highly important because it means that (i) if we can develop a theory to determine the value of the P/2e$^-$ ratio for any situation regarding the radiation field in the photobioreactor, we will then be able to obtain as a predictive mean both the composition of the produced biomass (in terms of the molar fraction of each intracellular-macromolecule class and in terms of the global C-molar formula) and the value of the mean quantum yield $\overline{\phi}_x$ involved in the thermokinetic law of coupling. (ii) All the dark reactions of the anabolism in the cells operate under the physical constraint of radiant light transfer controlled only by a tenth of light reactions in the Z-scheme. The first tentative attempt to calculate the P/2e$^-$ ratio in a predictive manner [using linear thermodynamics of irreversible processes (LTIP) to formulate a phenomenological model of the Z-scheme] will be presented later, in Section 5.4.

5.3 Calculation of Parameters Related to Dissipative Mechanisms in the Light-to-Chemical Energy Conversion Process

Regarding light as a wave or as photons, if we are reasoning either at the microscopic scale (reversibility) or at the macroscopic scale (irreversibility), then we can develop many theoretical approaches to analysis of the light-matter interaction and energy conversion. All these approaches are more-or-less tractable and could be used eventually to calculate the energy dissipation in the process of photon-exciton-electron generation at the early primary stages of photosynthesis. The authors of this chapter are currently working on reconciliation of different approaches in the fields of natural and artificial photosynthesis using materials or molecular complexes as photocatalysts. This reconciliation is necessary to better understand the link between the kinetic approach (flux of photons) and thermodynamic approach (energy flux) in formulation of the coupling. As a whole, this is a considerable amount of work, which is clearly beyond the scope of this chapter, especially because any theory would lead to the same value, with calculation of at least the maximum free chemical energy that can be extracted from a photon interacting with an exciton in a photosynthetic antenna.

We present later a classical approach (based on the macroscopic theory of radiant energy conversion) to such an analysis, which was recently reconciled with the non-equilibrium thermodynamic approach built on the definition of the chemical potential of a photon (Meszéna and Westerhoff, 1999).

The energetic yield ρ represents the dissipative part of the photonic energy absorbed by the antenna, ie, the part of the absorbed energy that does not lead to electron transfers in the carrier chain and to the reduction of $NADP^+$ (energy losses in the light-matter interaction process). First, generally speaking and as explained elsewhere (Cornet and Dussap, 2009), this local value decreases with the local specific absorption rate $\mathcal{A}$ according to a convenient relation postulated here as an approximation of a theoretical quantum mechanical study on the excitation transfer mechanisms in antennas (Paillotin, 1974):

$$\rho \simeq \rho_M \frac{1}{1+\dfrac{\mathcal{A}}{K}} = \rho_M \frac{K}{K+\mathcal{A}} \tag{128}$$

where ρ_M is the maximum value of the yield, obtained when the system operates under the optimal thermodynamic conditions, at a very low rate, near the compensation point of photosynthesis (the photon absorption rate $\mathcal{A}_c$ becomes negligible in regards to the half-saturation constant K). It must be pointed out at this stage that even if Eq. (128) remains a representation model, it adds some theoretical background to the well-known hyperbolic behavior of photosynthesis in relation to irradiance values or specific photon absorption rates (with respect to the use of ρ in the coupling law Eq. (118). Second, the maximum energetic yield ρ_M will be further discussed in the text later. This yield corresponds to the maximal thermodynamic efficiency (the minimal losses), for the photon conversion in terms of free chemical energy (excitons and electrons in photosystems) at the level of the reaction centers. This maximal efficiency for any radiant-energy conversion process has been extensively debated in the past and eventually clarified by Bejan (1987, 1988). We nevertheless propose here to use the simpler approach proposed by Jeter (used in photosynthesis analysis for a long time; Duysens, 1959). These authors simply consider an extension of the Carnot formula for maximal radiation conversion (Bejan, 1987), in the same form as that recently obtained independently from the thermodynamic definition of the chemical potential of photons (Meszéna and Westerhoff, 1999):

$$\rho_M = 1 - \frac{T}{T_R} \tag{129}$$

where the temperature T_R for the radiation is expressed as the ideal blackbody formula of Planck, with assignment to the spectral intensity $L_{c,\bar{\lambda}}$ (under optimal thermodynamic conditions) of the value at the compensation point for photosynthesis in the photobioreactor (Cornet and Dussap, 2009).

Evaluation of ρ_M by this approach in a wide range of radiation characteristics (such as angular distribution, frequency, or the value of the intensity) in prokaryotic and eukaryotic microorganisms (they have different optimal temperatures for functioning) generally yields numerical values ranging between 0.76 and 0.82 (in PAR). This result allows us to use in the first approximation the constant value $\rho_M = 0.8$ with less than 10% deviation for any photosynthetic microorganism considered in this study. It should be mentioned that this value is in good agreement with the theoretical values determined by a thorough analysis of the excitation transport in an antenna at the quantum mechanical level (Paillotin, 1974), demonstrating that near the compensation point, photosynthesis operates close to the optimal thermodynamic conditions.

The last kinetic parameter in Eq. (128), the half-saturation constant K, must be discussed at this point. Because Eq. (128) is only an approximate relation, today, there is no way to devise a predictive method producing a theoretical value of K as a function of a knowledge description of the OEC functioning. Accordingly, in the model, K appears to be the only parameter that needs to be identified, and this value is indeed specific for a given microorganism. It may be easily obtained, for example, by independent measurements of O_2 evolution as a function of the specific photon absorption rate (if the latter is rigorously quantified). Surprisingly, such experiments on different kinds of photosynthetic microorganisms (prokaryotic or eukaryotic) yield rather close values of K, within a range of $\pm$ 30%. Such variation indicates, however, that experimental determination in each microorganism under study can significantly increase the accuracy of the proposed approach. The corresponding value determined for *A. platensis* and used in this study has already been widely reported by the authors when they wrote the coupling kinetic formula as a function of irradiance (Cornet et al., 1998; Cornet and Dussap, 2009). The corresponding value in specific absorption rate units (see Eqs. (118) and (128)) is

$$K = (1 \pm 0.1) \times 10^4 \ \mu\text{mol}_{h\nu} \ \text{kg}_x^{-1} \ \text{s}^{-1} \tag{130}$$

5.4 The Use of Linear Thermodynamics of Irreversible Processes (LTIP) for Calculation of Parameters Related to Conservative Mechanisms in the Process of Light-to-Chemical Energy Conversion: P/2e$^-$ Calculation and Analysis

After the pioneering work of Prigogine (Glansdorff et al., 1971; Prigogine, 1967), the use of LTIP for modeling of different patterns of biological behavior was strongly debated in the 1980s and 1990s. When analyzing mainly the metabolism, and particularly the Z-scheme for photosynthesis (the light-driven primary metabolic reactions for photosynthetic microorganisms), it is very important to discuss applicability of such a phenomenological approach to situations involving photons and reactions operating far from equilibrium ($|A_i| > RT$). First, we have already shown that only the conserved and stoichiometric photons are considered in our model (calculation of the P/2e$^-$ ratio and stoichiometric quantum yield $\overline{\phi}_x$), and this approach enables proper use of the linear energy converter formalism. Second, it was demonstrated in the 1980s (Dussap, 1988; Stucki, 1979, 1980) that the mean variables describing biological systems and satisfying the non-asymptotic stability criterion (in the sense of Lyapunov) obey linear phenomenological relations. This state of affairs requires of course choosing a characteristic time point to perform time-averaged integrals for these variables and working after that with mean rates as multi-linear functions of mean affinity levels obeying the reciprocity Onsager relations. Consequently, LTIP seems to plausibly describe the mean functioning of processes far from equilibrium: unstable for instantaneous values but stable for time-averaged values during operation in a highly organized spatial biological structure (Dussap, 1988). The characteristic time point under consideration for variable observations must be consequently chosen at a level enabling the use of the pseudo-steady-state assumption for intermediate products of the metabolism (approximately 1 min). This period is clearly of the same order of magnitude as the mixing time inside a photobioreactor; consequently, we will assume later that because all the relations involved are linear, the mean time variables $\bar{B}$ are equivalent to the spatially averaged values $< B >$ inside the photobioreactor. Eventually, there are no additional restrictions on the use of LTIP to describe the Z-scheme for photosynthesis in the same form as it has been done in the seminal work of Stucki (1980, 1988) and Dussap (1988) for respiration.

As for the previous conditions of applicability, it is then possible to use linear phenomenological thermodynamics of irreversible processes to analyze the coupling between redox reactions leading to reducing $NADPH_2$ synthesis (specific molar rate J_{COF}) and the photo-phosphorylation mechanisms leading to ATP synthesis (specific molar rate J_{ATP}) according to the Z-scheme of photosynthesis (Fig. 28). The ratio of these two mean specific rates is defined as the P/2e$^-$ ratio (see Eq. (120)):

$$P_{2e} = \frac{\bar{J}_{ATP}}{\bar{J}_{COF}}$$

Because we work with "stoichiometric" photons (corresponding to the transfer of one electron), we can define the Z-scheme as two coupled reactions involving four photons for one molecule of water split at the OEC of photosystem II. This model requires that one equation be exergonic (positive affinity) and one endergonic (negative affinity), so that the partition among the four photons is unique and straightforward (Cornet et al., 1998):

$$3\,h\nu + \mathrm{ADP} + \mathrm{Pi} \overset{\bar{J}_{ATP}}{\rightarrow} \mathrm{ATP} + \mathrm{H_2O} \quad \text{where} \quad \bar{A}_{ATP} > 0 \tag{131}$$

$$h\nu + \mathrm{H_2O} + \mathrm{NADP^+} \overset{\bar{J}_{COF}}{\rightarrow} \frac{1}{2}\mathrm{O_2} + \mathrm{NADPH, H^+} \quad \text{where} \quad \bar{A}_{COF} < 0 \tag{132}$$

It must be noted here that this formulation remains purely phenomenological and must not be used in any case as a tentative mechanistic explanation of cyclic photo-phosphorylation. The affinity values A_i in this formula are defined by means of the chemical potential $\widetilde{\mu}_i$ calculated under cytosolic conditions from

$$\begin{aligned} \bar{A}_{ATP} &= 3\,h\nu + \widetilde{\mu}_{ADP} + \widetilde{\mu}_{Pi} - \widetilde{\mu}_{ATP} - \widetilde{\mu}_{H_2O} \\ \bar{A}_{COF} &= h\nu + \widetilde{\mu}_{H_2O} + \widetilde{\mu}_{NADP^+} - \widetilde{\mu}_{NADPH} - \widetilde{\mu}_{H^+} - \frac{1}{2}\widetilde{\mu}_{O_2} \end{aligned} \tag{133}$$

Assuming that 1 mol of photons at $\lambda = 680$ nm corresponds to enthalpy of 176 kJ/mol and using the models of calculation of thermodynamic properties developed in our lab (Ould Moulaye, 1998), we obtain the following result for the theoretical affinity values:

$$\begin{aligned} \bar{A}_{ATP,\ \max} &= 3\,h\nu_{\max} - 32,5 = 495,2\ \mathrm{kJ\ mol^{-1}} \\ \bar{A}_{COF,\ \max} &= h\nu_{\max} - 216,3 = -40,4\ \mathrm{kJ\ mol^{-1}} \end{aligned} \tag{134}$$

These affinity values are considered maximal because they are calculated with the enthalpy of a photon ($h\nu$), but the optimization procedure that is explained later will result in new affinity values at lower free enthalpy $h\nu_{eff}$. Therefore, it will become evident that energetic efficiency of the coupling decreases with the increasing specific photon absorption rate $<\mathcal{A}>$. By means of the partition of the entropy balance in the photobioreactor (Cornet et al., 1994), it is then possible to derive the expression for the rate of entropy production σ (or the dissipation function) in the reactive system under study (the Z-scheme). Assuming here the system as isotherm, we will soundly define the mean specific isotherm dissipation function at a given optimal temperature (T^*) as $\overline{\sigma}^* = \overline{\sigma}\, T^*/C_x$ in the following form:

$$\overline{\sigma}^* = \sum_{j=1}^{r} \overline{J}_j \overline{A}_j = \overline{J}_{ATP} \overline{A}_{ATP} + \overline{J}_{COF} \overline{A}_{COF} \geq 0 \tag{135}$$

Contrary to the complete expression established previously by the authors for photobioreactors (Cornet et al., 1994), the specific radiant light absorption rate here is ignored in the dissipation function because the radiant light energy in the photochemical process was taken into account directly in the affinity definitions (see Eqs. (131) to (134)).

The theory of linear energy converters then postulates a multi-linear relation between affinity values and specific rates, leading to the following equations, with the help of reciprocity relations of Onsager:

$$\overline{J}_{ATP} = L_{AA} \overline{A}_{ATP} + L_{AC} \overline{A}_{COF} \tag{136}$$

$$\overline{J}_{COF} = L_{AC} \overline{A}_{ATP} + L_{CC} \overline{A}_{COF} \tag{137}$$

where the phenomenological coefficients L_{ij} must be eliminated from convenient normalization (Dussap, 1988; Stucki, 1980) leading to definitions of the coupling coefficient:

$$q = \frac{L_{AC}}{\sqrt{L_{AA} L_{CC}}} \tag{138}$$

of the phenomenological stoichiometric coefficient:

$$\chi = \sqrt{\frac{L_{CC}}{L_{AA}}} \tag{139}$$

and finally, the ratio of generalized forces:

$$x = \chi \frac{\overline{A}_{COF}}{\overline{A}_{ATP}}, \quad x < 0 \tag{140}$$

Thus, the specific rates of cofactor or ATP synthesis can be expressed with the help of the previous variables as

$$\overline{J}_{ATP} = L_{AA} \overline{A}_{ATP} (1 + qx) \tag{141}$$

$$\bar{J}_{COF} = \chi L_{AA} \bar{A}_{ATP} (q + x) \tag{142}$$

yielding the values of chemical power as the key parameters involved in the dissipation function and in the energetic analysis of the photobioreactor (see Section 5.5):

$$\bar{J}_{ATP} \bar{A}_{ATP} = L_{AA} \bar{A}^2_{ATP} (1 + qx) \tag{143}$$

$$\bar{J}_{COF} \bar{A}_{COF} = L_{AA} \bar{A}^2_{ATP} \, x(q + x) \tag{144}$$

and eventually allowing us to establish a tractable expression for the specific source of entropy production rate:

$$\bar{\sigma}^* = L_{AA} \bar{A}^2_{ATP} (1 + 2qx + x^2) \tag{145}$$

The ultimate variable from which we intend to derive all the predictive information on the Z-scheme modeling is the P/2e$^-$ ratio rewritten from the normalized variables as

$$\mathrm{P}_{2e} = \frac{1 + qx}{\chi (q + x)} \tag{146}$$

The whole theoretical approach relies now on the ability to calculate, for any condition of radiation field in the photobioreactor, the variables q, χ, and x in a predictive manner. This calculation can be performed in two steps. First, it can be demonstrated (Dussap, 1988) that there are two non-adaptive conditions (ie, fixed conditions arising at the early stage of a cell's life cycle), allowing us to obtain constant values for q and χ. The first condition links the stoichiometric coefficient for the phosphorylation to the phenomenological stoichiometric coefficient in non-ideal coupling:

$$\chi = \upsilon_{COF-ATP} \, q \tag{147}$$

It is well established that there is only one site for phosphorylation in the electron carrier chain of photosynthesis; therefore, $\upsilon_{COF-ATP} = 1$, and consequently, according to Eq. (147):

$$\chi = q \tag{148}$$

The second non-adaptive condition means that the optimal thermodynamic functioning of the cell corresponds to maximization of chemical power (Eqs. (143) and (144)), leading to calculation of the coefficient q (Dussap, 1988; Stucki, 1980):

$$q = \sqrt{2\left(\sqrt{2}-1\right)} = 0.91 \tag{149}$$

Second, we must invoke the adaptive conditions of the cells experiencing a given mean radiation field inside the photobioreactor. As explained throughout this chapter, any photosynthetic microorganism operates under conditions of physical limitation by light. This means that the radiation field (or rather a spatial averaged function of the radiation field for the time constants considered in this analysis) is a constraint entirely determining the behavior of the Z-scheme functioning in the cells. If we assume that this behavior obeys the principle of minimum entropy production rate (Cornet et al., 1998; Dussap, 1988; Glansdorff et al., 1971; Stucki, 1980), then the ratio x of generalized forces can be determined by an optimization procedure confirming that

$$d\bar{\sigma}^* = 0 \tag{150}$$

This classical optimization problem under constraints requires introducing a Lagrange function $\mathcal{L}$ with Lagrange multipliers λ_k associated with the constraints g_k in the following form:

$$\bar{\mathcal{L}} = \bar{\sigma}^* - \sum_k \lambda_k \bar{g}_k \tag{151}$$

and requires calculating the partial derivatives

$$\left(\partial_{\bar{A}_j} \bar{\mathcal{L}}\right)_{\bar{A}_k} = 0 \quad , \quad (k \neq j) \tag{152}$$

Generally speaking, the main difficulty here is to formulate the constraints g_k linked to an averaged function of the radiation field $< f(\mathcal{A}) >$ in all the situations encountered in photobioreactors (eg, a kinetic regime, luminostat regime, and photo-limitation with dark zones). A comprehensive analysis of this complicated problem is still a work in progress by the authors and clearly beyond the scope of this chapter. In contrast, it is quite easy to examine two extreme situations for which the unique constraint on the Z-scheme functioning may be formulated as a limit, independently of knowledge about the function $< f(\mathcal{A}) >$.

The first situation that can be envisaged concerns functioning close to the compensation point for photosynthesis (very low specific photon absorption rates $< \mathcal{A} > \rightarrow 0$) corresponding to the strongest physical limitation by radiant light energy transfer. In this condition, the specific rate $\bar{J}_{ATP}$ is the constraint imposed on the photosynthetic functioning (Cornet et al., 1998; Dussap, 1988) and must be assumed to be a constant. Eq. (151) can then be written as

$$\bar{\mathcal{L}} = L_{AA}\bar{A}_{ATP}^{2}\,(1+2qx+x^{2}) - \lambda\, L_{AA}\bar{A}_{ATP}\,(1+qx) \tag{153}$$

Accordingly, the optimization (Eq. (152)) leads to calculation of the ratio of generalized forces:

$$x = 0 \tag{154}$$

With the previous values of q and χ, it is then possible to obtain the P/2e$^-$ value in this situation from Eq. (146): $P_{2e} = 1.21$. It is important to note that this value is very close to the value of $P_{2e} = 1.23$ obtained independently for the active biomass of *A. platensis* in a complete analysis of its metabolism (Cornet et al., 1998) and used in a previous stoichiometric equation for total biomass synthesis: Eq. (121) or (123).

The second simple situation is to consider, at the other extreme, the maximal saturation rate for photosynthesis (very high specific photon absorption rates $<\mathcal{A}> \rightarrow \infty$), ie, the functioning without the light transfer limitation. In this case, the constraint imposed on the Z-scheme metabolism is the chemical power output $\bar{J}_{ATP}\bar{A}_{ATP}$ (Cornet et al., 1998; Dussap, 1988), which is constant and independent of the radiation field. The Lagrange function takes the following form:

$$\bar{\mathcal{L}} = L_{AA}\bar{A}_{ATP}^{2}\,(1+2qx+x^{2}) - \lambda\, L_{AA}\bar{A}_{ATP}^{2}\,(1+qx) \tag{155}$$

and the optimization yields

$$x = \frac{\sqrt{1-q^{2}}-1}{q} = -0,643 \tag{156}$$

with the highest value for the P/2e$^-$ ratio: $P_{2e} = 1.71$.

Although general formulation of the problem is beyond the scope of this chapter as explained earlier, it is nevertheless possible to report here an important conclusion for a photobioreactor operating in optimal situations (Cornet, 2010) in terms of its kinetic or energetic performance (luminostat or photo-limitation). In this case, the maximum P/2e$^-$ value that can be reached with very high specific photon absorption rates [corresponding to incident photon flux density (PFD) of a full-sun AM 1.5 near 2000 $\mu mol_{h\nu}$ $m^{-2}\ s^{-1}$] is $P_{2e} = 1.5$. The highest values of the P/2e$^-$ ratio do not occur under natural outdoor sunlight conditions and in most of the artificially illuminated indoor photobioreactors. Considering now this optimal range of the $P/2e^-$ functioning for the Z-scheme in optimal situations for photobioreactor operation, ie, $1.2 < P_{2e} < 1.5$, and using Eqs. (124) and (127) for calculation of the mean stoichiometric quantum yield of the coupling law $\overline{\phi}_x$, we obtain a constant value equal to

$$\overline{\phi}_x = 7.8 \times 10^{-8}\ \mathrm{C-mol}_x\ \mu\mathrm{mol}_{h\nu}^{-1} \tag{157}$$

This surprising result is caused by the opposite effects in Eq. (127) where an increase in the $P/2e^-$ ratio is strictly compensated by a decrease in the stoichiometric coefficient υ_{NADPH,H^+-X} in the global stoichiometry for the quantum yield calculation. This phenomenon is important because it shows that the "stoichiometric" coupling may be considered linear coupling in Eqs. (118) and (119), independently of any function $<f(\mathcal{A})>$. It also shows that the energetic yield ρ remains the only radiation field-dependent parameter. In contrast, the biomass composition, ie, stoichiometry of the photosynthetic growth reaction and the C-molar formula of the biomass inside the photobioreactor are clearly dependent on the radiation field and $P/2e^-$.

The value of the quantum yield (Eq. (157)) was obtained with the preferred N source (nitrate) in the stoichiometry, but the calculation could be performed with ammonia as the N source. In this case, using Eq. (126) instead of Eq. (124) in Eq. (127), we also obtain the quasi-constant value:

$$\overline{\phi}_x = 1.0 \times 10^{-7}\ \mathrm{C-mol}_x\ \mu\mathrm{mol}_{h\nu}^{-1} \tag{158}$$

This interesting result shows that the efficiency of photosynthesis is 25% higher if ammonia is used instead of nitrate as the N source; this is a well-known phenomenon for cultivation of aerobic microorganisms as mentioned by Roels (1983).

5.5 Thermodynamic Efficiency and Energetic Coupling Analysis

As discussed in Introduction, the previous thermokinetic model of coupling (Sections 5.1–5.4) may be used to predictively calculate the mean averaged specific rates $< J_i >$, volumetric rates $< r_i >$, or surface rates $< s_i >$ by a volume integral for any given geometric structure of a photobioreactor. This approach requires first solving the radiative transfer equation using one-dimension-approximated or rigorous numerical approaches such as the Monte Carlo method discussed in Section 4 or the finite element method (proposed in Cornet et al., 1994) used for calculation of the local specific rate of photon absorption $\mathcal{A}$. In all these situations, it is easy to evaluate the local volumetric rate of radiant energy absorbed $\widetilde{\mathcal{A}}$ and its mean averaged integral over the total volume of the photobioreactor $<\widetilde{\mathcal{A}}>$. This parameter is derived from the data on the specific rate of photon absorption $\mathcal{A}$ from

$$\widetilde{\mathcal{A}} = \mathcal{A}\, C_x \aleph \quad (\text{in W m}^{-3}) \tag{159}$$

where $\aleph$ (in $J\,\mu\text{mol}_{h\nu}^{-1}$) is a conversion factor between micromoles of photons and joules (depending on the spectral nature of light in the incident PFD as a boundary condition). This is all that we need to calculate thermodynamic efficiency of the photochemical process as a whole. Considering first the thermodynamic efficiency of the photosynthesis inside the photobioreactor and using the entropic analysis proposed by Cornet et al. (1994), we obtain the following relation:

$$\langle \eta_{\text{th},\phi S} \rangle = \frac{\sum_{j=1}^{r}\sum_{p=1}^{n} \upsilon_{p,j} < r_j > \widetilde{\mu}_p}{<\widetilde{\mathcal{A}}> - \sum_{j=1}^{r}\sum_{s=1}^{m} \upsilon_{s,j} < r_j > \widetilde{\mu}_s} \simeq \frac{< r_x > \Delta g_x'^0}{<\widetilde{\mathcal{A}}>} \tag{160}$$

where all the parameters were determined by a predictive approach. Defining the total thermodynamic efficiency $< \eta_{\text{th}} >$ now requires working with the mean incident PFD $< q_{\cap} >$ arriving into the photobioreactor and then with surface rates rather than volumetric rates (see Section 1):

$$\langle \eta_{\text{th}} \rangle = \frac{\sum_{j=1}^{r}\sum_{p=1}^{n} \upsilon_{p,j} < s_j > \widetilde{\mu}_p}{<\widetilde{\mathcal{A}}> - \sum_{j=1}^{r}\sum_{s=1}^{m} \upsilon_{s,j} < s_j > \widetilde{\mu}_s} \simeq \frac{< s_x > \Delta g_x'^0}{< q_{\cap} >} \tag{161}$$

It has been previously demonstrated in both cases that these thermodynamic-efficiency levels strongly decrease with the increasing incident PFD $q_{\cap}$; this effect forms the basis of the light dilution concept for improvement of thermodynamic performance of photobioreactors (Cornet, 2010; Pruvost and Cornet, 2012). For example, in the DiCoFluV concept (for which simulations of the complete model are shown in Fig. 29 at the end of this section), the sunlight capture surface can be 50-fold smaller than the light distribution surface inside the photobioreactor (see chapter "Industrial Photobioreactors and Scale-up Concepts" by Pruvost et al.).

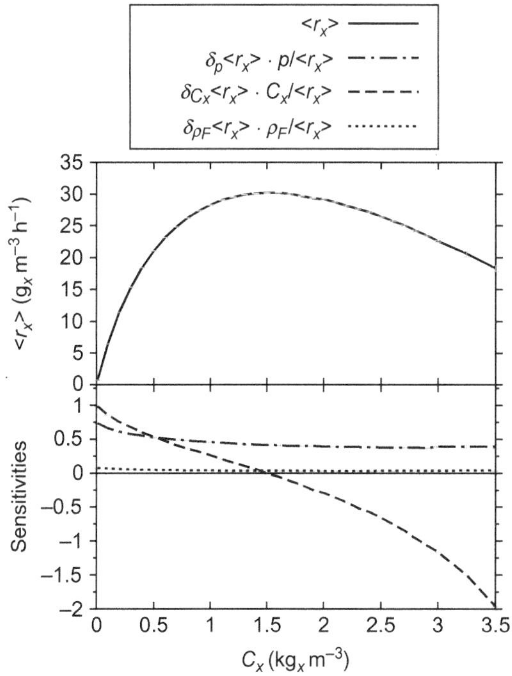

Figure 29 The volumetric biomass growth rate $< R_x >$ and its sensitivity parameters $\partial_p\langle r_x\rangle$, $\partial_{C_x}\langle r_x\rangle$, and $\partial_{\rho_{\mathcal{F}}}\langle r_x\rangle$ in a 25 L DiCoFluV photobioreactor (see Fig. 26) operating in continuous mode and cultivating *Chlamydomonas reinhardtii*. The results were obtained with the algorithm presented in Section 4.3 for 10^6 realizations, as a function of the dry-biomass concentration C_x. Statistical estimation of the numerical error is provided as error bars (gray). Relative types of sensitivity are shown, ie, $\partial_\pi\langle r_x\rangle \times \frac{\pi}{\langle r_x\rangle}$, where ∂_π is the partial derivative with respect to the parameter π under study: p is pigment content, C_x is the dry-biomass concentration, and $\rho^{\mathcal{F}}$ is reflectivity of the optical-fiber surface (see Fig. 26).

5.6 Experimental Validation of the Proposed Model for Different Simple Geometric Structures of a Photobioreactor

Experimental validation of the predictive knowledge model described in Sections 2–5 is shown later for simple geometric structures of photobioreactors. These results have already been published for analysis of validity of a simple and reliable engineering equation (Cornet and Dussap, 2009). They are used here again to analyze validity of the complete model, ie, for validation of the predictive approach as presented throughout this chapter. These experimental results have been obtained in eight very different, completely stirred photobioreactors, whose working liquid volume varied between 0.1 and 77.0 L, with the dark-volume fraction f_d ranging between

0 and 0.48 (ie, the volume fraction of photobioreactors that is not illuminated by design). Various kinds of geometric structures have been explored with different artificial illumination systems and with different methods of mixing the culture medium. In all the experiments, the microorganism was *Arthrospira (Spirulina) platensis* PCC 8005 grown axenically on the Cogne medium (Cogne et al., 2003). The temperature (35-36°C) and pH (between 8 and 10) were maintained at the levels optimal for growth. Incident hemispherical PFD ranged between 30 and 1600 $\mu\text{mol}_{h\nu}$ $\text{m}^{-2}\,\text{s}^{-1}$ (PAR, with multi-point measurements by means of the LI-COR cosine quantum sensor LI-190SA, with subsequent confirmation by actinometry). The main parameters influencing the biomass volumetric growth rates (geometric structure and characteristics of the illumination system), the culture conditions (mixing as well as pH and temperature control), and the operating conditions (batch mode or continuous culture) were described elsewhere (Cornet and Dussap, 2009) and are summarized here only briefly (see Table 2).

In all the experiments, the experimental mean volumetric growth rates were obtained according to the biomass balance in a well-mixed photobioreactor (defining the so-called residence time for continuous culture as $\tau = \frac{V_L}{Q_L}$):

$$< R_x > = \frac{C_x}{\tau} + d_t\, C_x \tag{162}$$

The resulting values of $< R_x >$, all obtained under optimal conditions of limitation by light (luminostat $\gamma = 1$, or photo-limitation $\gamma < 1$, see Cornet, 2010) are presented in Table 2. They are compared with the predictive model calculations presented in this chapter, where the radiative transfer equation was solved using the one-dimensional two-flux approximation for all the simple geometric structures of photobioreactors except for reactor PBR 2 (as indicated in Table 2), for which we used the three-dimensional finite element method developed by Cornet et al. (1994). As shown in the table, the mean deviation between the experimental results and the model calculation is less than 5% (ie, within the range of the experimental standard deviation), thus confirming the ability of the proposed predictive approach to quantify photobioreactor performance under many conditions of operation.

It must be noted that the new advances in the Monte Carlo method presented earlier would have led to the same $< R_x >$ values for the model

Table 2 Comparison Between Experimental Biomass Volumetric Growth Rates Obtained in Different Kinds of Photobioreactors Cultivating *Arthrospira platensis* and the Knowledge Model Presented in this Chapter

Geometry of the Reactor and Illuminating Characteristics	Reactor Type and Working Volume	Operating Cultivation Condition	Mean Incident Photon Flux Density q_0(PAR) (μ $mol_{h\nu} m^{-2} \cdot s^{-1}$)	Experimental Volumetric Growth Rate $< R_x >$ ($kg\ m^{-3}\ h^{-1}$)	Theoretical Volumetric Growth Rate Calculated by the Model $< R_x >$ ($kg\ m^{-3}\ h^{-1}$)	Deviation (%)
Rectangular, lightened by one side (1D) $a_{light} = 12.5\ m^{-1} (f_d = 0)$	**PBR 1** 4 L	Batch	40	$(1.6 \pm 0.2) \times 10^{-3}$	1.62×10^{-3}	+1
		Batch	50	$(2.1 \pm 0.2) \times 10^{-3}$	2.20×10^{-3}	+5
		Batch	85	$(3.2 \pm 0.2) \times 10^{-3}$	3.35×10^{-3}	+5
Cylindrical, lightened by one side (3D) $a_{light} = 12.5\ m^{-1} (f_d = 0)$	**PBR 2** 5 L	Batch	130	$(2.6 \pm 0.2) \times 10^{-3}$	2.65×10^{-3}	+2
		Batch	260	$(4.7 \pm 0.4) \times 10^{-3}$	4.78×10^{-3}	+2
		Batch	315	$(5.0 \pm 0.5) \times 10^{-3}$	4.95×10^{-3}	−1
		Batch	365	$(5.3 \pm 0.5) \times 10^{-3}$	5.24×10^{-3}	−1
		Batch	520	$(7.1 \pm 0.7) \times 10^{-3}$	6.93×10^{-3}	−1
		Batch	575	$(7.2 \pm 0.7) \times 10^{-3}$	7.42×10^{-3}	+3
		Batch	730	$(9.5 \pm 0.8) \times 10^{-3}$	9.23×10^{-3}	−3
		Batch	840	$(1.1 \pm 0.1) \times 10^{-2}$	1.11×10^{-2}	0
		Continuous	630	$(8.0 \pm 0.7) \times 10^{-3}$	7.85×10^{-3}	−2
		Continuous	1045	$(1.2 \pm 0.1) \times 10^{-2}$	1.18×10^{-2}	−2
		Continuous	1570	$(1.3 \pm 0.1) \times 10^{-2}$	1.35×10^{-2}	−4

Continued

Table 2 Comparison Between Experimental Biomass Volumetric Growth Rates Obtained in Different Kinds of Photobioreactors Cultivating *Arthrospira platensis* and the Knowledge Model Presented in this Chapter—cont'd

Geometry of the Reactor and Illuminating Characteristics	Reactor Type and Working Volume	Operating Cultivation Condition	Mean Incident Photon Flux Density q_0(PAR) (μ $mol_{hv}m^{-2}\cdot s^{-1}$)	Experimental Volumetric Growth Rate $< R_x >$ (kg m^{-3} h^{-1})	Theoretical Volumetric Growth Rate Calculated by the Model $< R_x >$ (kg m^{-3} h^{-1})	Deviation (%)
Cylindrical, radially lightened (1D) $a_{light} = 25\ m^{-1} (f_d = 0)$	**PBR 3** 5 L	Batch	245	$(1.3 \pm 0.1) \times 10^{-2}$	1.20×10^{-2}	−8
		Batch	620	$(1.9 \pm 0.2) \times 10^{-2}$	2.05×10^{-2}	+8
		Batch	1095	$(2.7 \pm 0.1) \times 10^{-2}$	2.70×10^{-2}	0
		Batch	1590	$(3.3 \pm 0.5) \times 10^{-2}$	3.40×10^{-2}	+3
Cylindrical, radially lightened (1D) $a_{light} = 40\ m^{-1} (f_d = 0.48)$	**PBR 4** 7 L	Continuous	235	$(1.0 \pm 0.1) \times 10^{-2}$	1.07×10^{-2}	+5
		Continuous	365	$(1.3 \pm 0.1) \times 10^{-2}$	1.31×10^{-2}	0
		Continuous	625	$(1.7 \pm 0.2) \times 10^{-2}$	1.78×10^{-2}	+5
		Continuous	780	$(1.9 \pm 0.2) \times 10^{-2}$	2.02×10^{-2}	+5
Oblate cylinder, lightened by one side (1D) $a_{light} = 43.5\ m^{-1} (f_d = 0)$	**PBR 5** 0.106 L	Batch	65	$(8.9 \pm 0.1) \times 10^{-3}$	9.01×10^{-3}	+1
Cylindrical, radially lightened (1D) $a_{light} = 26.7\ m^{-1} (f_d = 0.33)$	**PBR 6** 77 L	Batch	390	$(1.2 \pm 0.1) \times 10^{-2}$	1.18×10^{-2}	−2
		Continuous	525	$(1.4 \pm 0.2) \times 10^{-2}$	1.40×10^{-2}	0
		Continuous	840	$(1.7 \pm 0.2) \times 10^{-2}$	1.81×10^{-2}	+6

Annular and cylindrical, radially lightened (1D) $a_{light} = 40\ m^{-1} (f_d = 0)$	**PBR 7** 6 L	Batch	190	$(2.2 \pm 0.2) \times 10^{-2}$	2.08×10^{-2}	−5
		Batch	340	$(3.1 \pm 0.3) \times 10^{-2}$	3.02×10^{-2}	−3
		Batch	530	$(4.1 \pm 0.3) \times 10^{-2}$	3.90×10^{-2}	−5
Rectangular, lightened by one side (1D) $a_{light} = 25\ m^{-1} (f_d = 0)$	**PBR 8** 0.5 L	Batch and continuous	33	$(3.2 \pm 0.3) \times 10^{-3}$	3.23×10^{-3}	+0
		Continuous	135	$(1.1 \pm 0.1) \times 10^{-2}$	1.05×10^{-2}	−5

The photobioreactors' main characteristics and the experimental conditions are described in Cornet and Dussap (2009).

calculation, with numerous advantages such as shorter calculation time, assessment of standard deviations, model parameter sensibility analysis, evaluation of any local parameter without calculation of the whole field, possibility of increasing integral dimensionality of the problem without increasing the calculation time, and no need for a specific mesh grid. Additionally, as explained in Section 4, recent advances in this field allow researchers to separate the Monte Carlo algorithm formulation from analysis of complexity of a photobioreactor's geometric structure (Dauchet et al., 2013; Delatorre et al., 2014). These are the reasons why the authors today strongly recommend the Monte Carlo method, at least in the case of complex geometric structure or for development of a reference solution. Such Monte Carlo simulations are implemented in the next section for eukaryotic microalgae.

Finally, the reader will also find experimental validation of the proposed model in other interesting situations in chapter "Industrial Photobioreactors and Scale-up Concepts" by Pruvost et al.

5.7 Perspectives on Formulation of Thermokinetic Coupling for Eukaryotic Microalgae

The preliminary work on formulation of the thermokinetic coupling that we just presented earlier is limited to the metabolism of cyanobacteria or, in the case of eukaryotic microalgae, to their chloroplast functioning. Obviously, substantial additional work is needed to obtain such a general and predictive coupling law for eukaryotic photosynthetic microorganisms containing chloroplasts and mitochondria (which enable respiration at light). Although the LTIP approach that was described earlier has been strongly improved in relation to mitochondrial function (Dussap, 1988; Stucki, 1980, 1988), general formulation requires at least research into additional coupling between photosynthesis and respiration (including elucidation of the effect of light on respiration-specific rates). Evidently, this coupling must be formulated at the primary level, ie, taking into account the commutation between $NADPH_2$ and $NADH_2$ and its re-oxidation in the electron carrier chains of mitochondria. This commutation is possibly constrained by the radiation field inside the photobioreactor and the P/O ratio of mitochondria that is properly described by means of LTIP to analyze this new coupling. Unfortunately, we are today unable to provide a knowledge model for this thermokinetic coupling, and here, we will simply report tentative formulation of a representative law that seems consistent with numerous experimental observations (see chapter "Industrial Photobioreactors and Scale-up Concepts" by Pruvost et al.).

Because only primary stages are involved in photosynthesis and respiration (electron carrier chains and some key enzymes in the cofactor exchange reactions), it seems convenient to first formulate a kinetic coupling law related to the specific net oxygen production rate J_{O2}. Assuming that respiration rates are affected by the radiation field (as confirmed for some microalgae and reported for experimental measurements by specific analytical methods; Joliot, 1966), we can propose a rather symmetrical local law for respiration (this law is related to the already developed approach at the chloroplast level in the general form (Pruvost and Cornet, 2012)):

$$J_{O_2}(\mathbf{x}) = \rho_M \frac{K}{K + \mathcal{A}(\mathbf{x})} \overline{\phi}_{O_2} \mathcal{A}(\mathbf{x}) - J_r \frac{K_r}{K_r + \mathcal{A}(\mathbf{x})} \quad (\mathrm{mol_{O2}\, kg_x^{-1}\, s^{-1}}) \tag{163}$$

where the specific respiration rate is simply related to the rate of cofactor regeneration by respiration:

$$J_r = \frac{J_{NADH_2}}{\upsilon_{NADH_2-O_2}} = \frac{J_{COF}}{\upsilon_{NADH_2-O_2}} \text{and } \upsilon_{NADH_2-O_2} = 2 \tag{164}$$

The coupling equation above is still strongly linked to the radiation field, including the respiration term. At obscurity (anywhere in the photobioreactor) or at a very low value of $\mathcal{A}(\mathbf{x})$, the maximal respiration rate J_r must be considered constant and can be measured in independent experiments. In this case, the value of the new parameter K_r is not independent because it can be deduced directly from the data on the specific photon absorption rate at the compensation point $\mathcal{A}_c$ (the other parameters known in our knowledge model):

$$K_r = \frac{\mathcal{A}_c}{\dfrac{J_{NADH_2}}{\upsilon_{NADH_2-O_2}\, \rho_M \overline{\phi}_{O_2}} \left[\dfrac{1}{\mathcal{A}_c} + \dfrac{1}{K}\right] - 1} \tag{165}$$

In these equations, once $\mathcal{A}_c$ is specified, it is possible to use the predictive parameters described in the preceding sections (ρ_M and K). Then, the stoichiometric oxygen quantum yield is easily deduced from the data on the $P/2e^-$ ratio in the chloroplast from :

$$\overline{\phi}_{O_2} = \frac{10^{-6}}{4\,(1 + P/2e^-)} \simeq 1.1 \times 10^{-7} \mathrm{mol_{O2}\, \mu mol_{h\nu}^{-1}} \tag{166}$$

Finally, if necessary, the mean volumetric biomass growth rate in the photobioreactor $< R_x >$ is easily obtained by means of a spatial integral of the specific rate of net oxygen production $< J_{O2} >$ and again by means of

$$< R_x > = \frac{< J_{O_2} > C_x M_x}{\upsilon_{O_2-x}} \quad (\mathrm{kg}_x \, \mathrm{m}^{-3} \, \mathrm{s}^{-1}) \tag{167}$$

where υ_{O_2-x} and M_x are P/2e$^-$-dependent parameters.

Obviously, further work is necessary to devise a predictive and knowledge model of thermokinetic coupling for eukaryotic microalgae. This model may help to predict the specific compensation absorption rate $\mathcal{A}_c$ for any microorganisms (specific data explaining mainly the kinetic diversity of microalgae), which is perhaps linked to some hydrodynamic or physiological parameters. Such a model could also be used for predictive stoichiometric analysis, thus enabling assessment of quality of the produced biomass.

Now, the proposed kinetic coupling law (Eq. (163)), already used for example by Takache et al. (2012) for *C. reinhardtii* cultivation in a simple rectangular photobioreactor, is discussed in the text later as illustration of the whole methodology developed throughout this chapter, especially the use of the integral Monte Carlo formulation for easy calculation of the radiation field, volumetric biomass growth rate $< R_x >$, standard deviation of the results, and all the desired types of sensitivity (with the same calculation time).

Fig. 29 shows the results obtained from simulations of the whole predictive model (from calculations of optical and radiative properties to determination of the growth rate) presented in this chapter for *C. reinhardtii* cultivated in a geometrically complex photobioreactor with a thousand internal optical fibers (see Cornet, 2010; Dauchet et al., 2013 and Fig. 26). At each scale, the model was solved by the integral Monte Carlo method, including the volume integral, which implies formulation of the non-linear coupling law Eq. (163) (see Section 4.3). The CPU time ranged from 1 to 6 min for typical biomass concentrations varying between 0.5 and 3 g/L (2×10^5 realizations, standard deviation 1%, on Intel Core i7-2720QM). The DiCoFluV photobioreactor, with its complex geometric structure, obeys the same engineering rule as the rules classically developed for one-dimensional geometric structure (Cornet, 2010; Cornet and Dussap, 2009; Pruvost and Cornet, 2012): we identified an optimal biomass concentration that yields maximal productivity. Using the method presented in Section 4.4, we estimated the sensitivity of $< r_x >$ to different parameters (biomass concentration C_x, optical-fiber reflectivity $\rho^{\mathcal{F}}$, and pigment content p) simultaneously with $< r_x >$. The sensitivity to C_x is simply the slope of the curve $< r_x >$ vs C_x in the upper part of the figure. For optimal operation, when $\partial_{C_x} < r_x >$ is positive, the biomass concentration should be increased, but when $\partial_{C_x} < r_x >$ is negative, C_x should be decreased. Of course, the optimum corresponds to $\partial_{C_x} < r_x >= 0$. Estimation of $\partial_{C_x} < r_x >$ is especially relevant for maintenance of optimal operation without evaluating $< r_X >$ for the full range of biomass concentration. The sensitivity to $\rho^{\mathcal{F}}$ is low but positive, indicating that the fiber reflectivity should be increased to reduce photon losses at the boundary. Finally, $\partial_p < r_x >$ indicates that diminution of pigment content in microbial cells leads to homogenization of the radiation field, but the overall energy absorbed by the culture decreases, and losses at the fibers' and the photobioreactor's surface increase. This sensitivity is positive; therefore, the pigment content should be increased to improve performance of the photobioreactor.

ACKNOWLEDGMENTS

This work has been sponsored by the French government's research program "Investissements d'avenir" through the ANR programs PHOTOBIOH2 (2005-08), BIOSOLIS (2008-11), ALGOH2 (2011-15), PRIAM (2013-15), and the IMobS3 Laboratory of Excellence (ANR-10-LABX-16-01); by the European Union through the program "Regional competitiveness and employment" 2007–2013 (ERDF Auvergne region); and by the Auvergne region. This work was also funded by the CNRS through the PIE program PHOTORAD (2010-11) and the PEPS program "Intensification des transferts radiatifs pour le developpement de photobioreacteurs a haute productivite volumique" (2012-13). The authors wish to acknowledge the ESA/ESTEC for financial support through the MELiSSA project.

REFERENCES

Aas E: Refractive index of phytoplankton derived from its metabolite composition, *J Plankton Res* 18(12):2223–2249, 1996.

Aiba S: Growth kinetics of photosynthetic micro-organisms, *Adv Biochem Eng* 23:85–156, 1982.

Baran AJ: A review of the light scattering properties of cirrus, *J Quant Spectrosc Radiat Transf* 110(14-16):1239–1260, 2009.

Bejan A: Unification of three different theories concerning the ideal conversion of enclosed radiation, *Trans ASME J Sol Energy Eng* 109:46–51, 1987.

Bejan A: *Advanced engineering thermodynamics, 1997,* New York, NY, 1988, John Wiley and Sons Inc.

Berberoglu H, Gomez PS, Pilon L: Radiation characteristics of botryococcus braunii, chlorococcum littorale, and Chlorella sp. used for CO2 fixation and biofuel production, *J Quant Spectrosc Radiat Transf* 110(17):1879–1893, 2009.

Berberoglu H, Pilon L, Melis A: Radiation characteristics of Chlamydomonas reinhardtii CC125 and its truncated chlorophyll antenna transformants tla1, tlaX and tla1-CW+, *Int J Hydrog Energy* 33(22):6467–6483, 2008.
Berg MJ, Subedi NR, Anderson PA, Fowler NB: Using holography to measure extinction, *Opt Lett* 39(13):3993, 2014.
Bernard S, Probyn TA, Barlow RG: Measured and modelled optical properties of particulate matter in the southern Benguela, *S Afr J Sci* 97(9-10):410–420, 2001.
Bernard S, Probyn TA, Quirantes A: Simulating the optical properties of phytoplankton cells using a two-layered spherical geometry, *Biogeosci Discuss* 6(1):1497–1563, 2009.
Bi L, Yang P: Modeling of light scattering by biconcave and deformed red blood cells with the invariant imbedding T-matrix method, *Journal of biomedical optics* 18(5):055001-1–055001-13, 2013.
Bi L, Yang P: Impact of calcification state on the inherent optical properties of Emiliania huxleyi coccoliths and coccolithophores, *J Quant Spectrosc Radiat Transf* 155:10–21, 2015.
Bidigare RR, Ondrusek ME, Morrow JH, Kiefer DA: In-vivo absorption properties of algal pigments, *Ocean Opt* 1302(1):290–302, 1990.
Blanco S, Fournier R: An invariance property of diffusive random walks, *EPL (Europhys Lett)* 61(2):168, 2003.
Blanco S, Fournier R: Short-path statistics and the diffusion approximation, *Phys Rev Lett* 97(23):230604, 2006.
Bohren CF, Huffman DR: *Absorption and scattering of light by small particles,* New York, NY, 1983, Wiley-Interscience.
Bolton JR, Hall DO: The maximum efficiency of photosynthesis, *Photochem Photobiol* 53(4):545–548, 1991.
Bricaud A, Morel A: Light attenuation and scattering by phytoplanktonic cells: a theoretical modeling, *Appl Opt* 25(4):571–580, 1986.
Case KM, Zweifel PF: *Linear transport theory,* Reading, MA, 1967, Addison-Wesley.
Cassano AE, Martin CA, Brandi RJ, Alfano OM: Photoreactor analysis and design: fundamentals and applications, *Ind Eng Chem Res* 34:2155–2201, 1995.
Chami M, Thirouard A, Harmel T: Polvsm (polarized volume scattering meter) instrument: an innovative device to measure the directional and polarized scattering properties of hydrosols, *Opt Express* 22(21):26403, 2014.
Charon J, Blanco S, Cornet JF, et al: Monte Carlo implementation of Schiff's approximation for estimating radiative properties of homogeneous, simple-shaped and optically soft particles: application to photosynthetic micro-organisms, *J Quant Spectrosc Radiat Transf,* 2015. http://dx.doi.org/10.1016/j.jqsrt.2015.10.020 (in press).
Choi W, Yu CC, Fang-Yen C, Badizadegan K, Dasari R, Feld MS: Field-based angle-resolved light-scattering study of single live cells, *Opt Lett* 33(14):1596–1598, 2008.
Cogne G, Lehmann B, Dussap CG, Gros JB: Uptake of macrominerals and trace elements by the cyanobacterium Spirulina platensis (Arthrospira platensis PCC 8005) under photoautotrophic conditions: culture medium optimization, *Biotechnol Bioeng* 81(5):588–593, 2003.
Cornet JF: Calculation of optimal design and ideal productivities of volumetrically lightened photobioreactors using the constructal approach, *Chem Eng Sci* 65(2):985–998, 2010.
Cornet JF, Dussap C, Gros JB: Kinetics and energetics of photosynthetic micro-organisms in photobioreactors. In Bioprocess and algae reactor technology, apoptosis, *Advances in Biochemical Engineering Biotechnology*, vol. 59, Berlin, 1998, Springer, pp 153–224.
Cornet JF, Dussap CG: A simple and reliable formula for assessment of maximum volumetric productivities in photobioreactors, *Biotechnol Prog* 25(2):424–435, 2009.

Cornet JF, Dussap CG, Cluzel P, Dubertret G: A structured model for simulation of cultures of the cyanobacterium Spirulina platensis in photobioreactors: II. Identification of kinetic parameters under light and mineral limitations, *Biotechnol Bioeng* 40(7):826–834, 1992.
Cornet JF, Dussap CG, Dubertret G: A structured model for simulation of cultures of the cyanobacterium spirulina platensis in photobioreactors: I. Coupling between light transfer and growth kinetics, *Biotechnol Bioeng* 40(7):817–825, 1992.
Cornet JF, Dussap CG, Gros JB: Conversion of radiant light energy in photobioreactors, *AIChE J* 40(6):1055–1066, 1994.
Cornet JF, Dussap CG, Gros JB, Binois C, Lasseur C: A simplified monodimensional approach for modeling coupling between radiant light transfer and growth kinetics in photobioreactors, *Chem Eng Sci* 50(9):1489–1500, 1995.
Cornet JF, Favier L, Dussap CG: Modeling stability of photoheterotrophic continuous cultures in photobioreactors, *Biotechnol Prog* 19(4):1216–1227, 2003.
Csogor Z, Herrenbauer M, Schmidt K, Posten C: Light distribution in a novel photobioreactor–modelling for optimization, *J Appl Phycol* 13:325–333, 2001.
Dauchet J: *Analyse radiative des photobioréacteurs,* 2012. Ph.D. dissertation, Université Blaise Pascal de Clermont Ferrand, *n*° 2304.
Dauchet J, Blanco S, Cornet JF, El Hafi M, Eymet V, Fournier R: The practice of recent radiative transfer Monte Carlo advances and its contribution to the field of microorganisms cultivation in photobioreactors, *J Quant Spectrosc Radiat Transf* 128:52–59, 2013.
Dauchet J, Blanco S, Cornet JF, Fournier R: Calculation of the radiative properties of photosynthetic microorganisms, *J Quant Spectrosc Radiat Transf* 161:60–84, 2015.
De La Vara LG, Gomez-Lojero C: Participation of plastoquinone, cytochrome c 553 and ferrodoxin-NADP+ oxido-reductase in both photosynthesis and respiration in Spirulina maxima, *Photosynth Res* 8(1):65–78, 1986.
Delatorre J, Baud G, Bézian JJ, et al: Monte Carlo advances and concentrated solar applications, *Sol Energy* 103:653–681, 2014.
Dimov IT: *Monte carlo methods for applied scientists,* New Jersey, 2008, World Scientific.
Durian DJ: Influence of boundary reflection and refraction on diffusive photon transport, *Phys Rev E* 50:857–866, 1994.
Dussap CG: *Etude thermodynamique et cinetique de la production de polysaccharides microbiens par fermentation en limitation par le transfert d'oxygene,* 1988. Thèse d'Etat *n*° d'ordre 409, Université Blaise Pascal de Clermont Ferrand.
Duysens LNM: *Brookhaven symp. in biol.,* 1959.
Farges B, Laroche C, Cornet JF, Dussap CG: Spectral kinetic modeling and long-term behavior assessment of arthrospira platensis growth in photobioreactor under red (620 nm) light illumination, *Biotechnol Prog* 25(1):151–162, 2009.
Glansdorff P, Prigogine I: Thermodynamic theory of structure, stability and fluctuations, *Wiley-Interscience,* 1971.
Goody RM, Yung YL: *Atmospheric radiation: theoretical basis,* Oxford, 1964, Clarendon Press.
Hammersley JM, Handscomb DC: *Monte carlo methods,* London, 1964, Chapman and Hall.
Hellmers J, Schmidt V, Wriedt T: Improving the numerical stability of t-matrix light scattering calculations for extreme particle shapes using the nullfield method with discrete sources, *J Quant Spectrosc Radiat Transf* 112(11):1679–1686, 2011.
Hoepffner N, Sathyendranath S: Effect of pigment composition on absorption properties of phytoplankton, *Mar Ecol Prog Ser* 73(1):11–23, 1991.
Hottel HC, Sarofim AF: *Radiative transfer,* New York, NY, 1967, McGraw-Hill.
Howell JR: The Monte Carlo method in radiative heat transfer, *J Heat Transf* 120(3):547–560, 1998.
Irazoqui HA, Cerdá J, Cassano AE: The radiation field for the point and line source approximations and the three-dimensional source models: applications to photoreactions, *Chem Eng J* 11(1):27–37, 1976.

Ishimaru A: *Wave propagation and scattering in random media,* New York, NY, 1999, John Wiley & Sons.

Joliot P: Oxygen evolution in algae illuminated by modulated light, *Brookhaven symp Biol* 19:418, 1966.

Kahnert FM: Numerical methods in electromagnetic scattering theory, *J Quant Spectrosc Radiat Transf* 7980:775–824, 2003.

Kalos MH, Whitlock PA: *Monte carlo methods,* ed 2, Weinheim, 2008, Wiley-VCH.

Kolokolova L, Gustafsonm BAS: Scattering by inhomogeneous particles: microwave analog experiments and comparison to effective medium theories, *J Quant Spectrosc Radiat Transf* 70(4-6):611–625, 2001.

Liu L, Mishchenko MI, Arnott WP: A study of radiative properties of fractal soot aggregates using the superposition t-matrix method, *J Quant Spectrosc Radiat Transf* 109(15): 2656–2663, 2008.

Lucarini V: *Kramers-Kronig relations in optical materials research,* Berlin, 2005, Springer.

Marshak RE: The Milne problem for a large plane slab with constant source and anisotropic scattering, *Phys Rev* 72:47–50, 1947.

Meng MP, Viskanta R: Comparison of radiative transfer approximations for a highly forward scattering planar medium, *J Quant Spectrosc Radiat Transf* 29(5):381–394, 1983.

Mengüç MP: Characterization of fine particles via elliptically-polarized light scattering. In *Purdue Heat Transfer Celebration April 3-5, 2002, West Lafayette, IN,* 2002.

Meszéna G, Westerhoff HV: Non-equilibrium thermodynamics of light absorption, *J Phys A Math Gen* 32(2):301, 1999.

Metropolis N, Ulam S: The Monte Carlo method, *J Am Stat Assoc* 44(247):335–341, 1949.

Mishchenko MI, Berg MJ, Sorensen CM, van der Mee CVM: On definition and measurement of extinction cross section, *J Quant Spectrosc Radiat Transf* 110:323–327, 2009.

Mishchenko MI, Hovenier JW, Travis LD: *Light scattering by nonspherical particles: theory, measurements, and applications,* San Diego, 2000, Academic Press.

Mishchenko MI, Travis LD, Lacis AA: *Scattering, absorption, and emission of light by small particles,* 2002, Cambridge University Press.

Morel A, Bricaud A: Theoretical results concerning light absorption in a discrete medium, and application to specific absorption of phytoplankton, *Deep Sea Res A* 28(11): 1375–1393, 1981.

Moskalensky AE, Yurkin MA, Konokhova AI, et al: Accurate measurement of volume and shape of resting and activated blood platelets from light scattering, *J Biomed opt* 18(1):017001–017001, 2013.

Mouhim RF, Cornet JF, Fontane T, Fournet B, Dubertret G: Production, isolation and preliminary characterization of the exopolysaccharide of the cyanobacterium spirulina platensis, *Biotechnol Lett* 15(6):567–572, 1993.

Muoz O, Hovenier JW: Laboratory measurements of single light scattering by ensembles of randomly oriented small irregular particles in air. a review, *J Quant Spectrosc Radiat Transf* 112(11):1646–1657, 2011.

Naqvi KR, Merzlyak MN, Melo TB: Absorption and scattering of light by suspensions of cells and subcellular particles: an analysis in terms of Kramers-Kronig relations, *Photochem Photobiol Sci* 3:132–137, 2004.

Nousiainen T: Optical modeling of mineral dust particles: a review, *J Quant Spectrosc Radiat Transf* 110(14-16):1261–1279, 2009.

Ould Moulaye CB: Calcul des propriétés de formation en solution aqueuse des composés impliqués dans les procédés biologiques et alimentaires. prédiction et réconciliation de données; modélisation des équilibres chimiques et entre phases, 1998. (Ph.D. dissertation, Université Blaise Pascal de Clermont Ferrand, num. 1033).

Paillotin G: *Etude théorique des modes de création, de transport et d'utilisation de l'énergie d'excitation electronique chez les plantes superieures,* 1974 (Ph.D. dissertation, Université de Paris-Sud, centre d'Orsay).

Pharr M, Humphreys G: *Physically based rendering: from theory to implementation,* 2010, Morgan Kaufmann Pub.

Pilon L, Berberoglu H, Kandilian R: Radiation transfer in photobiological carbon dioxide fixation and fuel production by microalgae, *J Quant Spectrosc Radiat Transf* 112(17): 2639–2660, 2011.

Pottier L, Pruvost J, Deremetz J, Cornet JF, Legrand J, Dussap CG: A fully predictive model for one-dimensional light attenuation by chlamydomonas reinhardtii in a torus photobioreactor, *Biotechnol Bioeng* 91(5):569–582, 2005.

Prigogine I: *Introduction to thermodynamics of irreversible processes,* ed 3, New York, NY, 1967, Interscience.

Pruvost J, Cornet JF: Knowledge models for the engineering and optimization of photobioreactors. In Posten C, Walter C, editors: *Microalgal Biotechnolog, Potential Production,* vol. 1, Berlin, Germany, 2012, De Gruyter.

Rochatte V, Dauchet J, Cornet JF: *Experimental validation and modelling of a photobioreactor operating with diluted and controlled light flux,* 2015 (to be published in Journal of Physics: Conference Series, Proceedings of Eurotherm Conference 105: Computational Thermal Radiation in Participating Media V, France) (in press).

Roels JA: *Kinetics and energetics in biotechnology,* Amsterdam, 1983, Elsevier Biomedical Press.

Roger M, Blanco S, El Hafi M, Fournier R: Monte Carlo estimates of domain-deformation sensitivities, *Phys Rev Lett* 95(18):180601, 2005.

Schiff LI: Approximation method for high-energy potential scattering, *Phys Rev* 103(2):443–453, 1956.

Schuster A: Radiation through a foggy atmosphere, *Astrophys J* 21:1, 1905.

Siegel R, Howell JR: *Thermal radiation heat transfer,* Washington, DC, 1981, Hemisphere Pub. Corp.

Sihvola AH: *Electromagnetic mixing formulas and applications,* London, 1999, IEE.

Spadoni G, Bandini E, Santarelli F: Scattering effects in photosensitized reactions, *Chem Eng Sci* 33(4):517–524, 1978.

StarWest, n.d. Edstar development environment project. URL http://edstar.lmd.jussieu.fr/databases.

Stucki JW: Stability analysis of biochemical systems–a practical guide, *Prog Biophys Mol Biol* 33:99–187, 1979.

Stucki JW: The optimal efficiency and the economic degrees of coupling of oxidative phosphorylation, *Eur J Biochem* 109(1):269–283, 1980.

Stucki JW: Thermodynamics of energy conversion in the cells. In *From chemical to biological organization.* Berlin, 1988, Springer Verlag.

Takache H, Christophe G, Cornet JF, Pruvost J: Experimental and theoretical assessment of maximum productivities for the microalgae Chlamydomonas reinhardtii in two different geometries of photobioreactors, *Biotechnol Prog* 26(2):431–440, 2010.

Takache H, Pruvost J, Cornet JF: Kinetic modeling of the photosynthetic growth of chlamydomonas reinhardtii in a photobioreactor, *Biotechnol Prog* 28(3):681–692, 2012.

Thormählen I, Straub J, Grigull U: Refractive index of water and its dependence on wavelength, temperature, and density, *J Phys Chem Ref Data* 14:933–945, 1985.

Tuminello PS, Arakawa ET, Khare BN, Wrobel JM, Querry MR, Milham ME: Optical properties of Bacillus subtilis spores from 0.2 to 2.5 νm, *Appl Opt* 36(13):2818–2824, 1997.

Vaillon R, Geffrin JM, Eyraud C, Merchiers O, Sabouroux P, Lacroix B: A novel implementation of a microwave analog to light scattering measurement set-up, *J Quant Spectrosc Radiat Transf* 112(11):1753–1760, 2011.

van de Hulst HC: *Light scattering by small particules,* New York, 1981, Dover Publication, Inc.
Wozniak B, Dera J, Ficek D, et al: Model of the in vivo spectral absorption of algal pigments. Part 1. Mathematical apparatus, *Oceanologia* 42(2):177–190, 2000.
Wriedt T: Light scattering theories and computer codes, *J Quant Spectrosc Radiat Transf* 110(11):833–843, 2009.
Wu TT, Qu JY, Xu M: Unified Mie and fractal scattering by biological cells and subcellular structures, *Opt Lett* 32(16):2324–2326, 2007.
Wyatt PJ: Light scattering in the microbial world, *J Collo Interf Sci* 39(3):479–491, 1972.
Yang P, Zhang Z, Baum BA, Huang HL, Hu Y: A new look at anomalous diffraction theory (ADT): algorithm in cumulative projected-area distribution domain and modified ADT, *J Quant Spectrosc Radiat Transf* 89(1-4):421–442, 2004.
Yurkin MA, Hoekstra AG: The discrete-dipole-approximation code ADDA: capabilities and known limitations, *J Quant Spectrosc Radiat Transf* 112(13):2234–2247, 2011.

CHAPTER TWO

Interaction Between Light and Photosynthetic Microorganisms

Laurent Pilon[1], Razmig Kandilian
University of California, Los Angeles, Los Angeles, CA, United States
[1]Corresponding author: e-mail address: pilon@seas.ucla.edu

Contents

Abstract

This chapter aims to introduce the physical concepts and to provide the experimental and theoretical frameworks necessary to understand and to quantify the interaction between light and photosynthetic microorganisms. Indeed, light transfer is arguably the most critical aspect to consider in designing, optimizing, and operating photobioreactors of all sizes for the production of a wide range of value-added products. This chapter presents state-of-the art theoretical and experimental methods for determining

Advances in Chemical Engineering, Volume 48
ISSN 0065-2377
http://dx.doi.org/10.1016/bs.ache.2015.12.002

the scattering phase function and the absorption and scattering cross-sections of unicellular and multicellular microorganisms as well as of colonies. An extensive database of these so-called radiation characteristics over the photosynthetically active radiation region is presented for a wide variety of promising freshwater and marine microalgae, cyanobacteria, and nonsulfur purple bacteria with various shapes, sizes, pigments, and responses to stresses. The effects of photoacclimation and of progressive and sudden nitrogen starvation on the radiation characteristics are illustrated with *Nannochloropsis oculata*. Finally, limitations of current approaches are discussed and future research directions are suggested.

1. INTRODUCTION

Photosynthetic microorganisms use sunlight as their energy source and carbon dioxide as their carbon source. Some of them are capable of producing various value-added products including (i) nutritional supplements (Richmond, 2004), (ii) biofuels such as hydrogen (Das and Veziroğlu, 2001) or lipids, in particular triglycerides (TAGs), for biodiesel production (Chisti, 2007), as well as (iii) fertilizers (Richmond, 2004; Benemann, 1979). Other species are able to remove organic waste from effluent water (Richmond, 2004).

Due to the interest in the above-mentioned applications, the cultivation of photosynthetic microorganisms in photobioreactors (PBRs) exposed to artificial light (indoor) or to sunlight (outdoor) has been studied extensively. The economic viability of large-scale cultivation can be severely reduced by poor light penetration in dense microorganism cultures (Cornet et al., 1992; Pilon et al., 2011; Béchet et al., 2013; Pruvost et al., 2014). In fact, unlike nutrient concentrations, pH, and temperature, light intensity cannot be easily homogenized in the PBRs. As discussed in detail in other chapters of this book, it is essential to accurately predict light transfer in the culture in order to design, operate, monitor, and control PBRs with optimum light availability and maximum productivity and energy conversion efficiency (Pilon et al., 2011; Cornet and Dussap, 2009). To do so, understanding and quantifying the interactions between light and photosynthetic microorganisms are essential.

This chapter aims to provide the physical concepts needed to understand and to quantify the interaction between light and photosynthetic microorganisms from both experimental and theoretical points of view. It focuses on the optical phenomena taking place up to the moment when photons are absorbed. The subsequent biological processes and pathways involved in photosynthesis have been described in detail elsewhere (Ke, 2001; Blankenship, 2008) and need not be repeated. This chapter also discusses how the interaction between light and photosynthetic microorganisms are affected by stresses. Finally, it closes by offering a few prospects.

2. BACKGROUND

2.1 Photosynthetic Microorganisms: Shapes and Sizes

There are thousands of photosynthetic microorganism species classified as diatoms, green or red microalgae, eustigmatophytes, prymnesiophytes, and cyanobacteria (Canter-Lund and Lund, 1995; Rodolfi et al., 2009). While most diatoms and green microalgae exist in unicellular forms, cyanobacteria can be either unicellular or multicellular (Becker, 1994; Schirrmeister et al., 2013). This leads to photosynthetic microorganisms with a large variety of shapes and sizes. Fig. 1A shows a micrograph of unicellular green microalgae *Chlamydomonas reinhardtii* appearing spheroidal

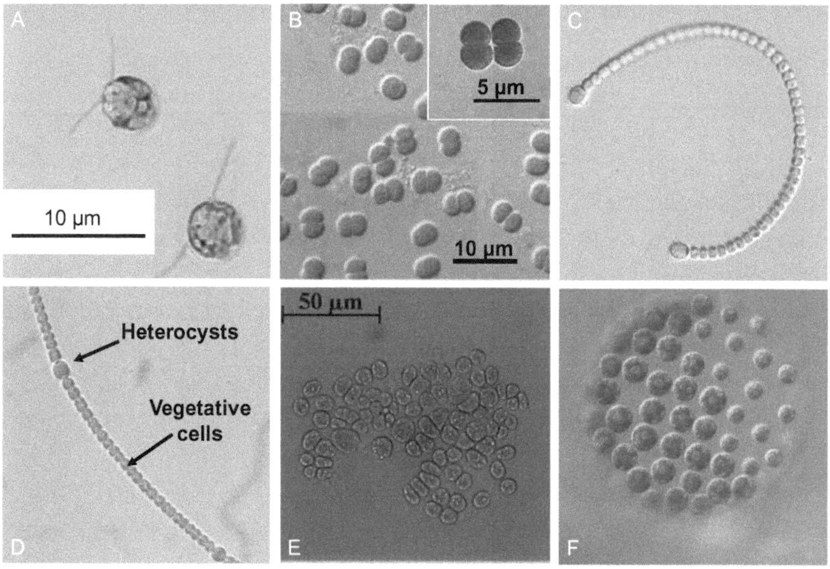

Figure 1 Micrographs of (A) *Chlamydomonas reinhardtii*, (B) dumbell-shaped *Synechocystis* sp. cell free floating and immediately after cell division (inset), (C) *Anabaenopsis* sp., (D) *Anabaena cylindrica*, (E) a colony of the microalgae *B. braunii*, and (F) *Pleodorina californica. Panels (B–D) and (F) are reproduced with permission from Prof. Yuuji Tsukii, Hosei University (http://protist.i.hosei.ac.jp/).*

with major and minor diameters around 7–10 μm. They have been considered for photobiological hydrogen (Benemann, 2000; Melis, 2002; Pilon and Berberoğlu, 2014) and lipids (Hu et al., 2008) production, and are often used as model systems. Fig. 1B shows a micrograph of a population of free-floating unicellular cyanobacterium *Synechocystis* sp. with a dumbbell shape whose lobs are about 3–5 μm in radius. They are also considered for biofuel production (Nakajima and Ueda, 1997). The inset of Fig. 1B shows a micrograph of *Synechocystis* sp. immediately after cell division into two morphologically identical daughter cells (Pinho et al., 2013). On the other hand, certain multicellular cyanobacteria such as *Anabaenopsis* sp., *elenkinii*, and *circularis* develop specialized cells called heterocysts that contain nitrogenase enzymes used for the biocatalytic reduction of atmospheric nitrogen into ammonia (Berman-Frank et al., 2003). This special ability to fix atmospheric nitrogen makes these cyanobacteria potential producers of fertilizers (Benemann, 1979). In addition, they are capable of producing hydrogen under certain conditions (Das and Veziroğlu, 2001; Tiwari and Pandey, 2012). Fig. 1C shows a micrograph of the cyanobacterium *Anabaenopsis* sp. consisting of spheroidal vegetative cells with 3–3.5 μm minor diameter and 4 μm major diameter and nearly spherical heterocysts 4–5 μm in diameter. Fig. 1D shows the filamentous heterocystous cyanobacterium *Anabaena cylindrica* consisting of connected and nearly spherical vegetative cells and fewer and larger heterocysts 2–4 μm in diameter. Filaments length varies widely but typically exceeds 100 μm.

Finally, several microalgae species of interest for various value-added products form colonies during their growth. For example, *Botryococcus braunii* secretes exopolysaccharides (EPS), a viscous substance coating the cell surface and causing their aggregation into colonies. EPS production is part of a protection mechanism activated in response to environmental conditions such as limited illumination (Dayananda et al., 2007), nonoptimal temperature (Demura et al., 2014), high salinity (Demura et al., 2014), and limited nutrient availability (Bayona and Garcés, 2014). In addition, a recent study demonstrated reversible cell aggregation in concentrated *Chlorella vulgaris* cultures (Souliés et al., 2013). Fig. 1E shows the colony formation of microalgae *Botryococcus braunii* consisting of tightly packed cells embedded in a semitransparent EPS matrix. These colonies resemble fractal aggregates formed by diffusion-limited aggregation (DLA). Finally, Fig. 1F illustrates how certain colonial green microalgae can form complex spherical aggregates containing a fixed number of distant cells including (i) eudorina (16, 32, or 64 cells), (ii) pleodorina (32–128 cells), or (iii) volvox (up to

50,000 cells). Note that large colonies are much easier to harvest that small free-floating cells which could prove practical and cost effective for industrial production (Lee et al., 2009).

2.2 Light Harvesting Antenna or Pigments

Photosynthetic microorganisms absorb photons in the photosynthetically active radiation (PAR) region ranging from 400 to 700 nm thanks to photosynthetic pigments, also referred to as light harvesting antenna. Each pigment absorbs light over different spectral bands of the solar spectrum enabling more efficient utilization of solar energy. Chlorophyll (Chl) *a*, *b*, and *c* molecules are the primary pigments responsible for absorbing visible photons and transferring the charges to the reaction center. Additionally, there exists a wide variety of accessory pigment carotenoids that can be divided into carotenes and xanthophylls (Ke, 2001). Carotenes are photosynthetic and absorb photons with wavelength corresponding to green and yellow colors and transfer the charges to chlorophyll molecules (Ke, 2001). They increase the solar light utilization efficiency of the microorganisms by broadening their absorption spectrum. On the other hand, xanthophylls act to protect the photosynthetic apparatus against excessive light (Ke, 2001). These photoprotective carotenoids shield the photosynthetic apparatus from photooxidation under large light intensities and convert excess radiant energy into heat (Lubián et al., 2000; Gentile and Blanch, 2001; Dubinsky and Stambler, 2009). In addition, phycobiliproteins are found in cyanobacteria and red algae (Madigan and Martinko, 2006). They include phycoerythrobilin (PEB) and phycourobilin (PUB), absorbing mainly around 500–550 nm, and phycocyanin (PCCN), absorbing strongly at 620 nm (Madigan and Martinko, 2006). Finally, bacteriochlorophylls absorb light mainly in the far to near infrared part of the electromagnetic spectrum (700–1000 nm) and are often found in purple bacteria (Ke, 2001).

Fig. 2 shows the specific absorption coefficient Ea (in m^2/mg) of Chl *a*, *b*, and *c*, photosynthetic (PSC) and photoprotective (PPC) carotenoids, as well as phycoerythrobilin (PEB) and phycourobilin (PUB) over the PAR region (Bidigare et al., 1990). It illustrates the two absorption peaks of Chl *a* and *b*, one in the blue and one in the red part of the visible spectrum (Ke, 2001). Chl *a* absorbs around 435, 630, and 676 nm while Chl *b* absorbs around 475 and 650 nm.

All photosynthetic species express Chl *a* but feature specific combination of other pigments. For example, marine eustigmatophycease *Nannochloropsis*

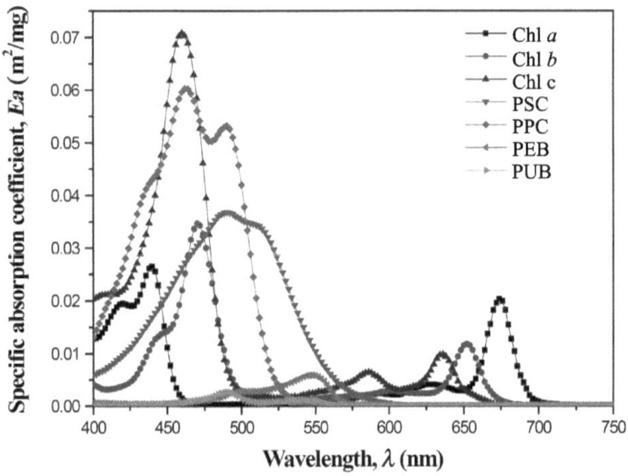

Figure 2 Specific absorption coefficient *Ea* of Chl *a*, *b*, *c*, photosynthetic (PSC) and photoprotective (PPC) carotenoids, as well as phycoerythrobilin (PEB) and phycourobilin (PUB) (Bidigare et al., 1990).

oculata contain the pigments Chl *a*, *β*-carotene, and the xanthophylls violaxanthin and vaucherxanthin but lack Chl *b* (Cohen, 1999). Note that advances in genetic engineering has enable the reduction of light harvesting antenna, i.e., the concentration of pigments (Nakajima and Ueda, 1997, 2000; Polle et al., 2003). For example, *Synechocystis* sp. has been genetically engineered with reduced light harvesting pigments (particularly PCCN), to increase their energetic yield per cell (Nakajima and Ueda, 1997). Similarly, *C. reinhardtii* has been genetically modified with truncated light antenna, i.e., with reduced Chl *a* and *b* pigment concentrations (Polle et al., 2003; Berberoğlu et al., 2008).

Finally, photosynthetic microorganisms not only absorb light but also scatter it due to the refractive index mismatch between the different cell compartments and between the cell wall and the surrounding growth medium (Jonasz and Fournier, 2007). Here, scattering refers to the elastic interaction between a photon and the microorganism resulting in the photon changing its direction while conserving its energy, i.e., its wavelength. Scattering depends mainly on the cell size and on the refractive index mismatch between the cell and the surrounding medium. The effective refractive index of the cell depends on their water content and their chemical composition (Jonasz and Fournier, 2007; Aas, 1996). The major cell constituents, namely, proteins, carbohydrates, and lipids do not absorb in the PAR region and have refractive indices larger than that of water. In addition, carbohydrates and proteins have larger refractive indices than lipids. All these constituents have refractive index nearly constant over the PAR region (Aas, 1996).

2.3 Light Transfer in Photobioreactors

Fig. 3 illustrates schematically the transport of photons in a PBR and their interaction with photosynthetic microorganisms including absorption and scattering. One can distinguish between *single scattering*, when photons are subjected to at most one scattering event, and *multiple scattering* when photons may be scattered more than once. In PBRs, microorganisms are typically uniformly distributed and randomly oriented thanks to stirring and/or bubble sparging, used as ways to keep them in suspension (Kumar et al., 2011). Thus, the PBR culture can be assumed to be homogeneous, absorbing, scattering, and nonemitting. Then, the spectral radiation intensity $I_\lambda(\mathbf{r}, \hat{\mathbf{s}})$ (in W/m^2 ·sr ·nm) along the direction $\hat{\mathbf{s}}$, at wavelength λ, and location $\mathbf{r}$ in the suspensions satisfies the radiative transfer equation (RTE) expressed as (Pilon et al., 2011)

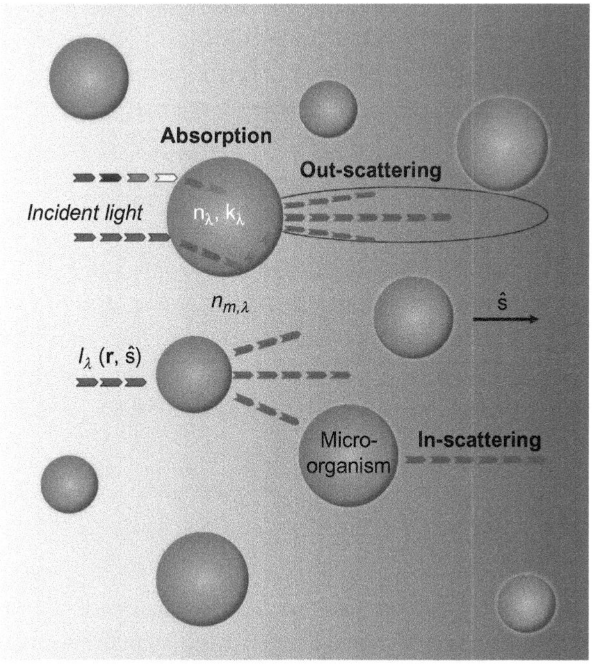

Figure 3 Illustration of light transfer in PBR including absorption and scattering of photons by photosynthetic microorganisms.

$$\hat{\mathbf{s}} \cdot \nabla I_\lambda(\mathbf{r}, \hat{\mathbf{s}}) = -\kappa_\lambda I_\lambda(\mathbf{r}, \hat{\mathbf{s}}) - \sigma_{s,\lambda} I_\lambda(\mathbf{r}, \hat{\mathbf{s}}) + \frac{\sigma_{s,\lambda}}{4\pi} \int_{4\pi} I_\lambda(\mathbf{r}, \hat{\mathbf{s}}_i) \Phi_{T,\lambda}(\hat{\mathbf{s}}_i, \hat{\mathbf{s}}) d\Omega_i. \quad (1)$$

Here, κ_λ and $\sigma_{s,\lambda}$ are the effective spectral absorption and scattering coefficients of the suspension (in m^{-1}), respectively. The extinction coefficient is defined as $\beta_\lambda = \kappa_\lambda + \sigma_{s,\lambda}$. The scattering phase function $\Phi_{T,\lambda}(\hat{\mathbf{s}}_i, \hat{\mathbf{s}})$ represents the probability that light propagating in the solid angle $d\Omega_i$ along direction $\hat{\mathbf{s}}_i$ be scattered into the solid angle $d\Omega$ along direction $\hat{\mathbf{s}}$. It is normalized such that

$$\frac{1}{4\pi} \int_{4\pi} \Phi_{T,\lambda}(\hat{\mathbf{s}}_i, \hat{\mathbf{s}}) d\Omega_i = 1. \quad (2)$$

The first and second terms on the right-hand side of Eq. (1) represent respectively the attenuation by absorption and out-scattering while the last term corresponds to the augmentation of radiation due to in-scattering (Fig. 3). This last term accounts for multiple scattering and vanishes when single scattering prevails, thus simplifying significantly the solution of the RTE.

Moreover, it is often interesting to define integral variables describing the scattering phase function $\Phi_{T,\lambda}$ in simpler terms. For example, the asymmetry factor, denoted by g_λ, for an axisymmetric phase function is defined as (Pilon et al., 2011)

$$g_\lambda = \frac{1}{2} \int_0^\pi \Phi_{T,\lambda}(\Theta) \cos\Theta \sin\Theta \, d\Theta \quad (3)$$

where Θ is the scattering angle between directions $\hat{\mathbf{s}}_i$ and $\hat{\mathbf{s}}$. The asymmetry factor varies between -1 and 1, corresponding to the limiting cases of purely backward and forward scattering, respectively. On the other hand, isotropic scattering features $\Phi_{T,\lambda}(\Theta) = 1$ and $g_\lambda = 0$. Similarly, the backward scattering ratio, denoted by b_λ, is defined as (Pottier et al., 2005)

$$b_\lambda = \frac{1}{2} \int_{\pi/2}^{\pi} \Phi_{T,\lambda}(\Theta) \sin\Theta d\Theta. \quad (4)$$

It is equal to 0, 1/2, and 1 for purely forward, isotropically, and purely backward scattering suspensions, respectively. Note that photosynthetic microorganism suspensions scatter visible light strongly in the forward direction due to their large dimensions compared with the wavelength. Then, g_λ approaches unity and b_λ tends to zero.

The average absorption $\bar{C}_{abs,\lambda}$ and scattering $\bar{C}_{sca,\lambda}$ cross-sections (in m^2) of a suspension of polydisperse microorganism cells can be related to its spectral absorption κ_λ and scattering $\sigma_{s,\lambda}$ coefficients according to (Pilon et al., 2011)

$$\bar{C}_{abs,\lambda} = \frac{\kappa_\lambda}{N_T} \quad \text{and} \quad \bar{C}_{sca,\lambda} = \frac{\sigma_{s,\lambda}}{N_T} \tag{5}$$

where N_T is the cell number density defined as the number of cells per m^3 of suspension. Alternatively, the biomass concentration of the microalgal suspension X, expressed in mass of dry weight per unit volume of suspension (g/L or kg/m^3), is often measured instead of N_T (Cornet et al., 1992; Takache et al., 2010, 2012; Kandilian et al., 2014a,b). Then, the average spectral mass absorption and scattering cross-sections $\bar{A}_{abs,\lambda}$ and $\bar{S}_{sca,\lambda}$, expressed in m^2/kg dry weight, can be expressed as

$$\bar{A}_{abs,\lambda} = \frac{\kappa_\lambda}{X} \quad \text{and} \quad \bar{S}_{sca,\lambda} = \frac{\sigma_{s,\lambda}}{X}. \tag{6}$$

Typically, the biomass concentration X in conventional PBR varies from 0.1 to 2.0 g/L (Takache et al., 2010). However, in closed intensified PBRs such as biofilm (Ozkan et al., 2012) or internally illuminated PBRs (Cornet, 2010) the biomass concentration X can reach up to 100 g/L.

2.4 Connection to Growth Kinetics and PBR Performance

From an energy point of view, the photosynthetic microorganisms "disregard" the direction of the incident photons. Then, instead of considering the directional intensity $I_\lambda(\mathbf{r}, \hat{\mathbf{s}})$, it is more appropriate to use the local spectral fluence rate $G_\lambda(\mathbf{r})$ defined as the irradiance incident from all directions, at location $\mathbf{r}$ in the PBR and expressed as

$$G_\lambda(\mathbf{r}) = \int_{4\pi} I_\lambda(\mathbf{r}, \hat{\mathbf{s}}) d\Omega. \tag{7}$$

Thanks to their different pigments, photosynthetic microorganisms can use photons with a wide variety of wavelengths. Then, the local fluence rate can be averaged over the PAR region to yield the PAR-averaged fluence rate $G_{PAR}(\mathbf{r})$ defined as (Pilon et al., 2011),

$$G_{PAR}(\mathbf{r}) = \int_{PAR} G_\lambda(\mathbf{r}) d\lambda. \tag{8}$$

Conveniently, simple analytical solutions of the RTE have been derived for $G_\lambda(\mathbf{r})$, based on the two-flux approximation, for one-dimensional flat-plate and tubular PBRs (Cornet et al., 1992, 1995). In fact, they have been shown to predict $G_\lambda(\mathbf{r})$ and $G_{PAR}(\mathbf{r})$ accurately for outdoor open ponds and flat plate PBRs exposed to both collimated and diffuse sunlight (Lee et al., 2014). Such analytical solutions bypass the need to solve the RTE numerically. However, they still requires knowledge of the radiation characteristics of the microorganism suspensions, namely κ_λ, $\sigma_{s,\lambda}$, and b_λ.

Finally, microalgae are in suspension and move quickly through the PBR. Then, the average fluence rate G_{ave} over the entire PBR of volume V can be estimated from the local PAR-averaged fluence rate as,

$$G_{ave} = \frac{1}{V}\int_V G_{PAR}(\mathbf{r})dV. \tag{9}$$

The average fluence rate G_{ave} has been used in growth kinetics models such as the Haldane-type model (Andrews, 1968; Sukenik et al., 1991; Grima et al., 1996; Acien Fernandez et al., 1997; Chen et al., 2011; Béchet et al., 2013). It can be used for optically thin PBRs where the PAR-averaged fluence rate does not vary significantly within the PBR (Fernandes et al., 2010; Lee et al., 2013; Kong and Vigil, 2014). However, when the PBR features strong gradient in $G_{PAR}(\mathbf{r})$, a more general approach is to relate the local growth rate $\mu(\mathbf{r})$ to the local fluence rate $G_{PAR}(\mathbf{r})$ and average $\mu(\mathbf{r})$ over the volume of the PBR (Cornet et al., 1998; Yun and Park, 2003; Pruvost et al., 2008; Cornet and Dussap, 2009; Murphy and Berberoğlu, 2011; Takache et al., 2012; Lee et al., 2014).

An alternative approach, based on thermodynamic and biochemical considerations, consists of defining the specific local rate of photon absorption (LRPA), $\mathscr{A}$ expressed in $\mu\text{mol}_{h\nu}/\text{kg}\cdot\text{s}$ represents the amount of photons in the PAR region absorbed per unit weight of biomass and per unit time (Cornet et al., 1992; Pruvost and Cornet, 2012). The LRPA depends on the mass spectral absorption cross-section of the species and on the spectral fluence rate in the PBR. It can be expressed as (Cornet et al., 1992)

$$\mathscr{A}(\mathbf{r}) = \int_{PAR} \bar{A}_{abs,\lambda} G_\lambda(\mathbf{r})d\lambda. \tag{10}$$

It has been used to predict the growth kinetics and biomass or lipid productivities of the PBR (Pruvost and Cornet, 2012; Takache et al., 2012; Kandilian et al., 2014a).

The specific mean rate of photon absorption (MRPA) can be determined by averaging the LRPA over the volume of the PBR according to (Cornet et al., 1992)

$$\langle \mathscr{A} \rangle = \frac{1}{V}\int_V \mathscr{A}(\mathbf{r})dV. \tag{11}$$

The MRPA $\langle \mathscr{A} \rangle$ accounts for the cumulative effects of (i) biomass concentration, (ii) the spectral mass absorption cross-section $\bar{A}_{abs,\lambda}$, and (iii) the local spectral fluence rate $G_\lambda(\mathbf{r})$ inside the PBR.

Overall, the process variables $G_{PAR}(\mathbf{r})$, G_{ave}, $\mathscr{A}(\mathbf{r})$, and $\langle \mathscr{A} \rangle$ are strongly associated with growth kinetics of microorganisms and with the productivity and efficiency of the PBRs. In order to determine these variables, it is necessary to know the radiation characteristics $\Phi_{T,\lambda}(\Theta)$, κ_λ, and $\sigma_{s,\lambda}$ of the suspension. This can be achieved numerically or experimentally, as discussed in the next sections.

3. THEORETICAL PREDICTIONS

3.1 Introduction

Theoretical predictions of the radiation characteristics $\Phi_{T,\lambda}(\Theta)$, ($\bar{C}_{abs,\lambda}$, $\bar{C}_{sca,\lambda}$) or ($\bar{A}_{abs,\lambda}$, $\bar{S}_{sca,\lambda}$) of a suspension of polydisperse photosynthetic microorganisms can be obtained by solving Maxwell's equations of electromagnetic wave theory based on the cells' shapes, size distribution, and complex index of refraction. Lorenz–Mie theory refers to the analytical solution of Maxwell's equations for homogeneous and spherical particles (Mie, 1908; Bohren and Huffman, 1998). Analytical solutions also exist for homogeneous concentric spheres or coated spheres (Aden and Kerker, 1951; Bohren and Huffman, 1998) and randomly oriented and infinitely long cylinders (Wait, 1955; Kerker, 1969; Bohren and Huffman, 1998). Note, however, that all these analytical expressions require the use of a computer program.

For more complex shapes, Maxwell's equations can be solved numerically. However, given the complexity and variations in the morphology of the photosynthetic microorganisms and despite the increasing available computing resources, simplifications of the shape and/or of the optical properties of the photosynthetic microorganisms are necessary. This section presents the different theoretical approaches used to predict the radiation characteristics of microorganisms and discusses their advantages and limitations.

3.2 Heterogeneous vs Homogeneous

Despite the heterogeneous nature of photosynthetic microorganism cells (Fig. 1), they have typically been treated as homogeneous with some effective complex index of refraction (Quirantes and Bernard, 2004; Pottier et al., 2005; Jonasz and Fournier, 2007; Berberoğlu et al., 2007; Gordon, 2011; Lee et al., 2013; Dauchet et al., 2015). This assumption can be justified by the often small mismatch in complex index of refraction between the different cell compartments. In addition, it was validated by Quirantes and Bernard (2004) who modeled single cell microalgae as homogeneous spheres and as coated spheres. The outer coating was assumed to be nonabsorbing and represented the cellular cytoplasm. By contrast, the inner core, representing the organelles and chloroplasts, was absorbing and featured a larger refractive index than the outer coating. The authors found that light absorption and scattering cross-sections of a homogeneous sphere with volume-averaged complex index of refraction were similar to those of the coated sphere for representative wavelengths, cell dimensions, and optical properties (Quirantes and Bernard, 2004).

Alternatively, one could treat microorganisms as heterogeneous cells. Advanced numerical tools can solve Maxwell's equations for very complex heterogeneous structures (Waterman, 1965; Mackowski, 1994; Mishchenko et al., 2002, 1995). However, the number of input parameters would be very large to account for the shapes, dimensions, volume fractions, and spectral complex index of refraction of the various cell compartments (e.g., nucleus, chloroplast, cell wall, mitochondria, cytoplasm, starch grains). The latter is difficult to measure *in vivo* and usually is not precisely known, in particular as a function of wavelength over the PAR region (Jonasz and Fournier, 2007). In addition, the computational cost to predict spectral radiation characteristics while accounting for the different organelles as well as the polydispersity of the cell population seems quite prohibitive for currently available computing resources. However, for some species, a compromise could be to model cells as coated spheres without adding significant complexity. For example, *Chlorella*, feature a relatively thick (~130 nm; Gerken et al., 2013) and nonabsorbing but strongly refracting ($n \sim 1.5$; Atkinson Jr et al., 1972; Traverse, 2007) cell wall that could be modeled as the coating of a homogeneous core.

3.3 Effective Optical Properties of Photosynthetic Microorganisms

Jonasz and Fournier (2007) reviewed various methods used to predict or measure (i) the average refractive index n, (ii) the spectral refractive index n_λ,

(iii) the spectral absorption index k_λ, or (iv) the complex index of refraction m_λ for phytoplankton and bacteria in water. For example, Aas (1996) used the Lorenz–Lorenz effective medium approximation (EMA) to determine n from the various components of the phytoplankton cells. The authors pointed out that uncertainty in the water content had a significant effect on the predictions, even more than the choice of EMA. Alternatively, Bricaud, Morel, and Stramski (Bricaud and Morel, 1986; Bricaud et al., 1988; Stramski et al., 2001) used the Helmholtz–Ketteler theory (Jonasz and Fournier, 2007) for n_λ and k_λ to predict the complex index of refraction of various phytoplanktons. The parameters of the model were retrieved by fitting theoretical predictions with experimental measurements of absorption spectra for various phytoplankton suspensions.

Pottier et al. (2005) predicted the radiation characteristics of *C. reinhardtii* using the Lorenz–Mie theory assuming that (i) the cells were homogeneous and spherical, (ii) the refractive index was constant over the PAR and equal to 1.55, and (iii) the absorption index was given by

$$k_\lambda = \frac{\lambda}{4\pi}\sum_j C_j Ea_j = \frac{\lambda}{4\pi}\rho_{dry}(1-x_w)\sum_j w_j Ea_j \tag{12}$$

where C_j is the concentration of jth pigment in the cell (in kg/m^3) while ρ_{dry} is the density of the dry biomass (in kg/m^3), x_w is the average water mass fraction in the cells, and $w_j = C_j/X$ is the concentration of jth pigment on a dry mass basis. Moreover, Ea_j (in m^2/kg) is the specific absorption cross-section of individual pigments, as reported by Bidigare et al. (1990) and reproduced in Fig. 2.

Recently, Dauchet et al. (2015) relaxed the assumption of constant refractive index. Instead, they predicted the refractive index n_λ of microalgae cells using the subtractive Kramers–Kronig relation based on the absorption index k_λ of the microorganism, estimated by Eq. (12) (Pottier et al., 2005). Then, the refractive index of the cell was estimated according to (Dauchet et al., 2015)

$$n_\nu = n_{\nu_p} + 2\frac{(\nu^2-\nu_p^2)}{\pi}P\int_{\nu_{\min}}^{\nu_{\max}} \frac{\nu' k_{\nu'}}{(\nu'^2-\nu^2)(\nu'^2-\nu_p^2)}d\nu'. \tag{13}$$

where $\nu = c/\lambda$ is the frequency of radiation, c is the speed of light in vacuum, and P is the Cauchy principal value. The anchor frequency denoted by ν_p was chosen such that the cells did not absorb at that frequency, i.e., $k_{\nu_p} = 0$. On the other hand, the value for n_{ν_p} must be known or retrieved experimentally.Dauchet et al. (2015) chose an anchor wavelength λ_p as 820 nm for *C. reinhardtii* as green microalgae do not absorb at $\lambda \geq 750$ nm (Dauchet et al., 2015). The authors retrieved a value of $n_{\nu_p} = 1.44$ for *C. reinhardtii* using an inverse method that minimized the difference between the measured and the predicted normal–hemispherical transmittance at 820 nm. The latter was estimated by solving the RTE using the Monte Carlo method and the predicted radiation characteristics of the microorganisms.

Finally, Pilon and coworkers (Lee et al., 2013; Kandilian et al., 2013; Heng et al., 2014) retrieved the spectral complex index of refraction m_λ over the PAR region for various microalgae and cyanobacteria from the experimental measurements of their absorption and scattering cross-sections. The authors used an inverse method based on genetic algorithm and a forward method based on one of the theoretical models described in the next sections.

3.4 Radiation Characteristics of Unicellular Spheroidal Microorganisms

The radiation characteristics of axisymmetric spheroidal microorganisms, such a *C. reinhardtii* (Fig. 1A), with major and minor diameters a and b can be predicted numerically using (i) the T-matrix method (Waterman, 1965; Mackowski, 1994; Mishchenko et al., 2002, 1995), (ii) the discrete-dipole approximation (Draine, 1988), and (iii) the finite-difference time-domain method (Liou, 2002). Most often, however, they have been approximated as homogeneous spheres with some equivalent radius r_{eq} and some effective complex index of refraction $m_\lambda = n_\lambda + ik_\lambda$ (Pottier et al., 2005; Berberoğlu et al., 2007; Dauchet et al., 2015), as discussed in Section 3.6.1.

In general, the size-averaged absorption $\bar{C}_{abs,\lambda}$ and scattering $\bar{C}_{sca,\lambda}$ cross-sections of polydisperse spheroidal microalgae suspension with size distribution $f(a, b)$ can be estimated as (Pilon et al., 2011)

$$\bar{C}_{abs,\lambda} = \int_0^\infty \int_0^\infty C_{abs,\lambda}(a,b) f(a,b) da db$$
$$\text{and} \quad \bar{C}_{sca,\lambda} = \int_0^\infty C_{sca,\lambda}(a,b) f(a,b) da db. \tag{14}$$

Note that the same expressions apply to the mass cross-sections $\bar{A}_{abs,\lambda}$ and $\bar{S}_{sca,\lambda}$. Similarly, the total scattering phase function $\Phi_{T,\lambda}(\Theta)$ of the suspension is expressed (Modest, 2013)

$$\Phi_{T,\lambda}(\Theta) = \frac{1}{\bar{C}_{sca,\lambda}} \int_0^\infty \int_0^\infty C_{sca,\lambda}(a,b)\Phi_\lambda(a,b,\Theta) f(a,b)\, da\, db \quad (15)$$

where $\Phi_\lambda(a, b, \Theta)$ is the scattering phase function of a single spheroidal scatterer with major and minor diameters a and b. Similar averaging over the cell population can be formulated for particles with other shapes as long as the geometry can be parameterized with one or more parameters.

Finally, the average absorption $\bar{C}_{abs,\lambda}$ and scattering $\bar{C}_{sca,\lambda}$ cross-sections of the suspension are related to the mass absorption $\bar{A}_{abs,\lambda}$ and scattering $\bar{S}_{sca,\lambda}$ cross-sections by (Pottier et al., 2005)

$$\bar{A}_{abs,\lambda} = \frac{\bar{C}_{abs,\lambda}}{V_{32}\rho_{dry}(1-x_w)} \quad \text{and} \quad \bar{S}_{sca,\lambda} = \frac{\bar{C}_{sca,\lambda}}{V_{32}\rho_{dry}(1-x_w)}. \quad (16)$$

Here, V_{32} (in m^3) is the Sauter mean diameter of the cells, x_w is the average mass fraction of water in the cells, and ρ_{dry} is the density of dry material in the biomass. This relationship is often useful when comparing theoretical predictions and experimental measurements or for retrieving the effective complex index of refraction of microorganisms from the measurements of $\bar{A}_{abs,\lambda}$ and $\bar{S}_{sca,\lambda}$ (Lee et al., 2013; Heng et al., 2014). However, it requires knowledge of x_w whose measurement is often affected by large experimental uncertainty (Jonasz and Fournier, 2007).

3.5 Multicellular Microorganisms and Colonies

Several numerical methods exist to estimate the radiation characteristics of (i) multicellular microorganisms, such as filamentous cyanobacteria, and (ii) aggregates consisting of spherical cells, such as microalgae colonies. They include the superposition T-matrix method (Mackowski, 1994; Mackowski and Mishchenko, 1996, 2011; Mishchenko, 2015), the generalized multiparticle-Mie theory (Xu, 1997), and the volume integral method (Iskander et al., 1989), to name a few. The superposition T-matrix method is based on the superposition solutions of Maxwell's equations for single spherical monomers or cells. The electromagnetic (EM) field scattered by the entire aggregate of cells is the sum of the EM fields scattered by each of the constituent cells (Mackowski and Mishchenko, 1996). The EM field incident onto a monomer takes into account not only the incident EM field but also the scattered fields from all the other cells in the aggregate (Mackowski and Mishchenko, 1996). The interacting fields are transformed into a system of sphere-centered equations for the scattering coefficients and inverted to obtain the T-matrix (Mackowski and Mishchenko, 1996). Then, the absorption and scattering cross-sections of the randomly oriented aggregate of spherical cells can be obtained from operations on the T-matrix (Mackowski and Mishchenko, 1996).

The use of the T-matrix method for nonspherical particles and for aggregates of spherical particles has been popularized by Mishchenko and Mackowski thanks to the availability of regularly updated computer programs and their user's manual (Mishchenko et al., 2002; Mackowski and Mishchenko, 2011; Mishchenko, 2015). However, depending on the number of cells and/or on the size of the aggregate, calculations can be time consuming and often require large computational resources (Kimura et al., 2003).

3.6 Equivalent Scattering Particles

Efforts have been made to approximate the radiation characteristics of (i) nonspherical unicellular microorganisms, (ii) multicellular microorganisms, and (iii) aggregates of cells with complex morphologies by those of particles with simple shapes such as spheres, coated spheres, or infinitely long cylinders. The radiation characteristics of such scatterers with simple shapes can be computed relatively rapidly (Kerker, 1969; Bohren and Huffman, 1998; Kahnert et al., 2014) compared with predictions by the T-matrix method, for example. This is particular important for real-time monitoring and control of PBRs as well as for inverse method aiming to retrieve the spectral complex index of refraction of microorganisms from their measured cross-sections (Heng et al., 2014).

3.6.1 Nonspherical Unicellular Microrganisms

As previously mentioned, nonspherical cells have been modeled as spheres with equivalent radius and effective complex index of refraction. This approximation can be justified by the fact that they are typically well mixed and randomly oriented in the PBRs. Then, the equivalent radius r_{eq} can be approximated such that either the volume or the surface area of the equivalent sphere is identical to that of the actual cell. The radius r_v of the volume-equivalent sphere can be expressed as

$$r_v = a\left(\frac{3}{2\epsilon}\right)^{1/3} \tag{17}$$

where, ϵ is the spheroid aspect ratio defined as $\epsilon = a/b$. Alternatively, the radius r_s of the surface area-equivalent sphere is given by

$$r_s = \frac{1}{4}\left(2a^2 + 2ab\frac{\sin^{-1}e}{e}\right)^{1/2} \quad \text{where} \quad e = \frac{(\epsilon^2 - 1)^{1/2}}{\epsilon}. \tag{18}$$

Lorenz–Mie theory predicts the absorption $C_{abs,\lambda}$ and scattering $C_{sca,\lambda}$ cross-sections of a single homogeneous spherical cell based on (i) the equivalent cell radius r_{eq} (e.g., r_V or r_s), (ii) the wavelength λ of interest, (iii) the refractive index $n_{m,\lambda}$ of the nonabsorbing medium at λ, and (iv) the complex index of refraction of the microorganism $m_\lambda = n_\lambda + ik_\lambda$. In fact, the cross-sections depend only on the size parameter $\chi_{eq} = 2\pi r_{eq}/\lambda$ and on the relative index of refraction $m_{r,\lambda} = m_\lambda / n_{m,\lambda}$, i.e., $C_{abs/sca,\lambda} = C_{abs/sca,\lambda}(\chi_{eq}, m_{r,\lambda})$.

In addition, the anomalous diffraction approximation can also be used to predict the cross-sections of spherical cells based on the facts that (i) their relative complex index of refraction $m_{r,\lambda}$ is such that $|m_{r,\lambda} - 1| \ll 1$ and (ii) the size parameter $\chi_{eq} = 2\pi r_{eq}/\lambda$ satisfies $\chi_{eq}|m_{r,\lambda} - 1| \ll 1$ (van de Hulst, 2012; Jonasz and Fournier, 2007). This approximation offers simple analytical expressions for the cross-sections expressed as $C_{abs/sca,\lambda} = C_{abs/sca,\lambda}(\chi_{eq}, m_{r,\lambda})$. It has been widely used in the ocean optics community to predict the radiation characteristics of phytoplanktons (Bricaud and Morel, 1986; Stramski et al., 1988; Bricaud et al., 1988; Jonasz and Fournier, 2007).

Nonspherical cells can also be easily modeled as coated spheres. For example, Quirantes and Bernard (2006) modeled *Aureococcus anophagefferens* cells as coated spheres with a shell volume fraction of 15%. The inner core and outer coating corresponded to the cytoplasm and chloroplast and their complex index of refraction was equal to 1.36 and 1.4+i0.005, respectively. The authors compared theoretical predictions of algal bloom reflectance to measurements by a tethered surface radiometer. They found better agreement between measurements and prediction when the cells were modeled as coated spheres compared to when they were modeled as homogeneous spheres. This was attributed to the larger backscattering ratio of the coated spheres compared to homogeneous spheres of the same outer radius and effective volume-averaged complex index of refraction.

3.6.2 Multicellular Microorganisms and Colonies

Recently, Lee and Pilon (2013) demonstrated that the absorption and scattering cross-sections per unit length of randomly oriented linear chains of spheres, representative of filamentous cyanobacteria (Fig. 1D), can be approximated as those of randomly oriented infinitely long cylinders with equivalent volume per unit length. Then, for linear chains of monodisperse cells of diameter d_s, the diameter $d_{c,V}$ of the volume-equivalent infinitely long cylinder is given by $d_{c,V} = \sqrt{2/3}d_s$. This approximation was used to retrieve the spectral complex index of refraction of *Anabeana cylindrica* over the PAR region (Heng et al., 2014).

Heng et al. (2015) demonstrated that the absorption $\bar{C}_{abs,\lambda}$ and scattering $\bar{C}_{sca,\lambda}$ cross-sections and the asymmetry factor g_λ of bispheres, quadspheres, and rings of up to 20 spherical cells could be approximated as those of coated spheres such that (i) the coating has the same total volume V_T and complex index of refraction m_λ as the cells, (ii) the inner core has the same index of refraction $n_{m,\lambda}$ as the surrounding medium, and (iii) the projected area of the equivalent coated sphere is the same as the average projected area $\bar{A}_p$ of the multicellular microorganisms. Kandilian et al. (2015) proved that this volume and average projected area equivalent coated sphere approximation can also be used for fractal aggregates of up to 1000 spherical microalgae with a wide range of size parameter and relative complex index of refraction. In fact, this approximation was able to capture the effects of both multiple scattering and shading among constituent cells on the integral radiation characteristics of the aggregates. Then, the equivalent coated sphere has inner $r_{i,V+\bar{A}_p}$ and outer $r_{o,V+\bar{A}_p}$ radii expressed as (Heng et al., 2015)

$$r_{i,V+\bar{A}_p} = \left(r_{o,V+\bar{A}_p}^3 - \frac{3}{4\pi}V_T\right)^{1/3} \quad \text{and} \quad r_{o,V+\bar{A}_p} = \left(\frac{\bar{A}_p}{\pi}\right)^{1/2} \tag{19}$$

The total volume V_T of an arbitrary aggregate made of N_s polydisperse spherical cells of radius $(r_j)_{1\leq j\leq N_s}$ can be written as (Heng et al., 2015)

$$V_T = \sum_{j=1}^{N_s} \frac{4\pi}{3} r_j^3 \tag{20}$$

The average projected area $\bar{A}_p$ of randomly oriented multicellular microorganisms and of colonies of spherical cells can be estimated numerically, as described in Heng et al. (2015). For monodisperse spherical cells, it was found to be proportional to the square of the constitutive cell radius r_s such that (Heng et al., 2015)

$$\bar{A}_p = \alpha(N_s) r_s^2 \tag{21}$$

where α is a constant depending on the number of cells N_s in the multicellular microbe. For bispheres and quadspheres, $\alpha(2)$ and $\alpha(4)$ were found to be equal to 5.35 and 9.70, respectively. For a circular ring of N_s monodisperse cells $\alpha(N_s)$ was such that $\alpha(N_s) = 2.42N_s$ for $N_s \geq 5$. Similarly, for fractal aggregates, $\alpha(N_s)$ was given by $\alpha(N_s) = \pi N_s^\gamma$ where the exponent γ was a function of the aggregate's fractal dimension D_f. It was fitted with numerically generated data for different values of D_f varying from the limiting cases of $D_f = 1.0$ corresponding to linear chains of spheres and $D_f = 3.0$ for spheres aggregated in a simple cubic packing so that (Kandilian et al., 2015)

$$\gamma = 0.73 + 0.19\left[1 + \left(\frac{D_f - 1}{2}\right)^{1.8}\right]^{1/1.8} \tag{22}$$

Note that the coated sphere approximation can be used in other fields as different from light transfer in PBR as combustion systems (Drolen and Tien, 1987; Mengüç et al., 1994) and atmospheric science (Latimer and Wamble, 1982; Latimer, 1985). In addition, $\bar{A}_p$ can be measured using image analysis of two-dimensional micrographs of freely suspended microorganisms (Brown and Vickers, 1998).

Overall, the main challenges of the theoretical approach for predicting the radiation characteristics of photosynthetic microorganisms reside in (i) predicting accurately both their effective refractive index n_λ and the absorption index k_λ as functions of wavelength and of the cell's biochemical composition and in (ii) accounting for their complex shape and their polydispersity. To date, these challenges have not been fully addressed and/or the state of the art models have not been rigorously validated. Experimental measurements offer an alternative to determine the microorganisms' radiation characteristics without relying on assumption difficult to verify.

4. EXPERIMENTAL MEASUREMENTS

This section presents a versatile method to measure directly the complete set of radiation characteristics $\Phi_{T,\lambda}(\Theta)$, and ($\bar{C}_{abs,\lambda}$, $\bar{C}_{sca,\lambda}$) or ($\bar{A}_{abs,\lambda}$, $\bar{S}_{sca,\lambda}$) of photosynthetic microorganisms of various shapes and sizes.

4.1 Assumptions

The following assumptions are necessary in the data analysis of experimental measurements: (1) the photosynthetic microorganisms are well mixed and randomly oriented. (2) For all measurements, the pathlength and cell concentration of the samples are relatively small such that single scattering prevails, i.e., photons undergo one scattering event at most as they travel through the suspension. (3) The scattering phase function $\Phi_{T,\lambda}(\Theta)$ has azimuthal symmetry and is only a function of the polar angle. This can be satisfied by ensuring that the microorganisms are randomly oriented (Jonasz and Fournier, 2007). In addition, (4) $\Phi_{T,\lambda}$ is assumed to be time invariant and constant over the PAR region. Finally, (5) the suspension is scattering in the forward direction, i.e., the scatterers are large compared with the wavelength of interest.

4.2 Scattering Phase Function

The scattering phase function $\Phi_{T,\lambda}(\Theta)$ of the microalgae can be measured using a polar nephelometer, as illustrated in Fig. 4. A typical nephelometer is comprised of a probe with a small acceptance angle such that it can measure the scattered radiation as a function of the polar angle. A laser provides a continuous beam narrowly centered around wavelength λ in the PAR region. It is modulated by a beam chopper at constant frequency. The modulated beam is collimated and reduced in size by a set of collimating lenses and a pinhole. The collimated beam enters a sample holder dish containing

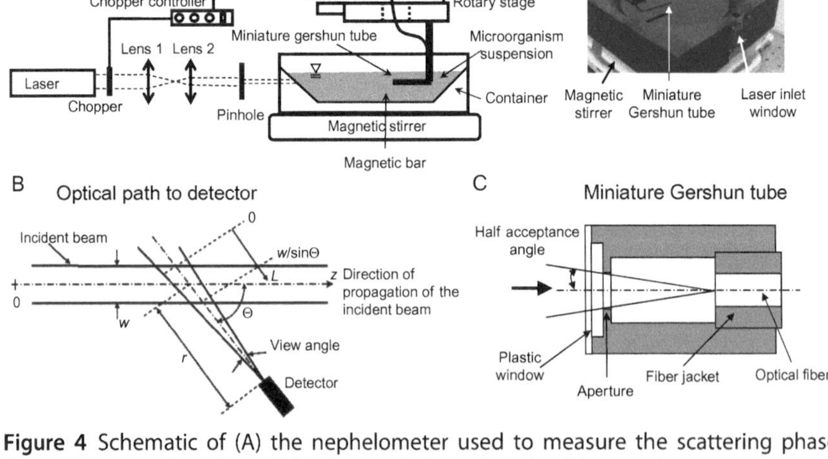

Figure 4 Schematic of (A) the nephelometer used to measure the scattering phase function at wavelength λ of the laser. (B) Optical path with coordinate system used in recovering the scattering phase function from the measured intensity distribution (Privoznik et al., 1978). (C) The miniaturized Gershun tube (drawings not to scale).

the microorganism suspension through a transparent glass window. The microorganisms are kept in suspension and randomly oriented with the aid of a black magnetic stirring bar and a magnetic stirrer. The scattered light is collected with a custom made fiber-optic probe immersed in the suspension and consisting of (i) a miniaturized Gershun tube with a small half acceptance angle and (ii) a UV–IR fiber-optic cable. The probe is mounted on a computer controlled motorized rotary stage (Fig. 4A). The collected light is detected with a photomultiplier tube (PMT) and amplified with a lock-in amplifier. The PMT is powered with a variable high voltage power supply. The latter enables the sensitivity of the PMT to be varied so that the input to the lock-in amplifier is within its detection range. Use of the lock-in amplifier together with the beam chopper enables the detection of noisy signals otherwise difficult to detect.

The nephelometer measures the scattered intensity I_λ in $\mathrm{Wm^{-2}sr^{-1}}$ as a function of the polar angle Θ. Then, the scattering phase function can be obtained based on the analysis derived by Privoznik et al. (1978) and leading to

$$\Phi_\lambda(\Theta) = \frac{2 I_\lambda(\Theta)[U_\lambda(\Theta)]^{-1}}{\int_0^\pi I_\lambda(\Theta)[U_\lambda(\Theta)]^{-1} \sin\Theta d\Theta} \tag{23}$$

The geometrical correction term $U_\lambda(\Theta)$ accounts for the variation of the scattering volume and the pathlength with detection angle and is given by (Privoznik et al., 1978)

$$U_\lambda = \int_0^{w/\sin\Theta} \left(1 + \beta_\lambda \frac{w}{2} c\tan\Theta - \beta_\lambda L \cos\Theta\right)\left[1 - \beta_\lambda\left(r - \frac{w}{2\sin\Theta}\right)\right] \times \left[1 - \beta_\lambda\left(\frac{w}{\sin\Theta} - L\right)\right] dL \tag{24}$$

where w is the beam diameter, r is the radius of rotation of the fiber-optic probe, and L is the coordinate direction along the line of sight of the detector, marking the length of the scattering volume (Fig. 4B).

Moreover, the extinction coefficient $\beta_\lambda = \kappa_\lambda + \sigma_{s,\lambda}$ of the suspension can be determined with the nephelometer by measuring the radiation flux $F_\lambda(z)$, expressed in $\mathrm{Wm^{-2}}$, at two different locations z_1 and z_2 along the path of a divergent incident beam. Then, the extinction coefficient is given by

$$\beta_\lambda = \frac{\ln|F_\lambda(z_2)/F_\lambda(z_1) + \ln(z_2^2/z_1^2)}{z_1 - z_2} \tag{25}$$

where z is the distance between the detector and the virtual image of the last lens in the optical setup.

Finally, the above measurements can be performed for different wavelengths by employing different types of lasers emitting at different wavelengths. However, there exists a limited number of options in the PAR region. Alternatively, one can assume that the scattering phase function is independent of wavelength as done in the literature for *A. variabilis* (Merzlyak and Naqvi, 2000). This has been corroborated with analysis and experimental measurements for cyanobacteria *Synechococcus* (Stramski and Mobley, 1997) and green microalgae *N. occulata* (Kandilian et al., 2013).

4.3 Absorption and Scattering Cross-Sections

The average absorption and scattering cross-sections $\bar{C}_{abs,\lambda}$ and $\bar{C}_{sca,\lambda}$ of microorganisms suspensions can be experimentally measured using a spectrometer equipped with an integrating sphere. First, the spectral normal–normal $T_{nn,\lambda}$ and normal–hemispherical $T_{nh,\lambda}$ transmissions of several dilute suspensions with different known concentrations are measured, as illustrated in Fig. 5. Here, the scattering phase function $\Phi_{T,\lambda}(\Theta)$ previously measured for the same suspension is used to correct for various optical effects.

The apparent extinction coefficient β_λ^* can be obtained from normal–normal transmittance measurements of cuvettes, of pathlength t containing either the microalgae suspension $T_{nn,\lambda,X}$ or the reference medium $T_{nn,\lambda,ref}$ (Pilon et al., 2011)

$$\beta_\lambda^* = -\frac{1}{t}\ln\left(\frac{T_{nn,\lambda,X}}{T_{nn,\lambda,ref}}\right). \tag{26}$$

Similarly, the apparent absorption coefficient κ_λ^* can be defined from the normal–hemispherical transmittance $T_{nh,\lambda}$ as (Pilon et al., 2011)

$$\kappa_\lambda^* = -\frac{1}{t}\ln\left(\frac{T_{nh,\lambda,X}}{T_{nh,\lambda,ref}}\right) \tag{27}$$

In addition, the apparent extinction coefficient β_λ^* can also be expressed as a function of the actual absorption κ_λ and scattering $\sigma_{s,\lambda}$ coefficients (Pilon et al., 2011)

$$\beta_\lambda^* = \kappa_\lambda + (1-\epsilon_n)\sigma_{s,\lambda}. \tag{28}$$

Here, ϵ_n represents the fraction of light scattered in the forward direction and detected by the spectrometer. Ideally, ϵ_n is equal to zero and $\beta_\lambda^* = \beta_\lambda$.

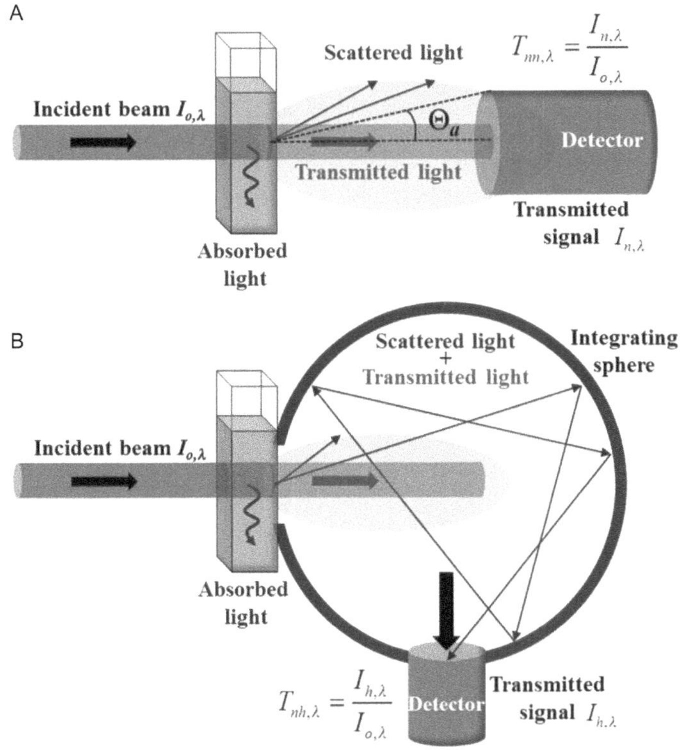

Figure 5 Schematic of experimental setup used to determine (A) the extinction coefficient β_λ from normal–normal spectral transmittance and (B) the absorption coefficient κ_λ from normal–hemispherical spectral transmittance.

However, due to the finite size of the acceptance angle of the detector, ϵ_n is larger than zero and is assumed to be constant over the PAR region. It can be defined from the suspension's scattering phase function $\Phi_{T,\lambda}(\Theta)$ previously measured as (Pilon et al., 2011)

$$\epsilon_n = \frac{1}{2}\int_0^{\Theta_a} \Phi_{T,\lambda}(\Theta)\sin\Theta d\Theta \tag{29}$$

where Θ_a is the half acceptance angle of the spectrometer's detector (Fig. 5A). The actual extinction coefficient $\beta_\lambda = \kappa_\lambda + \sigma_{s,\lambda}$ can then be determined according to

$$\beta_\lambda = \frac{\beta_\lambda^* - \epsilon_n \kappa_\lambda}{1 - \epsilon_n}. \tag{30}$$

Similarly, the apparent absorption coefficient κ_λ^* is related to the actual absorption κ_λ and scattering $\sigma_{s,\lambda}$ coefficients according to (Pilon et al., 2011)

$$\kappa_\lambda^* = \kappa_\lambda + (1 - \epsilon_h)\sigma_{s,\lambda}. \tag{31}$$

Here, ϵ_h is the fraction of the scattered light detected by the detector connected to the integrating sphere. Ideally, when all the scattered light is accounted for, ϵ_h is equal to unity. Moreover, at $\lambda = 750$ nm green microalgae are assumed to be nonabsorbing, i.e., $\kappa_{750} = 0\ \text{m}^{-1}$. Then, Eqs. (28) and (31) at 750 nm simplify to

$$\beta_{750}^* = (1 - \epsilon_n)\sigma_{s,750} \quad \text{and} \quad \kappa_{750}^* = (1 - \epsilon_h)\sigma_{s,750}. \tag{32}$$

Combining Eqs. (30) to (32) yields

$$\kappa_\lambda = \kappa_\lambda^* - \kappa_{750}^* \left(\frac{\beta_\lambda^* - \kappa_\lambda^*}{\beta_{750}^* - \kappa_{750}^*} \right) \quad \text{and} \quad \sigma_{s,\lambda} = \frac{\beta_\lambda^* - \epsilon_n \kappa_\lambda}{1 - \epsilon_n} - \kappa_\lambda. \tag{33}$$

Then, the average absorption $\bar{C}_{abs,\lambda}$ and scattering $\bar{C}_{sca,\lambda}$ cross-sections of the microorganism suspension can be estimated as

$$\bar{C}_{abs,\lambda} = \kappa_\lambda / N_T \quad \text{and} \quad \bar{C}_{sca,\lambda} = \sigma_{s,\lambda} / N_T. \tag{34}$$

Similarly, κ_λ and $\sigma_{s,\lambda}$ can be divided by the samples' respective dry mass concentration X to obtain the average mass absorption and scattering cross-sections $\bar{A}_{abs,\lambda}$ and $\bar{S}_{sca,\lambda}$.

Finally, in this method, the pathlength and concentration of the samples are to be chosen such that single scattering prevails, ie, photons undergo at most one scattering event as they travel through the suspension (Assumption 1). To verify this important assumption, van de Hulst (2012) suggested that "a simple and conclusive test for the absence of multiple scattering" consists of demonstrating that the scattered intensity is directly proportional to the particle concentration. In other words, the spectral cross-sections $\bar{C}_{abs,\lambda}$ and $\bar{C}_{sca,\lambda}$ (or $\bar{A}_{abs,\lambda}$ and $\bar{S}_{sca,\lambda}$) for different values of cell density N_T (or X) should collapse onto a single line if single and independent scattering prevailed. This provides further validation of the experimental procedure and data analysis.

4.4 Validation of the Experimental Procedure

Before measuring the radiation characteristics of photosynthetic microorganisms, the experimental setups, procedures, and data analysis should be rigorously validated. To do so, experimental results for $\Phi_{T,\lambda}(\Theta)$, κ_λ, and $\sigma_{s,\lambda}$ of scatterers of known shape, size distribution, and complex index of refraction can be compared with theoretical predictions based on the exact solution of Maxwell's equations (see Section 3). Examples of such well characterized and commercially available suspensions include polystyrene latex or glass microspheres and long glass fibers.

4.4.1 Validation of the Scattering Phase Function Measurements

Fig. 6A compares the experimentally determined scattering phase function at 632.8 nm of polystyrene latex microspheres with predictions by the Lorenz–Mie theory. The microspheres had a Gaussian size distribution of mean diameter 19 μm and standard deviation 3.56 μm. The particle complex index of refraction (in air) at 633 nm was $m_\lambda = 1.5823 + i4 \times 10^{-4}$ (Ma et al., 2003). Note that experimental measurements describe a smooth line whereas theoretical predictions show strong oscillations in $\Phi_{T,\lambda}(\Theta)$. This difference was due to the fact experimentally, the microspheres were polydisperse and in constant motion in the stirred suspension. In fact, several of them may pass by the probing volume (Fig. 4B) during the finite acquisition time of the PMT. By contrast, the theoretical predictions considered a single spherical particle of diameter 19 μm at rest in the electromagnetic (EM) field. Similarly, Fig. 6B compares the experimentally measured scattering phase function of long glass fibers and theoretical prediction for randomly oriented infinitely long cylinders of diameter 15–20 μm with complex index of refraction of $1.4567 + i\,10^{-7}$ at 632.8 nm (Malitson, 1965; Kang et al., 2001). Similarly successful validation has been obtained with monodisperse

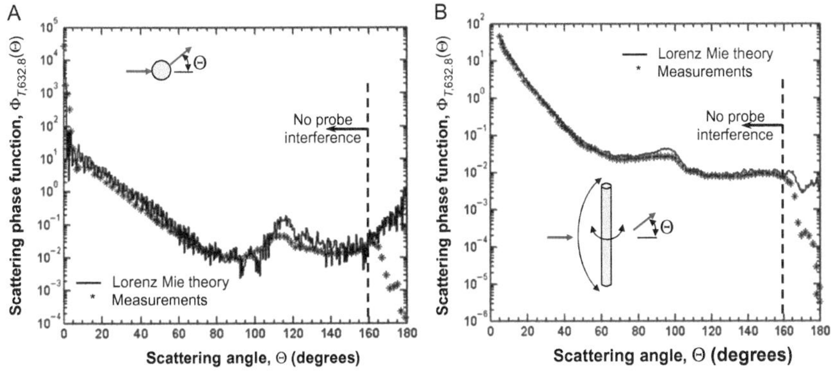

Figure 6 Comparison of the scattering phase functions at 632.8 nm measured experimentally for (A) polydisperse polystyrene latex microspheres with mean diameter 19 μm and (B) randomly oriented infinitely long glass fibers 15–20 μm in diameter along with the corresponding theoretical predictions (Berberoğlu and Pilon, 2007; Berberoğlu et al., 2008).

polystyrene latex microspheres 5 μm in diameter (Kandilian, 2014). Note that measurements beyond scattering angle of 160° should be disregarded because of the interference of the rotating fiber optic probe with the incident laser beam (Berberoğlu and Pilon, 2007). This has little consequence on the determination of the asymmetry factor g_λ since only a small amount of energy is present in these backscattering angles beyond 160° due to the large size of the microorganisms compared with the wavelength.

Overall, for both polydisperse microspheres and randomly oriented long cylinders, very good agreement was found between experimental measurements and theoretical predictions. These results demonstrate the capability of the nephelometer to measure the scattering phase function of scatterers of various shapes and sizes.

4.4.2 Validation of the Cross-Section Measurements

Fig. 7 compares the experimentally measured (A and C) absorption $\bar{C}_{abs,\lambda}$ and (B and D) scattering $\bar{C}_{sca,\lambda}$ cross-sections between 400 and 700 nm of monodisperse latex spheres 2.02 and 4.5 μm diameter with Lorenz–Mie theory predictions using the complex index of refraction of latex reported by Ma et al. (2003). Here also, the good agreement between theoretical and experimental results successfully validated the experimental setup and the data analysis. Similar validation has been performed with the same polydisperse polystyrene latex microspheres and randomly oriented and infinitely long glass fibers considered for validating the scattering phase function measurements, as illustrated in Fig. 6 (Berberoğlu and Pilon, 2007).

4.4.3 Validation of Single Scattering Assumption

Fig. 8A and B respectively show the spectral absorption κ_λ and scattering $\sigma_{s,\lambda}$ coefficients measured in the PAR region for dilute solutions of cyanobacteria *A. cylindrica* with mass concentrations X equals to 0.202, 0.296, and 0.431 kg/m^3. Each data point represents the arithmetic mean of κ_λ and $\sigma_{s,\lambda}$ measured three times for each concentration and the error bars correspond to 95% confidence interval. It is evident that the scattering and absorption coefficients increased with increasing mass concentration X. In addition, *A. cylindrica* absorbed mainly in the spectral region from 400 to 700 nm with peaks (i) at 435 and 676 nm corresponding to absorption by Chl *a* (Bidigare et al., 1990), (ii) at 630 nm corresponding to PCCN (Wolk and Simon, 1969), and (iii) a shoulder around 480 nm corresponding to absorption by PSC and PPC (Bidigare et al., 1990). In addition, scattering

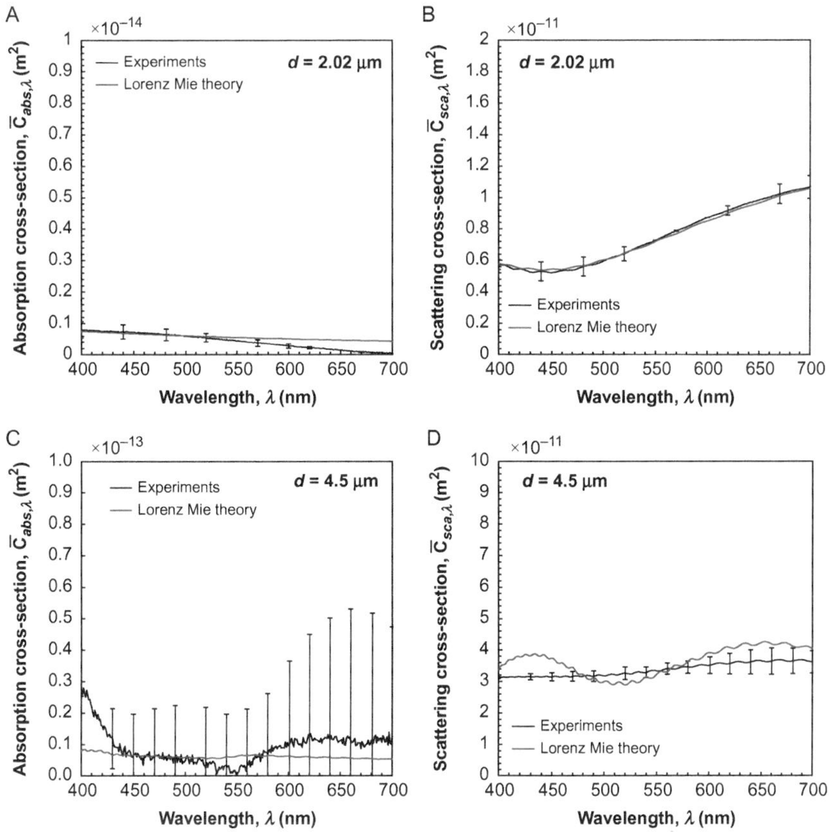

Figure 7 Experimental measurement and Lorenz–Mie theory predictions of the average absorption $\bar{C}_{abs,\lambda}$ and scattering $\bar{C}_{sca,\lambda}$ cross-sections between 400 and 700 nm of monodisperse polystyrene latex microspheres with diameters d equal to (A and B) 2.02 μm and (C and D) 4.5 μm, respectively (Kandilian, 2014).

dominated over absorption at all wavelengths between 400 and 750 nm, i.e., $\sigma_{s,\lambda} \gg \kappa_\lambda$, due to the fact that the cells were optically soft, i.e., $|m_{r,\lambda} - 1| \ll 1$.

Fig. 8C and D show the average mass absorption $\bar{A}_{abs,\lambda}$ and scattering $\bar{S}_{sca,\lambda}$ cross-sections after normalizing κ_λ and $\sigma_{s,\lambda}$ by X according to Eq. (6). It is evident that the three datasets collapsed on a single line. This confirms that single scattering prevailed and that absorption and scattering were linear processes. It is interesting to note the small dips in the scattering cross-section $\bar{S}_{sca,\lambda}$ coincided with the peaks in the absorption cross-section. This "cross-talk" between absorption and scattering can be attributed to resonance behavior in the real part (or refractive index) of the complex index of refraction of the microalgae at wavelengths when the imaginary

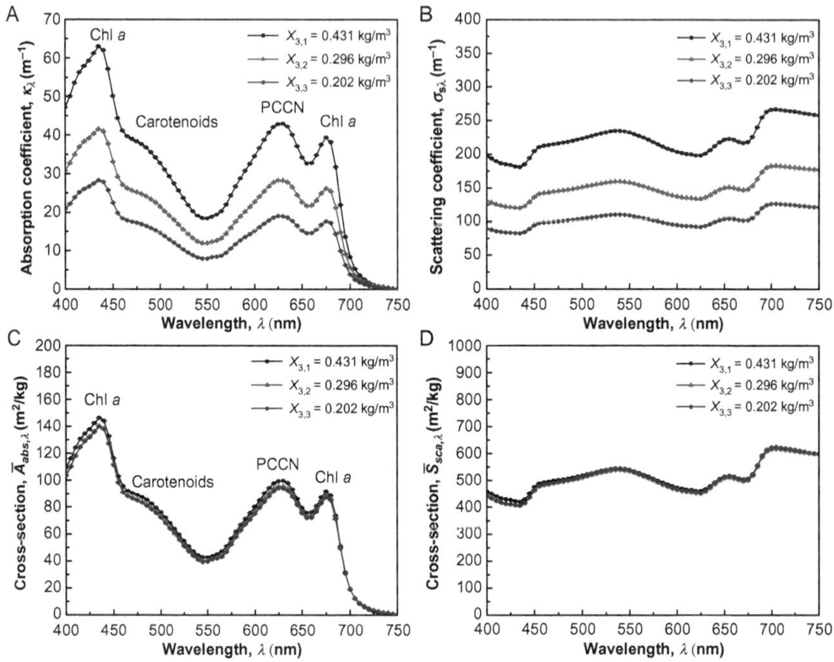

Figure 8 Experimental measurements of the (A) absorption κ_λ, and (B) scattering $\sigma_{s,\lambda}$ coefficients and of the average (C) absorption $\bar{A}_{abs,\lambda}$, and (D) scattering $\bar{S}_{sca,\lambda}$ cross-sections of dilute suspensions of *A. cylindrica* over the PAR region with different dry mass concentrations *X* (Heng et al., 2014).

part (or absorption index) features strong absorption peaks. Such resonance can be predicted by the Ketteler–Helmholtz theory (Jonasz and Fournier, 2007), among others.

Overall, these different results demonstrate the validity and the versatility of the described experimental method. Note that it can also be used for other absorbing and/or scattering particles as long as they can be suspended in a liquid.

5. RADIATION CHARACTERISTICS UNDER VARIOUS CONDITIONS

5.1 Exponential Growth

Experimental measurements have been performed on a wide variety of photosynthetic microorganism species. These species include (i) the green microalgae *C. reinhardtii* CC125 and its truncated chlorophyll antenna transformants tla1, tla1-CW+, and tlaX (Berberoğlu et al., 2008), (ii) the freshwater green microalgae *Botryococcus braunii* (Berberoğlu et al., 2009) and (iii) *Chlorella* sp. (Berberoğlu et al., 2009), (iv) the marine microalgae *Chlorococcum littorale* (Berberoğlu et al., 2009) and (v) *Nannochloropsis oculata* (Kandilian et al., 2013; Heng and Pilon, 2014), (vi) the purple nonsulfur bacteria *Rhodobacter sphaeroides* (Berberoğlu and Pilon, 2007), and (vii) the filamentous cyanobacteria *Anabaena variabilis* (Berberoğlu and Pilon, 2007) and (viii) *A. cylindrica* (Heng et al., 2014). Unless otherwise noted, these measurements were performed under replete conditions during the culture's exponential growth phase. Heng and Pilon (2014) demonstrated that light transfer in PBRs can be predicted using constant radiation characteristics measured during the exponential growth phase with reasonable accuracy provided that the cultures were not nitrogen-limited. Indeed, during nitrogen starvation, pigment concentrations and radiation characteristics evolved rapidly and irreversibly with time, as discussed later in this Section.

Fig. 9 shows the scattering phase function at 632.8 nm measured experimentally for various species previously mentioned. It indicates that all these microorganisms scatter light strongly in the forward direction. In fact, in all

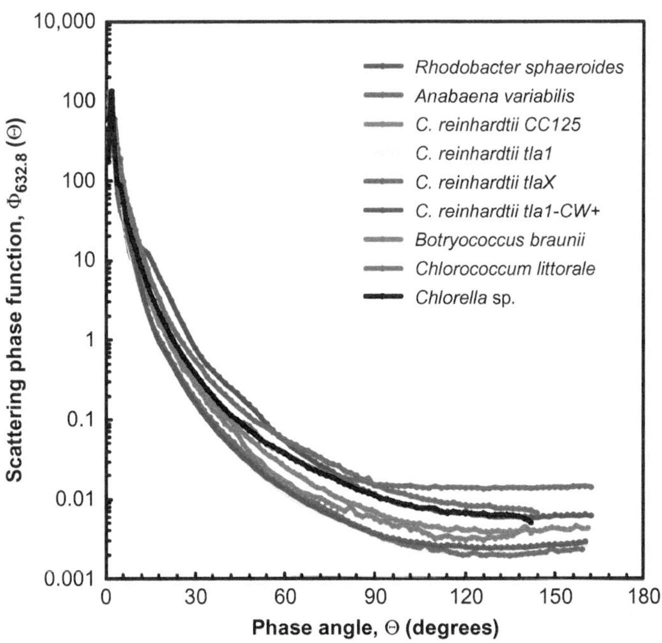

Figure 9 Scattering phase function $\Phi_{T,632.8}(\Theta)$ at 632.8 nm reported for various freshwater and marine green microalgae, cyanobacteria, and nonsulfur purple bacteria (Berberoğlu and Pilon, 2007; Berberoğlu et al., 2008, 2009; Kandilian et al., 2013).

cases, the associated asymmetry factor g_λ was larger than 0.95 and did not change significantly with wavelength (Kandilian et al., 2013).

Similarly, Fig. 10 compares the average mass spectral absorption $\bar{A}_{abs,\lambda}$ and scattering $\bar{S}_{sca,\lambda}$ cross-sections of the same species. It indicates that microorganisms presented different absorption peaks depending on the species. *C. reinhardtii* and its truncated light-harvesting antenna (tla) transformants featured absorption peaks at 435 and 676 nm corresponding to Chl *a* while the peak at 475 nm and the peak broadening around 650 nm can be attributed to Chl *b*. Note that genetic engineering led to a reduction in the absorption cross-sections across the PAR ranked by decreasing order as tla1-CW+ (with cell wall), tla1 and tlaX (without cell wall) (Berberoğlu et al., 2008). On the other hands, all *C. reinhardtii* strains had similar scattering cross-section. It is also interesting to note that filamentous cyanobacteria *A. variabilis* presented the same Chl *a* absorption peaks at 435 and 676 nm but also a peak at 621 nm corresponding to phycocyanin (PCCN) (Madigan et al., 2006). The scattering cross-section of *A. variabilis* was the largest of all microorganisms considered, most likely due to its long filaments. Finally, the nonsulfur purple bacteria *R. sphaeroides* stands out for its absorption peaks around 370, 480, 790, and 850 nm associated to the presence of bacteriochlorophyll (BChl) *b* and cartenoids (Madigan et al., 2006; Broglie et al., 1980).

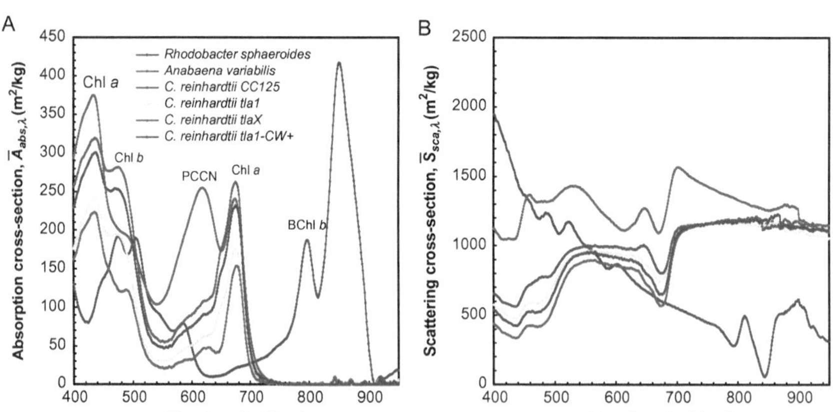

Figure 10 Average spectral mass (A) absorption $\bar{A}_{abs,\lambda}$ and (B) scattering $\bar{S}_{sca,\lambda}$ cross-sections in the spectral range from 400 to 750 nm for various green microalgae, cyanobacteria, and nonsulfur purple bacteria (Berberoğlu and Pilon, 2007; Berberoğlu et al., 2008, 2009; Kandilian et al., 2013).

The radiation characteristics shown in Figs. 9 and 10 for various species are available directly online (Pilon, 2015) or from the corresponding authors upon request.

5.2 Effect of Stresses

5.2.1 Definitions

5.2.1.1 Photoacclimation and Chromatic Acclimation

Photosynthetic microorganisms may experience photoacclimation and chromatic acclimation in response to different incident irradiance and spectrum, respectively (Fisher et al., 1996; Gentile and Blanch, 2001; Dubinsky and Stambler, 2009). Photoacclimation refers to the ability of photosynthetic microorganisms to adjust their light harvesting capacity, on the time scale of hours to days, based on the amount of light energy available to carry out photosynthesis. On the other hand, chromatic adaptation refers to the adjustment of pigment composition based on the spectral composition of the incident light.

In practice, photosynthetic microorganisms tend to increase their pigment concentrations in light-limited conditions and to reduce them under strong light illumination. For example, Fisher et al. (1996) found that *Nannochloropsis* sp. grown under 30 μmol/m^{2}s, in continuous cultures, had a steady-state chlorophyll concentration 4.5 times larger than when grown under 650 μmol/m^{2}s. Gentile and Blanch (2001) observed an 80% and 60% decrease in Chl *a* and vioxanthin, respectively, in batch grown *Nannochloropsis gaditana* when the incident irradiance was increased from 70 to 880 μmol/m^{2}s. However, low incident light may not always lead to significant changes in the microorganisms radiation characteristics as increasing the concentration of chlorophylls also decreases their *in vivo* specific absorption coefficient due to mutual shading of pigment molecules (Dubinsky and Stambler, 2009). The latter is partially responsible for what is known as the package effect corresponding to the nonlinear relationship between cell pigment concentrations and cell absorption cross-section (Jonasz and Fournier, 2007).

Moreover, microalgae may increase their *photoprotective* carotenoid concentration in response to large irradiance while reducing the amount of *photosynthetic* carotenoids through the so-called xanthophyll cycle (Lubián et al., 2000; Gentile and Blanch, 2001; Dubinsky and Stambler, 2009). The latter does not usually lead to changes in the overall carotenoid concentration as changes in the two types of carotenoids compensate each other (Dubinsky and Stambler, 2009; Lubián et al., 2000).

Finally, photoacclimation and chromatic acclimation depend on the microalgae species. For example, Lubián et al. (2000) demonstrated that *N. oculata* had lower concentrations of carotenoids and larger Chl *a* concentration per cell compared with *N. gaditana* and *N. salina* for cultures grown under the same conditions.

5.2.1.2 Photoinhibition

Exposing microalgae to large irradiance causes photooxidative damage in some of their photosystem units. This so-called photoinhibition leads to a decrease in the photosynthetic efficiency. This is primarily due to the destruction of reaction center proteins (Ke, 2001). The chloroplast repairs such damage by destroying the affected proteins and synthesizing new ones and integrating them into the affected photosystems. In fact, the cells continuously perform a damage repair cycle to repair the damaged photosystems (Baroli and Melis, 1996; Neidhardt et al., 1998). However, when the damage rate exceeds the repair rate, photoinhibition prevails and the overall efficiency of the cells decreases (Ke, 2001). In addition, the overall chlorophyll content can also decrease during the growth due to intense incident light. This is sometimes referred to as chlorophyll bleaching (Baroli and Melis, 1996). As a result, the absorption cross-section decreases over the PAR region and particularly at the chlorophylls absorption peaks (Fig. 2).

5.2.1.3 Nitrogen Starvation

Several strategies can be used to enhance microalgal lipid productivity (Williams and Laurens, 2010). For example, nitrogen starvation triggers large amounts of neutral lipid accumulation in various species mainly in the form of triglyceride fatty acids (TAGs) (Hu et al., 2008; Van Vooren et al., 2012). The latter are believed to serve as carbon and energy storage compound for the cells (Hu et al., 2008). TAGs are also the main feedstock for lipid to biodiesel conversion through transesterification reaction with methanol to produce methyl esters of fatty acids that are essentially biodiesel (Chisti, 2007). Kandilian et al. (2014a) demonstrated that TAG synthesis and productivity from microalgae *N. oculata* was limited by the mean rate of photon absorption $\langle \mathcal{A} \rangle$ (Eq. (11)) in the PBR during nitrogen starvation.

Nitrogen starvation can be achieved by either sudden or progressive starvation. *Sudden starvation* consists of two steps: first, microalgae are grown in nitrogen replete conditions. Then, they are transferred into a nitrogen-free medium. *Progressive starvation* consists of initially adding a small amount of nitrogen to the culture medium, in the form of nitrate, for example.

After inoculating the PBR, the microalgae grow and multiply until they consume all the nitrates in the medium and the culture medium becomes deprived of nitrogen.

In general, nitrogen limitation results in a decrease in pigment concentrations and in a significant change of color of the suspension (Kandilian et al., 2014a). In addition, the cells increase their carotenoid to Chl *a* concentration ratio (Heath et al., 1990). This ratio is related to the so-called stress index defined as the ratio of the optical densities (OD) of the cells' pigment extract at wavelengths 480 and 665 nm (Heath et al., 1990). It is an indicator of the "nutrient status" of the cells and is inversely correlated to the C/N ratio of the cells (Heath et al., 1990). In fact, Flynn et al. (1993) reported that nitrogen replete *N. oculata* cells had a carbon to nitrogen ratio (C/N) of 6 while NH_4^+ deprived cells featured C/N ratio of nearly 26 (Flynn et al., 1993).

5.2.2 Photoacclimation and Progressive Nitrogen Starvation

Fig. 11A and B presents the average spectral absorption cross-sections $\bar{C}_{abs,\lambda}$ of *N. oculata* at different times during their growth in a flat-plate PBR operated in batch mode under 7500 and 10,000 lux, respectively (Heng and Pilon, 2014). The incident light was provided by red LEDs emitting at 630 nm. The average absorption cross-section $\bar{C}_{abs,\lambda}$ displayed peaks at 435, 630, and 676 nm corresponding to *in vivo* absorption peaks of Chl *a* and at 485 nm corresponding to that of carotenoids. It also varied significantly with time in response to changes in light and nutrients availability. Similar trends were observed for both incident irradiances.

Fig. 11C and D plots $\bar{C}_{abs,\lambda}$ at wavelengths 485 and 676 nm with respect to time. Similarly, Fig. 11E and F shows the measured Chl *a* and total carotenoids (PSC + PPC) concentrations as functions of time. It is evident that the trends in the absorption peaks $\bar{C}_{abs,676}$ and $\bar{C}_{abs,485}$ closely follow the trends in Chl *a* and PSC + PPC concentrations, respectively. In fact, $\bar{C}_{abs,676}$ and $\bar{C}_{abs,485}$ and the corresponding pigment concentration reach their maximum and minimum at the same times. The initial downregulation of pigments was caused by exposure to excessive amounts of light when the cell concentration was relatively small. It contributed to reducing the energy absorbed per cell in order to prevent photodamage to their light-harvesting antenna. It is interesting to note that the duration of the initial downregulation of pigments closely coincided with the duration of the lag phase observed in the growth curves (see Fig. 2 of Heng and

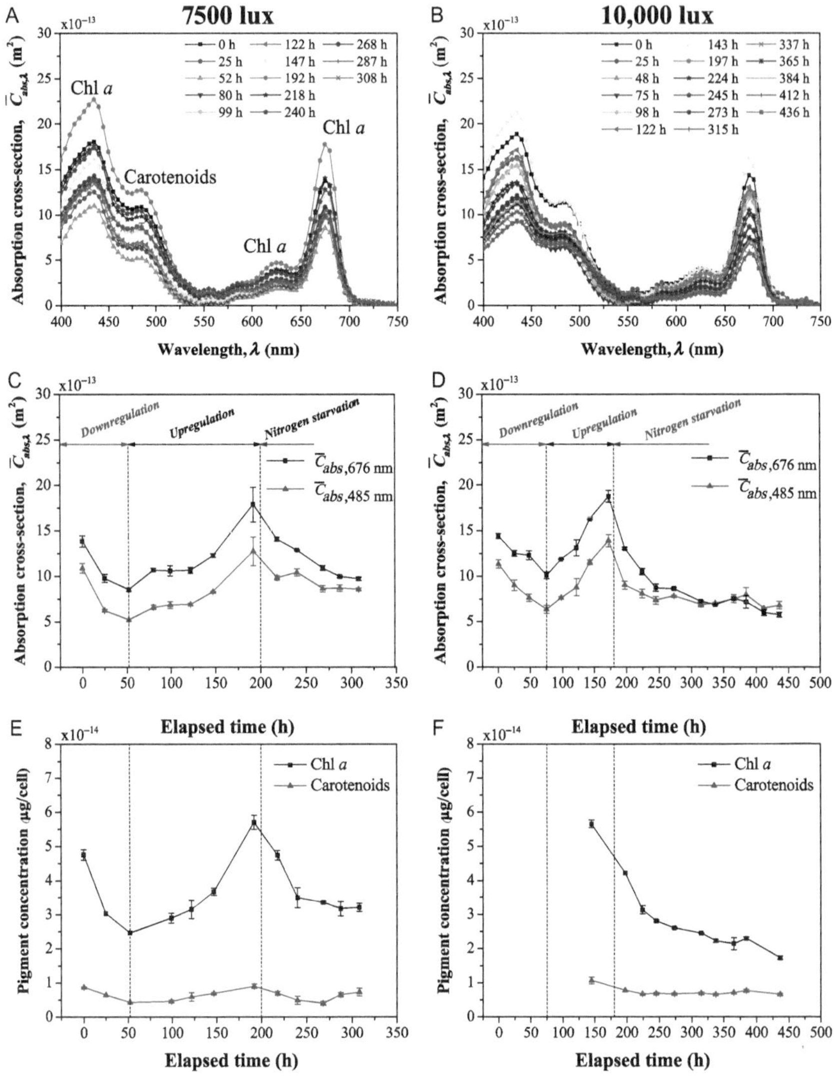

Figure 11 (A and B) Average spectral absorption cross-section $\bar{C}_{abs,\lambda}$ of *N. oculata*, (C and D) temporal evolutions (C and D) of average absorption cross-sections at 485 and 676 nm, and (E and F) of pigment Chl *a* and carotenoids for *N. oculata* grown under 7500 and 10,000 lux, respectively (Heng and Pilon, 2014).

Pilon, 2014). Then, Chl *a* and carotenoids concentrations increased between times 50 and 200 h for the culture grown under 7500 lux and between 75 and 180 h for those grown under 10,000 lux. This was due to upregulation of pigments by microalgae during the exponential growth phase to avoid photolimitation. Finally, elemental analysis predicted that the cultures grown under 7500 and 10,000 lux became nitrogen-limited after about 200 and 180 h, respectively. Interestingly, Fig. 11E and F shows that pigment concentrations decreased sharply around those times.

5.2.3 Sudden Nitrogen Starvation

Fig. 12A and B shows the temporal evolution of the average mass absorption $\bar{A}_{abs,\lambda}$ and scattering $\bar{S}_{sca,\lambda}$ cross-sections over the PAR region for *N. oculata* during sudden nitrogen starvation of a batch culture with an initial biomass concentration X_0 of 0.23 kg/m^3 (Kandilian et al., 2014a). It illustrates how $\bar{A}_{abs,\lambda}$ decreased sharply by nearly one order of magnitude across the PAR region within 96 h. This was accompanied by a decrease in Chl *a* concentration from 3–3.5 to 0.25–0.75 wt.% and in carotenoid from 0.45–0.6 to 0.1–0.2 wt.% over the same time period (Kandilian et al., 2014a). The stress index also increased continuously during that time, as did the TAG concentration from 5–10 to 30–45 wt.% (Kandilian et al., 2014a). On the other hand, Fig. 12B indicates that the scattering cross-section $\bar{S}_{sca,\lambda}$ did not change significantly during sudden nitrogen starvation. Using these radiation characteristics, Kandilian et al. (2014a) demonstrated that (i) TAG productivity correlated with light absorption rate by cells and (ii) a critical light absorption rate was needed to achieve large TAG accumulation.

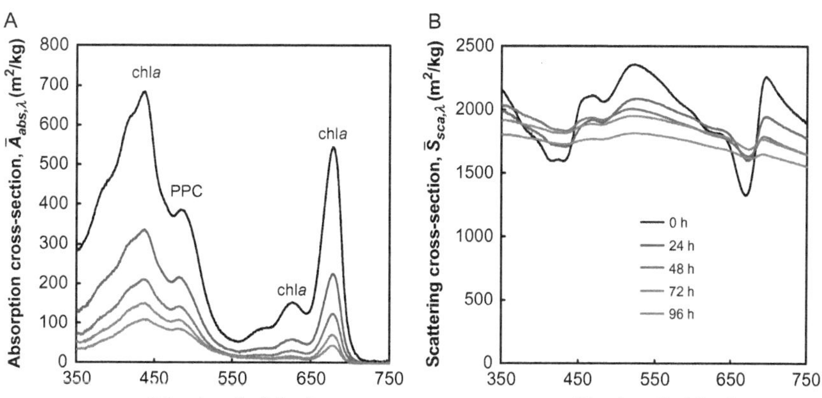

Figure 12 Average mass (A) absorption and (B) scattering cross-sections of *N. oculata* after 0, 24, 48, 72, and 96 h of cultivation during sudden nitrogen starvation of batch culture exposed to 250 $\mu mol_{h\nu}/m^2$ s with initial biomass concentration $X_0 = 0.23$ kg/m^3.

6. CONCLUSIONS AND PROSPECTS

This chapter has emphasized the importance of understanding and quantifying the interaction between light and photosynthetic microorganisms in designing, optimizing, monitoring, and operating PBRs of all sizes for the production of value-added products. To do so, knowing the radiation characteristics of photosynthetic microorganisms and their evolution with time and various stresses is essential. In fact, they are directly related to growth kinetics and to lipid production. This chapter has presented theoretical and experimental methods to determine the radiation characteristics of a wide variety of promising microorganism species with various shapes, sizes, and responses to stresses.

First, the theoretical methods for predicting the radiation characteristics of photosynthetic microorganisms are relatively fast and could be used for simulating microalgae growth under various operating conditions. It could also be employed for real-time monitoring and model-based control of PBRs to achieve their maximum productivity. However, existing models make simplifications on the shape of the cells or may require several input parameters difficult to obtain in practice in order to predict the spectral complex index of refraction. In fact, they have only been validated indirectly by considering the normal–hemispherical transmittance measurements of *Chlamydomonas reinhardtii* suspension grown under optimal conditions (Pottier et al., 2005; Dauchet et al., 2015). Therefore, one should perform direct comparison between experimental measurements and numerical predictions of the radiation characteristics and of the effective complex index of refraction for selected representative microalgae grown under various conditions. To do so, their size distribution, pigment concentrations, cell composition, and radiation characteristics should be measured simultaneously.

On a more fundamental level, contradicting arguments appear in the literature on the validity of approximating cells as homogeneous (Quirantes and Bernard, 2004, 2006). In particular, it remains unclear (i) how the chloroplast spatial distribution within the cell affects its absorption cross-section, (ii) how to accurately model the package effect, (iii) how the cell organelles participate to light scattering, and (iv) how this should be accounted for. Addressing these questions is made even more difficult by the fact that "very little is known about the optical properties of these organelles" (Jonasz and Fournier, 2007). Experimental determination of *in vivo* organelles' optical properties present major challenges. However, it could ultimately help determine the conditions under which microorganisms could be treated as homogeneous with some effective optical properties.

Moreover, the experimental measurements presented in Section 4 can faithfully capture the effect of the microorganisms' size, shape, and polydispersity. However, the experimental setup can be costly and the experimental procedure is time consuming. Thus, it may be difficult to implement in actual production systems. In addition, measurements are valid only for specific growth conditions and need to be repeated each time conditions change including pH, temperature, illumination, medium composition, etc. Thus, it would be beneficial to develop a simplified experimental method to determine the radiation characteristics and in particular the absorption cross-section which is the most influence on light transfer in PBRs (Kandilian, 2014).

Finally, many photosynthetic microorganisms have highly nonspherical and sometimes very complex shapes such as *Scenedesmus*, *Spirunila*, or *Golenkinia*. Similarly, many genus or species of interest are colonial (e.g., *Scenedesmus*, *Botryococcus*) forming complex and sometimes large colonies. Experimental measurements and theoretical methods or approximations to determine their radiation characteristics are still lacking, for the most part.

REFERENCES

Aas E: Refractive index of phytoplankton derived from its metabolite composition, *J Plankton Res* 18(12):2223–2249, 1996.

Acien Fernandez FG, Garcia Camacho F, Sanchez Perez JA, Fernandez Sevilla JM, Molina Grima E: A model for light distribution and average solar irradiance inside outdoor tubular photobioreactors for the microalgal mass culture, *Biotechnol Bioeng* 55(5):701–714, 1997.

Aden AL, Kerker M: Scattering of electromagnetic waves from two concentric spheres, *J Appl Phys* 22(10):1242–1246, 1951.

Andrews JF: A mathematical model for the continuous culture of microorganisms utilizing inhibitory substrates, *Biotechnol Bioeng* 10(6):707–723, 1968.

Atkinson AW Jr, Gunning BES, John PCL: Sporopollenin in the cell wall of *Chlorella* and other algae: ultrastructure, chemistry, and incorporation of 14C-acetate, studied in synchronous cultures, *Planta* 107(1):1–32, 1972.

Baroli I, Melis A: Photoinhibition and repair in *Dunaliella salina* acclimated to different growth irradiances, *Planta* 198(4):640–646, 1996.

Bayona KCD, Garcés LA: Effect of different media on exopolysaccharide and biomass production by the green microalga *Botryococcus braunii*, *J Appl Phycol* 26(5):2087–2095, 2014.

Béchet Q, Shilton A, Guieysse B: Modeling the effects of light and temperature on algae growth: state of the art and critical assessment for productivity prediction during outdoor cultivation, *Biotechnol Adv* 31(8):1648–1663, 2013.

Becker EW: *Microalgae: biotechnology and microbiology*, Cambridge, UK, 1994, Cambridge University Press.

Benemann JR: Production of nitrogen fertilizer with nitrogen-fixing blue-green algae, *Enzym Microb Technol* 1(2):83–90, 1979.

Benemann JR: Hydrogen production by microalgae, *J Appl Phycol* 12:291–300, 2000.

Berberoğlu H, Gomez PS, Pilon L: Radiation characteristics of *Botryococcus braunii*, *Chlorococcum littorale*, and *Chlorella* sp. used for fixation and biofuel production, *J Quant Spectrosc Radiat Transf* 110(17):1879–1893, 2009.
Berberoğlu H, Pilon L: Experimental measurement of the radiation characteristics of *Anabaena variabilis* ATCC 29413-U and *Rhodobacter sphaeroides* ATCC 49419, *Int J Hydrog Energy* 32(18):4772–4785, 2007.
Berberoğlu H, Pilon L, Melis A: Radiation characteristics of *Chlamydomonas reinhardtii* CC125 and its truncated chlorophyll antenna transformants *tla1*, *tlaX* and *tla1-CW+*, *Int J Hydrog Energy* 33(22):6467–6483, 2008.
Berberoğlu H, Yin J, Pilon L: Light transfer in bubble sparged photobioreactors for H_2 production and CO_2 mitigation, *Int J Hydrog Energy* 32(13):2273–2285, 2007.
Berman-Frank I, Lundgren P, Falkowski P: Nitrogen fixation and photosynthetic oxygen evolution in cyanobacteria, *Res Microbiol* 154(3):157–164, 2003.
Bidigare R, Ondrusek M, Morrow J, Kiefer D: *In vivo* absorption properties of algal pigments, *Proc SPIE Ocean Opt X* 1302:290–301, 1990.
Blankenship RE: *Molecular mechanisms of photosynthesis*, Hoboken, NJ, 2008, Wiley-Blackwell.
Bohren CF, Huffman DR: *Absorption and scattering of light by small particles*, New York, NY, 1998, John Wiley & Sons.
Bricaud A, Bédhomme AL, Morel A: Optical properties of diverse phytoplanktonic species: experimental results and theoretical interpretation, *J Plankton Res* 10(5):851–873, 1988.
Bricaud A, Morel A: Light attenuation and scattering by phytoplanktonic cells: a theoretical modeling, *Appl Opt* 25(4):571–580, 1986.
Broglie RM, Hunter CN, Delepelaire P, Niederman RA, Chua NH, Clayton RK: Isolation and characterization of the pigment-protein complexes of *Rhodopseudomonas sphaeroides* by lithium dodecyl sulfate/polyacrylamide gel electrophoresis, *Proc Natl Acad Sci* 77(1):87–91, 1980.
Brown DJ, Vickers GT: The use of projected area distribution functions in particle shape measurement, *Powder Technol* 98(3):250–257, 1998.
Canter-Lund H, Lund JWG: *Freshwater algae: their microscopic world explored*, Somerset, UK, 1995, Biopress Limited.
Chen X, Goh QY, Tan W, Hossain I, Chen WN, Lau R: Lumostatic strategy for microalgae cultivation utilizing image analysis and chlorophyll *a* content as design parameters, *Bioresour Technol* 102(10):6005–6012, 2011.
Chisti Y: Biodiesel from microalgae, *Biotechnol Adv* 25(3):294–306, 2007.
Cohen Z: *Chemicals from microalgae*, London, UK, 1999, Taylor & Francis.
Cornet JF: Calculation of optimal design and ideal productivities of volumetrically lightened photobioreactors using the constructal approach, *Chem Eng Sci* 65(2):985–998, 2010.
Cornet JF, Dussap CG: A simple and reliable formula for assessment of maximum volumetric productivities in photobioreactors, *Biotechnol Prog* 25(2):424–435, 2009.
Cornet JF, Dussap CG, Dubertret G: A structured model for simulation of cultures of the cyanobacterium *Spirulina platensis* in photobioreactors: I. Coupling between light transfer and growth kinetics, *Biotechnol Bioeng* 40(7):817–825, 1992.
Cornet JF, Dussap CG, Gros JB: Kinetics and energetics of photosynthetic micro-organisms in photobioreactors. In *Bioprocess and algae reactor technology, apoptosis*, vol. 59, Advances in Biochemical Engineering Biotechnology, Berlin, 1998, Springer, pp 153–224.
Cornet JF, Dussap CG, Gros JB, Binois C, Lasseur C: A simplified monodimensional approach for modeling coupling between radiant light transfer and growth kinetics in photobioreactors, *Chem Eng Sci* 50(9):1489–1500, 1995.
Das D, Veziroğlu TN: Hydrogen production by biological processes: a survey of literature, *Int J Hydrog Energy* 26(1):13–28, 2001.

Dauchet J, Blanco S, Cornet JF, Fournier R: Calculation of the radiative properties of photosynthetic microorganisms, *J Quant Spectrosc Radiat Transf* 161:60–84, 2015.

Dayananda C, Sarada R, Rani MU, Shamala TR, Ravishankar GA: Autotrophic cultivation of *Botryococcus braunii* for the production of hydrocarbons and exopolysaccharides in various media, *Biomass Bioenergy* 31(1):87–93, 2007.

Demura M, Ioki M, Kawachi M, Nakajima N, Watanabe M: Desiccation tolerance of *Botryococcus braunii* (trebouxiophyceae, chlorophyta) and extreme temperature tolerance of dehydrated cells, *J Appl Phycol* 26(1):49–53, 2014.

Draine BT: The discrete-dipole approximation and its application to interstellar graphite grains, *Astrophys J* 333:848–872, 1988.

Drolen BL, Tien CL: Absorption and scattering of agglomerated soot particulate, *J Quant Spectrosc Radiat Transf* 37(5):433–448, 1987.

Dubinsky Z, Stambler N: Photoacclimation processes in phytoplankton: mechanisms, consequences, and applications, *Aquat Microb Ecol* 56:163–176, 2009.

Fernandes BD, Dragone GM, Teixeira JA, Vicente AA: Light regime characterization in an airlift photobioreactor for production of microalgae with high starch content, *Appl Biochem Biotechnol* 0273-2289. 161(1-8):218–226, 2010.

Fisher T, Minnaard J, Dubinsky Z: Photoacclimation in the marine alga *Nannochloropsis* sp. (eustigmatophyte): a kinetic study, *J Plankton Res* 18(10):1797–1818, 1996.

Flynn KJ, Davidson K, Cunningham AJ: Relations between carbon and nitrogen during growth of *Nannochloropsis oculata* (droop) Hibberd under continuous illumination, *New Phytol* 125(4):717–722, 1993.

Gentile MP, Blanch HW: Physiology and xanthophyll cycle activity of *Nannochloropsis gaditana*, *Biotechnol Bioeng* 75(1):1–12, 2001.

Gerken HG, Donohoe B, Knoshaug EP: Enzymatic cell wall degradation of *Chlorella vulgaris* and other microalgae for biofuels production, *Planta* 237(1): 239–253, 2013.

Gordon HR: Light scattering and absorption by randomly-oriented cylinders: dependence on aspect ratio for refractive indices applicable for marine particles, *Opt Express* 19(5):4673–4691, 2011.

Grima EM, Sevilla JMF, Perez JAS, Camacho FG: A study on simultaneous photolimitation and photoinhibition in dense microalgal cultures taking into account incident and averaged irradiances, *J Biotechnol* 45:59–69, 1996.

Heath MR, Richardson K, Kiørboe T: Optical assessment of phytoplankton nutrient depletion, *J Plankton Res* 12(2):381–396, 1990.

Heng RL, Lee E, Pilon L: Radiation characteristics and optical properties of filamentous cyanobacterium *Anabaena cylindrica*, *J Opt Soc Am A* 31(4):836–845, 2014.

Heng RL, Pilon L: Time-dependent radiation characteristics of *Nannochloropsis oculata* during batch culture, *J Quant Spectrosc Radiat Transf* 144:154–163, 2014.

Heng RL, Sy KC, Pilon L: Absorption and scattering by bispheres, quadspheres, and circular rings of spheres and their equivalent coated spheres, *J Opt Soc Am A* 32(1):46–60, 2015.

Hu Q, Sommerfeld M, Jarvis E, et al: Microalgal triacylglycerols as feedstocks for biofuel production: perspectives and advances, *Plant J* 54(4):621–639, 2008.

Iskander MF, Chen HY, Penner JE: Optical scattering and absorption by branched chains of aerosols, *Appl Opt* 28(15):3083–3091, 1989.

Jonasz M, Fournier G: *Light scattering by particles in water: theoretical and experimental foundations*, San Diego, CA, 2007, Academic Press.

Kahnert M, Nousiainen T, Lindqvist H: Review: model particles in atmospheric optics, *J Quant Spectrosc Radiat Transf* 146:41–58, 2014.

Kandilian R: *Optimization and control of light transfer in photobioreactors for biofuel production*, Ph.D. thesis, Los Angeles, USA, 2014, University of California.

Kandilian R, Heng RL, Pilon L: Absorption and scattering by fractal aggregates and by their equivalent coated spheres, *J Quant Spectrosc Radiat Transf* 151:310–326, 2015.
Kandilian R, Lee E, Pilon L: Radiation and optical properties of *Nannochloropsis oculata* grown under different irradiances and spectra, *Bioresour Technol* 137:63–73, 2013.
Kandilian R, Pruvost J, Legrand J, Pilon L: Influence of light absorption rate by *Nannochloropsis oculata* on triglyceride production during nitrogen starvation, *Bioresour Technol* 163:308–319, 2014a.
Kandilian R, Tsao TC, Pilon L: Control of incident irradiance on a batch operated flat-plate photobioreactor, *Chem Eng Sci* 119:99–108, 2014b.
Kang RJ, Shi DJ, Cong W, Cai ZL, Ouyang F: Dispersion of the optical constants of quartz and polymethyl methacrylate glasses, *Opt Commun* 188:129–139, 2001.
Ke B: Photosynthesis: photobiochemistry and photobiophysics, *Advances in Photosynthesis,* Dordrecht, The Netherlands, 2001, Kluwer Academic Publishers.
Kerker M: *The scattering of light, and other electromagnetic radiation,* New York, NY, 1969, Academic Press.
Kimura H, Kolokolova L, Mann I: Optical properties of cometary dust: constraints from numerical studies on light scattering by aggregate particles, *Astron Astrophys* 407(1): L5–L8, 2003.
Kong B, Vigil RD: Simulation of photosynthetically active radiation distribution in algal photobioreactors using a multidimensional spectral radiation model, *Bioresour Technol* 158:141–148, 2014.
Kumar K, Dasgupta CN, Nayak B, Lindblad P, Das D: Development of suitable photobioreactors for CO_2 sequestration addressing global warming using green algae and cyanobacteria, *Bioresour Technol* 102(8):4945–4953, 2011.
Latimer P: Experimental tests of a theoretical method for predicting light scattering by aggregates, *Appl Opt* 24(19):3231–3239, 1985.
Latimer P, Wamble F: Light scattering by aggregates of large colloidal particles, *Appl Opt* 21(13):2447–2455, 1982.
Lee AK, Lewis DM, Ashman PJ: Microbial flocculation, a potentially low-cost harvesting technique for marine microalgae for the production of biodiesel, *J Appl Phycol* 21(5):559–567, 2009.
Lee E, Heng RL, Pilon L: Spectral optical properties of selected photosynthetic microalgae producing biofuels, *J Quant Spectrosc Radiat Transf* 114:122–135, 2013.
Lee E, Pilon L: Absorption and scattering by long and randomly oriented linear chains of spheres, *J Opt Soc Am A* 30(9):1892–1900, 2013.
Lee E, Pruvost J, He X, Munipalli R, Pilon L: Design tool and guidelines for outdoor photobioreactors, *Chem Eng Sci* 106:18–29, 2014.
Liou KN: *An introduction to atmospheric radiation,* Waltham, MA, 2002, Academic Press.
Lubián LM, Montero O, Moreno-Garrido I, et al: *Nannochloropsis* (eustigmatophyceae) as source of commercially valuable pigments, *J Appl Phycol* 12:249–255, 2000.
Ma X, Lu JQ, Brock RS, Jacobs KM, Yang P, Hu HH: Determination of complex refractive index of polystyrene microspheres from 370 to 1610 nm, *Phys Med Biol* 48:4165–4172, 2003.
Mackowski DW: Calculation of total cross sections of multiple-sphere clusters, *J Opt Soc Am A* 11(11):2851–2861, 1994.
Mackowski DW, Mishchenko MI: Calculation of the T-matrix and the scattering matrix for ensembles of spheres, *J Opt Soc Am A* 13(11):2266–2278, 1996.
Mackowski DW, Mishchenko MI: A multiple sphere T-matrix Fortran code for use on parallel computer clusters, *J Quant Spectrosc Radiat Transf* 112(13):2182–2192, 2011.
Madigan MT, Martinko JM: *Biology of microorganisms,* Upper Saddle River, NJ, 2006, Pearson Prentice Hall.

Madigan MT, Martinko JM, Parker J, Brock TD: *Biology of microorganisms,* Upper Saddle River, NJ, 2006, Pearson Prentice Hall.

Malitson IH: Interspecimen comparison of the refractive index of fused silica, *J Opt Soc Am* 55(10):1205–1209, 1965.

Melis A: Green alga hydrogen production: process, challenges and prospects, *Int J Hydrog Energy* 27:1217–1228, 2002.

Mengüç MP, Manickavasagam S, D'Sa DA: Determination of radiative properties of pulverized coal particles from experiments, *Fuel* 73(4):613–625, 1994.

Merzlyak MN, Naqvi KR: On recording the true absorption spectrum and scattering spectrum of a turbid sample: application to cell suspensions of cyanobacterium *Anabaena variabilis, J Photochem Photobiol B Biol* 58:123–129, 2000.

Mie G: Beiträge zur Optik trüber Medien, speziell kolloidaler Metallösungen, *Ann Phys* 25(3):377–445, 1908.

Mishchenko MI: *Electromagnetic scattering by particles and surfaces,* 2015. http://www.giss.nasa.gov/staff/mmishchenko/t_matrix.html.

Mishchenko MI, Mackowski DW, Travis LD: Scattering of light by bispheres with touching and separated components, *Appl Opt* 34(21):4589–4599, 1995.

Mishchenko MI, Travis LD, Lacis AA: *Scattering, absorption, and emission of light by small particles,* Cambridge, UK, 2002, Cambridge University Press.

Modest M: *Radiative heat transfer,* ed 3, Oxford, UK, 2013, Elsevier.

Murphy TE, Berberoğlu H: Effect of algae pigmentation on photobioreactor productivity and scale-up: a light transfer perspective, *J Quant Spectrosc Radiat Transf* 112(18): 2826–2834, 2011.

Nakajima Y, Ueda R: Improvement of photosynthesis in dense microalgal suspension by reduction of light harvesting pigments, *J Appl Phycol* 9(6):503–510, 1997.

Nakajima Y, Ueda R: The effect of reducing light-harvesting pigment on marine microalgal productivity, *J Appl Phycol* 12(3-5):285–290, 2000.

Neidhardt J, Benemann JR, Zhang L, Melis A: Photosystem-II repair and chloroplast recovery from irradiance stress: relationship between chronic photoinhibition, light-harvesting chlorophyll antenna size and photosynthetic productivity in *Dunaliella salina* (green algae), *Photosynth Res* 56:175–184, 1998.

Ozkan A, Kinney K, Katz L, Berberoğlu H: Reduction of water and energy requirement of algae cultivation using an algae biofilm photobioreactor, *Bioresour Technol* 114:542–548, 2012.

Pilon L: *Radiation characteristics of microalgae for CO_2 fixation and biofuel production,* 2015. http://www.seas.ucla.edu/pilon/downloads.htm.

Pilon L, Berberoğlu H, Kandilian R: Radiation transfer in photobiological carbon dioxide fixation and fuel production by microalgae, *J Quant Spectrosc Radiat Transf* 112(17):2639–2660, 2011.

Pilon L, Berberoğlu H: Chap. 11: Photobiological hydrogen production. In Sherif SA, Goswami DY, Stefanakos EKL, Steinfeld A, editors: *Handbook of hydrogen energy,* Boca-Raton, FL, 2014, CRC Press, pp 369–418.

Pinho MG, Kjos M, Veening JW: How to get (a) round: mechanisms controlling growth and division of coccoid bacteria, *Nat Rev Microbiol* 11(9):601–614, 2013.

Polle JE, Kanakagiri SD, Melis A: tla1, a DNA insertional transformant of the green alga *Chlamydomonas reinhardtii* with a truncated light-harvesting chlorophyll antenna size, *Planta* 217:49–59, 2003.

Pottier L, Pruvost J, Deremetz J, Cornet JF, Legrand J, Dussap CG: A fully predictive model for one-dimensional light attenuation by *Chlamydomonas reinhardtii* in a torus photobioreactor, *Biotechnol Bioeng* 91(5):569–582, 2005.

Privoznik KG, Daniel KJ, Incropera FP: Absorption, extinction and phase function measurements for algal suspensions of *Chlorella pyrenoidosa, J Quant Spectrosc Radiat Transf* 20(4):345–352, 1978.

Pruvost J, Cornet JF: Knowledge models for engineering and optimization of photobioreactors. In Posten C, Walter C, editors: *Microalgal biotechnology*, Berlin, Germany, 2012, Walter De Gruyter, pp 181–224.

Pruvost J, Cornet JF, Legrand J: Hydrodynamics influence on light conversion in photobioreactors: an energetically consistent analysis, *Chem Eng Sci* 63(14):3679–3694, 2008.

Pruvost J, Cornet JF, Pilon L: Large scale production of algal biomass: photobioreactors. In Chisti Y, Bux F, editors: *Algae biotechnology: products and processes*, UK, 2014, Springer.

Quirantes A, Bernard S: Light scattering by marine algae: two-layer spherical and nonspherical models, *J Quant Spectrosc Radiat Transf* 89(1):311–321, 2004.

Quirantes A, Bernard S: Light-scattering methods for modelling algal particles as a collection of coated and/or nonspherical scatterers, *J Quant Spectrosc Radiat Transf* 100(1-3):315–324, 2006.

Richmond A: *Handbook of microalgal culture: biotechnology and applied phycology,* Oxford, UK, 2004, Blackwell Science Ltd.

Rodolfi L, Zittelli GC, Bassi N, et al: Microalgae for oil: strain selection, induction of lipid synthesis and outdoor mass cultivation in a low-cost photobioreactor, *Biotechnol Bioeng* 102(1):100–112, 2009.

Schirrmeister BE, de Vos JM, Antonelli A, Bagheri HC: Evolution of multicellularity coincided with increased diversification of cyanobacteria and the Great Oxidation Event, *Proc Natl Acad Sci* 110(5):1791–1796, 2013.

Souliés A, Pruvost J, Legrand J, Castelain C, Burghelea TI: Rheological properties of suspensions of the green microalga *Chlorella vulgaris* at various volume fractions, *Rheol Acta* 52(6):589–605, 2013.

Stramski D, Bricaud A, Morel A: Modeling the inherent optical properties of the ocean based on the detailed composition of the planktonic community, *Appl Opt* 40(18):2929–2945, 2001.

Stramski D, Mobley CD: Effects of microbial particles on oceanic optics: a database of single-particle optical properties, *Limnol Oceanogr* 42(3):538–549, 1997.

Stramski D, Morel A, Bricaud A: Modeling the light attenuation and scattering by spherical phytoplanktonic cells: a retrieval of the bulk refractive index, *Appl Opt* 27(19):3954–3956, 1988.

Sukenik A, Levy R, Levy Y, Falkowski P, Dubinsky Z: Optimizing algal biomass production in an outdoor pond: a simulation model, *J Appl Phycol* 0921-89713:191–201, 1991.

Takache H, Christophe G, Cornet JF, Pruvost J: Experimental and theoretical assessment of maximum productivities for the microalgae *Chlamydomonas reinhardtii* in two different geometries of photobioreactors, *Biotechnol Prog* 26(2):431–440, 2010.

Takache H, Pruvost J, Cornet JF: Kinetic modeling of the photosynthetic growth of *Chlamydomonas reinhardtii* in a photobioreactor, *Biotechnol Prog* 28(3):681–692, 2012.

Tiwari A, Pandey A: Cyanobacterial hydrogen production–a step towards clean environment, *Int J Hydrog Energy* 37(1):139–150, 2012.

Traverse A: Paleopalynology, Topics in geobiology (Dordrecht, The Netherlandds, 2007, Springer.

van de Hulst HC: *Light scattering by small particles,* Mineola, NY, 2012, Courier Dover Publications.

Van Vooren G, Grand FL, Legrand J, Cuine S, Peltier G, Pruvost J: Investigation of fatty acids accumulation in *Nannochloropsis oculata* for biodiesel application, *Bioresour Technol* 124:421–432, 2012.

Wait JR: Scattering of a plane wave from a circular dielectric cylinder at oblique incidence, *Can J Phys* 33(5):189–195, 1955.

Waterman PC: Matrix formulation of electromagnetic scattering, *Proc Inst Electr Electron Eng* 53(8):805–812, 1965.

Williams PJleB, Laurens LML: Microalgae as biodiesel and biomass feedstocks: review and analysis of the biochemistry, energetics and economics, *Energy Environ Sci* 3:554–590, 2010.

Wolk CP, Simon RD: Pigments and lipids of heterocysts, *Planta* 86:92–97, 1969.

Xu YL: Electromagnetic scattering by an aggregate of spheres: far field, *Appl Opt* 36(36):9496–9508, 1997.

Yun YS, Park JM: Kinetic modeling of the light-dependent photosynthetic activity of the green microalga *Chlorella vulgaris*, *Biotechnol Bioeng* 83(3):303–311, 2003.

CHAPTER THREE

Modeling of Microalgae Bioprocesses

Matthias Schirmer, Clemens Posten[1]
Institute of Process Engineering in Life Sciences, Karlsruhe Institute of Technology (KIT), Karlsruhe, Germany
[1]Corresponding author: e-mail address: Clemens.Posten@kit.edu

Contents

Abstract

Modeling of photobioprocesses is a powerful tool for process development and understanding. Nevertheless, this tool is nowadays not exhaustively employed due to lack of clear insights into the specific structure and small data basis. The following chapter gives an introduction on modeling phototrophic processes. Emphasis lays on simplification to be applicable to process development based on measurable data. The first step is to structure the problem into a reactor level, the level of metabolic fluxes, and the intracellular control level. Second, the general approach of lumping consecutive metabolic steps to metabolic fluxes and setting up appropriate balance equations and kinetics is outlined. Combining linear dependent parameters to observable yield coefficients is another approach for model reduction. To consider complexity on the control level an optimization approach is explained which has been successfully applied already to heterotrophs and phototrophs. Starting from a generic simulation example, third, the specific features are outlined in more detail. These are especially photosynthesis,

Advances in Chemical Engineering, Volume 48
ISSN 0065-2377
http://dx.doi.org/10.1016/bs.ache.2015.12.001

carbon uptake, and carbon partitioning. For dynamic description of the complex reactions of the cells to environmental changes, examples are listed, some of them supported with data where available. The idea of this chapter is to give the basic biological background, to deduce step by step the model equations, and to give simulation results with quantitative parameter values. The chapter shall give the basis for a straight start into own modeling projects of the reader.

ABBREVIATIONS

abs. absorbed
act active
app apparent
ATP adenosine triphosphate
c concentration (g/L; mol/L)
$C_6H_{12}O_6$ glucose
carb carboxylase
CH, $(CH_2O)_i$ carbohydrate
CO_2 carbon dioxide
CTR carbon dioxide transfer rate
diss dissipation
D_R diameter (reactor)
E energy (kJ)
GAP glyceraldehyde-3-phosphate
Gluc glucose
h Planck's constant (6.626068×10^{-34} J/s)
H heat of combustion (kJ)
H_2O water
I light intensity (μmol/m^2/s)
k_La volumetric mass transfer coefficient (1/h)
L lipid
m mass (g/L)
max maximum
M molar weight (g/mol)
Macro macroscopic
n_f number of metabolic fluxes
NADH nicotinamide adenine dinucleotide (reduced)
NH_3 ammonia
NuAc nucleic acid
O_2 oxygen
opt optimum
oxyg oxygenase
p_i partial pressure
ps photosynthesis
P phosphor
PCE photoconversion efficiency (kJ/kJ)
PI photosynthesis irradiance

PQ photosynthetic quotient
Prot protein
q volumetric flow rate (mL(L)/min(s))
q molar specific turnover rate (mol/g/h)
r vector of specific turnover rates
r_{Comp} specific turnover rate of compound (g/g/h)
resp subscript, respiration
R volumetric reaction rate (g/L/h)
RuBisCO ribulose-1,5-bisphosphate carboxylase/oxygenase
RuBP ribulose-1,5-bisphosphate
S sulfur
t time (h)
t_f fermentation/process time (d)
T temperature (°C)
T superscript in matrix, transposed
TCC tricarboxylic acid cycle
V_R reaction volume (L)
X biodrymass
Y stoichiometric matrix
$y_{C1/C2}$ yield coefficient of compounds C1 from C2 (g/g; g/mol; mol/mol)
α, $y_{X,I}$ photosynthetic efficiency (m^2/s/μmol/d)
ε molar absorption coefficients (L/m/g)
σ absorption cross section (m^2)
μ specific growth rate (1/d)
ν frequency of the radiation (Hz)

1. INTRODUCTION

Modeling of bioprocesses has already been developed into a tool supporting engineering tasks like control design, process optimization with respect of process strategies and debottlenecking, scale up, and last but not least better understanding of the internal processes inside the cell and the reactor. However, for photobioprocesses modeling is still not so far developed due to specific qualities of such processes. The first reason is of course the light. It is not miscible and forms strong gradients inside the reactor. Photosynthesis as a unique feature is the energy source for the microalgae and is a complex reaction system on the molecular level. Nevertheless, lumped stoichiometric models can be set up using well-known stoichiometries. Other ingredients of modeling can be translated from modeling of heterotrophs like energy and reductant generation in TCC and respiratory chain.

Microalgae react in complex ways to environmental conditions and form a diversity of different products making modeling quite specific for different production processes. Many production processes are running under outdoor conditions with day/night cycles for light or changing temperatures what is another specific issue. The following paragraphs give examples for phototrophic models. One red thread is the concept of lumping consecutive metabolic steps to metabolic fluxes and then to condense unknown parameters to macroscopically measurable ones. The biological background given so far is necessary for setting up respective equations for metabolic fluxes. Special emphasis is laid on model bricks unique for phototrophic microorganisms. Simulation results and numerical values of parameters shall give support for model understanding and starting with modeling on own processes.

It is recommended to study the literature given in "Further Reading" section.

2. BASIC CONSIDERATIONS AND GENERAL APPROACH

The principles of metabolic modeling can be consequently applied also for photobioprocesses to have a first structured model which has to be amended with specific submodels for photosynthesis and product formation. These principles include thinking on different process levels basically reactor and cell level and application of chemical reaction principles like balances, stoichiometry, and kinetics. In the following paragraphs this approach will be outlined and explained introducing a simple generic example.

2.1 Model Hierarchy and System Boundaries

Many systems in nature and technology can be structured hierarchically in different levels with own dominant aspects, variables, and a characteristic degree of simplification. In Fig. 1 this is shown for a typical photobioprocess.

The photoreactor has to transform external conditions into an artificial environment for the cells. The processes on the reactor level are mainly transport processes. Light and dissolved nutrients are transported into and out of the reactor, distributed in its volume, and finally taken up by the cells. Reactions take place predominately inside the cells but are counted on the reactor level as volumetric mass flows and reaction rates R, reading for a specific compound:

$$R_{\text{Comp}} = \frac{\Delta m_{\text{Comp}}}{V_{\text{R}} \cdot \Delta t} = \frac{\Delta c_{\text{Comp}}}{\Delta t} \left(\frac{\text{g}}{\text{L} \cdot \text{h}}\right) \tag{1}$$

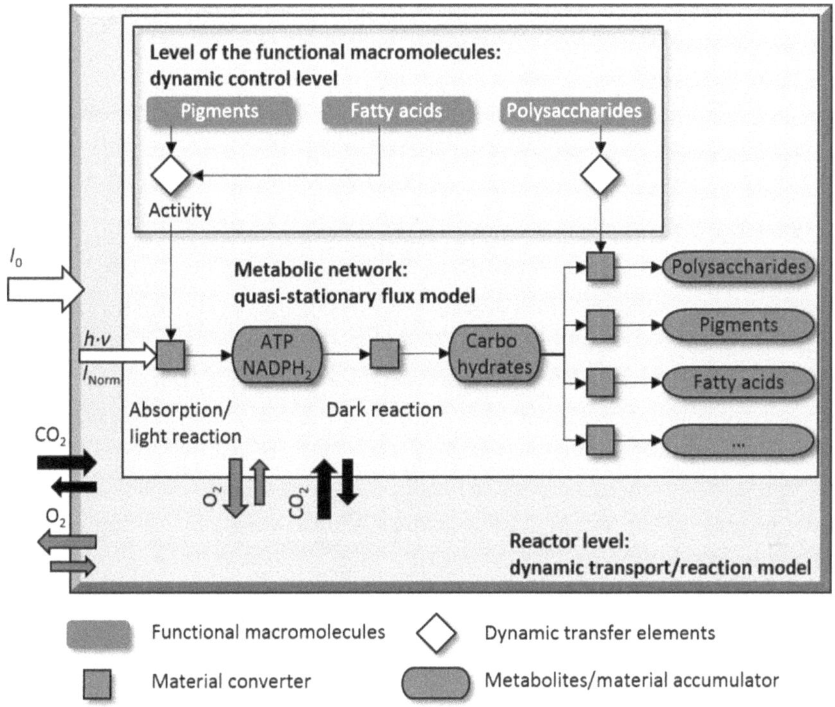

Figure 1 Hierarchical structure of photobioprocesses. The outer layer is the reactor itself being the link between the environment and the biomass. The cells can be understood from the quasi-stationary metabolic network. Acclimation to different environmental situations is mainly done on the level of functional macromolecules controlling the fluxes in the framework of stoichiometric constraints.

On this level the performance of a process with respect to volumetric productivity can be assessed.

Of special interest is the energetic analysis on the basis of photoconversion efficiency (PCE), what is the chemical energy formed as biomass (heat of combustion H_X) per light energy absorbed in the reactor volume

$$\mathrm{PCE} = \frac{\Delta H_X}{\Delta E_{\mathrm{Light,\,abs}}} \left(\frac{\mathrm{kJ}}{\mathrm{kJ}}\right). \tag{2}$$

To give this value a clear meaning the system boundaries, here the reactor volume, have to be defined, what is not always the case in literature. Looking from the next higher level, the environment, to the reactor, eg, also light reflection on the surface, has to be considered.

Starting from the reactor level material fluxes into and inside the cell level are modeled as specific turnover rates being the converted amount of a given compound per biodrymass per time

$$r_{\text{Comp}} = \frac{R_{\text{Comp}}}{m_X} = \frac{\Delta m_{\text{Comp}}}{m_X \cdot \Delta t} \left(\frac{\text{g}}{\text{g} \cdot \text{h}}\right). \tag{3}$$

The specific turnover rate for the autocatalytic biomass formation itself r_X is classically referred to as μ. Results on the biomass level give insight into the performance of a given strain and its physiological behavior under the specific conditions in the reactor. The fluxes are driven by light and substrate availability and coupled by enzyme reactions. The cells can adapt to the environment on a control level, where enzyme activities are regulated or the cell composition changes with respect to functional macromolecules.

In the following paragraphs the hierarchical modeling levels will be outlined in more detail giving a look at the different ways of thinking and the related methods employed.

2.2 Modeling Metabolic Fluxes

Many systems in nature and technology exhibit typical structures which can be used for modeling considerations via analogy. One of those structures are networks. Examples include the vascular system of animals, rivers or streets in a given region, or electrical networks. The features and related scientific questions are

- The system variables are regarded as fluxes in case of metabolic networks.
- Each variable in the system interacts only with a few other variables in distinct knots.
- Nevertheless, the system as a whole shows specific features.
- How do sources and perturbations spread over the network?
- Can we find information about the network from peripheral measurements?
- Where are limiting steps or bottlenecks?

A cell comprises several thousands of enzymatic steps. For process models it is not possible and not necessary to know all of them with their kinetic data. Consecutive enzymatic steps can be summarized as metabolic pathways. Background is the assumption that metabolites entering a specific pathway leave it at the same molar amount and that metabolites will not accumulate and are therefore considered as quasi-stationary. On this point the modeling work consists of defining the most important metabolic fluxes, necessary to make quantitative statements about the behavior of the cells, and the process as a whole under the light of a given question. Modeling being an iterative process, the list of considered fluxes can be amended in a later stage of modeling.

Besides ordering enzymatic steps along metabolic pathways, cofactors like ATP or NADH can be regarded as a pool to reach further simplification. This is accompanied by lumping-related yield parameters like $y_{ATP,I}$ and $y_{X,ATP}$ to $y_{X,I}$.

The direction of the counting arrows is usually chosen in a way to result in positive fluxes for the normal physiological state of the organism.

For complex metabolic networks we can set up a structured approach for finding suitable balance boundaries and setting up the balance equations from network analysis. Background knowledge is given by Roels (1983) and Stephanopoulos et al (1998).

- Each knot has to be inside (at least) one balance boundary. Otherwise the stoichiometry of this knot is not employed for the model.
- Each path has to be cut (at least) by one balance boundary. Otherwise it will not appear in the model.
- The maximum number of linear independent equations for one knot with n fluxes is $<n$. Otherwise the knot would determine itself completely and would be decoupled from the remaining part of the network and the outer world.

The maximum number of balances equals the number of metabolic knots. An additional balance around the whole cell may be useful for later model validation, but does not give additional information besides the balances around the individual knots.

Besides, also weaker linear relations are observed in living cells. These are not directly stoichiometrically determined but have their origin in intracellular energy or material balances which are subject to change by the cells according to the environmental conditions.

Metabolic fluxes interact in specific ways corresponding to specific approaches finding model equations. A first approach is setting up material balances and stoichiometric relations. Elemental balances are a first choice because they can easily be formulated.

In the simple example shown in Fig. 2 the list of $n_f = 7$ metabolic fluxes including the flux of light energy is given as the column vector

$$\mathbf{r} = \left[r_{I,abs}, r_{CO_2}, r_{H_2O}, r_{O_2}, r_{Gluc}, r_{NH_3}, r_X\right]^T \quad (4)$$

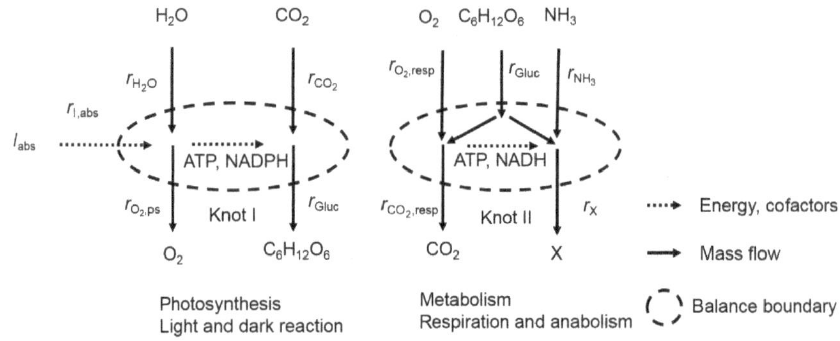

Figure 2 Simplified example of a generic phototrophic organism for setting up balance equations.

For this example setting up the linear equations is outlined. For knot I we obtain three elemental balance equations:

$$+1 \cdot \frac{1}{M_{CO_2}} \cdot r_{CO_2} + 0 \cdot \frac{1}{M_{H_2O}} \cdot r_{H_2O} - 0 \cdot \frac{1}{M_{O_2}} \cdot r_{O_2} - 6 \cdot \frac{1}{M_{Gluc}} \cdot r_{Gluc} = 0 \quad \text{(Molar C-balance)} \tag{5}$$

$$+2 \cdot \frac{1}{M_{CO_2}} \cdot r_{CO_2} + 1 \cdot \frac{1}{M_{H_2O}} \cdot r_{H_2O} - 2 \cdot \frac{1}{M_{O_2}} \cdot r_{O_2} - 6 \cdot \frac{1}{M_{Gluc}} \cdot r_{Gluc} = 0 \quad \text{(Molar O-balance)} \tag{6}$$

$$+0 \cdot \frac{1}{M_{CO_2}} \cdot r_{CO_2} + 2 \cdot \frac{1}{M_{H_2O}} \cdot r_{H_2O} - 0 \cdot \frac{1}{M_{O_2}} \cdot r_{O_2} - 12 \cdot \frac{1}{M_{Gluc}} \cdot r_{Gluc} = 0 \quad \text{(Molar H-balance)} \tag{7}$$

Now we have three equations for four unknown variables concerning fluxes in photosynthesis. The balance equations represent the well-known gross reaction equations for photosynthesis

$$6 \cdot H_2O + 6 \cdot CO_2 \rightarrow C_6H_{12}O_6 + 6 \cdot O_2. \tag{8}$$

These stoichiometric relations could of course be used directly to set up three linear equations.

From the example in Fig. 2 a nitrogen balance can be obtained for knot II:

$$e_{N,NH_3} \cdot r_{NH_3} - e_{N,X} \cdot r_X = 0 \quad \text{(N-balance knot II)} \tag{9}$$

expressing that growth, especially protein and nucleic acid formation, directly depends on nitrogen availability. It can be understood as a "weak" stoichiometry. This holds of course only as long as the cell does not change its molecular composition towards carbohydrates. Nevertheless, this equation is very useful to predict growth or ammonia consumption. The elemental composition of the biomass $e_{N,X}$ has to be verified by independent measurements. Similarly, a P balance could be applied, not shown here. As the parameters e_{N,NH_3} and $e_{N,X}$ are linearly dependent, a more lumped yield parameter $y_{X,NH_3} = e_{N,NH_3}/e_{N,X}$ is often introduced in references

$$y_{X,NH_3} \cdot r_{NH_3} - r_X = 0. \tag{10}$$

This lumped parameter has to be estimated for different nitrogen sources or deduced from the elemental equation.

As we are allowed to set up further equations for this knot we chose the stoichiometric relation between $r_{O_2,resp}$ and $r_{CO_2,resp}$

$$\frac{1}{M_{O_2}} \cdot r_{O_2} - \frac{1}{M_{CO_2}} \cdot r_{CO_2} = 0 \quad \text{stoichiometry} \tag{11}$$

A third equation could be the mass balance for the knot

$$\sum r_i = 0 \tag{12}$$

Of course other compounds are used by the cell in minor amounts and have to be considered in a more precise model. The same holds for hydration and carboxylation conversion taking place in several steps of the metabolism. For knot II one degree of freedom remains, what is the split of glucose flux to respiration and anabolism. To avoid going from the macroscopic level to the intracellular ATP flux for the moment, an a priori unknown yield parameter $y_{X,Gluc}$ is introduced leading to:

$$y_{X,Gluc} \cdot r_{Gluc} - r_X = 0 \tag{13}$$

The exact value of this yield coefficient has to be determined by parameter estimation on a given data set.

At the end of this process looking at intracellular stoichiometry we obtain the linear set of equations

$$\mathbf{Y} \cdot \mathbf{r} = 0, \tag{14}$$

where **Y** is the system (stoichiometric, yield) matrix of the network.

In the example we have now seven equations for eight unknown fluxes. In this example no linear dependency between the equations occurs, what has to be nevertheless validated, so one additional equation is necessary to set up a complete model of the cell. These equations come usually from kinetics, where further biological knowledge can be included.

Similarly as in other networks, cell models have to contain sources to keep the fluxes running. These are represented by kinetics, especially substrate uptake kinetics. Kinetic equations are basically the link between the reactor model and the cell model and represent one way for the cell to react to environmental conditions. In the case of the simple example, CO_2, NH_3, and of course light harvesting I_{abs} are worth considering. As shown above, only one additional equation is necessary to come to a full model description. Biologically speaking, the cell is either light, or carbon, or nitrogen limited at a time. Under light limitation for example, NH_3 and CO_2 are taken up only to an extent which can be used by the cell in accordance with the given stoichiometry. In references sometimes connected expressions like the product of different kinetics can be found. However, this is in normal cases not a structured approach. The limitation may of course change during a cultivation process

$$\mathbf{r} \leq f(\mathbf{c}). \tag{15}$$

The equal sign is valid where a kinetics is fully employed by the cell; the less than sign indicates that the cell has to reduce some of the uptake rates because of stoichiometric constraints.

As the ammonia uptake rate is an enzymatic step, a Michaelis–Menten-type equation is chosen:

$$r_{NH_3} \leq r_{NH_3,\max} \cdot \frac{c_{NH_3}}{k_{NH_3} + c_{NH_3}} \tag{16}$$

CO_2 enters the cell via diffusion, but the enzymatic carbon fixation at the RuBisCo is potentially limiting, leading to Michaelis–Menten-type kinetics as well

$$r_{CO_2} \leq r_{CO_2,\max} \cdot \frac{c_{CO_2}}{k_{CO_2} + c_{CO_2}}. \tag{17}$$

The actual meaning of equal or less than is calculated during simulation from the stoichiometric model. For simulation it is advisable to extend the kinetic equations for negative concentrations like: if $c_S < 0$, then $r_S = -k_S / r_{S,\max} \cdot c_S$. That avoids numerical problems of function evaluation for negative values, what can happen during iteration of the integrator.

For the moment we assume that ATP production for growth depends on the absorbed light with a given yield and energy demand for growth is constant as well.

In our generic example we consider light limitation in its most simple form:

$$r_{\mathrm{Gluc}} \leq y_{\mathrm{Gluc,I}} \cdot \frac{I_{\mathrm{abs}}}{c_{\mathrm{X}}} \tag{18}$$

The impact of light is represented by the lumped yield coefficient. This represents the first linear part in the so-called PI curve, see below. Of course this does not go at infinitum, but ends at a specific maximum rate represented by the second more or less constant part of the PI curve. This maximum value can be determined either by maximum capacity of the light reaction in photosynthesis or by another limiting step, may be capacity of RuBisCo for CO_2 fixation or nutrient availability. Also other intracellular bottlenecks in metabolism cannot be excluded a priori.

The actual values are calculated during simulation, especially which of the kinetic equations is active at its maximum and which one is calculated from the stoichiometric model. Knowing the experimental conditions, the limiting step seems often to be clear. But for a structured model development, rules are needed to decide during simulation about the active kinetic equation.

Assumptions for finding the limiting kinetic steps:

- The material and energy uptake rates given by kinetics cannot be exceeded.
- The uptake rates can be reduced by the cell in case further processing of the respective compound is not possible.
- The cell will keep the turnover rates as high as possible under the kinetic and stoichiometric constraints. This leads as an example to the two branches of the PI curve.
- Choosing infeasible kinetic limitations during simulation will result in violating one of the others in the sense of predicting a value higher than r_{max} or less than 0.

2.3 Modeling the Intracellular Control Level

For process modeling it is not possible to understand all intracellular control loops. But in many cases we can formulate model hypothesis on the final effect of intracellular control. Some assumptions have already been done implicitly, like the assumption that the organism will not waste ATP in normal growth modus or that the cell exploits the kinetic constraints as much as possible.

Possible model hypotheses concerning flux control for photobioprocesses:

- The cell optimizes metabolic fluxes subject to a defined criterion. To find and formulate this criterion is part of the model-building process.
- Shift of ATP/NADPH ratio in the light reaction is possible by the ratio of cyclic and linear electron flow and will happen according to the needs of metabolism, eg, with respect to degree of reduction of the biomass.
- Light respiration depends on RuBisCo kinetics for oxygen turnover but can be modified by the cell, as it is a reduction of available metabolic energy.
- Metabolic cycles and anaplerotic sequences can partially circumvent stoichiometric and kinetic limitations.
- Partitioning of the produced glucose to different functional macromolecules depends on the mentioned constrains but also on recently not exactly understood intracellular control goals.

The next step is to look at the rank of the stoichiometric matrix. Rank deficiency means biologically speaking that the organism has additional degrees of freedom to adjust metabolic pathways. As a possible control goal maximization of the specific growth rate has been proposed. The idea was first formulated for yeast, where ethanol formation occurs only, if maximum respiratory capacity is reached but glucose uptake still possible.

For a given working point where light intensity and nutrient concentrations have distinct values $\mathbf{c}$ this optimization criterion reads:

$$\max r_X \text{ subject to } \mathbf{Y} \cdot \mathbf{r} = 0 \quad \text{and} \quad 0 \leq \mathbf{r} \leq \mathbf{r}(\mathbf{c})$$

model hypothesis of maximum specific growth rate (19)

As the optimization criterion, the stoichiometric matrix, and the values of the kinetics for given concentrations are linear, this optimization problem can be solved by means of linear programming. The solution lies always at one of the constraints and especially at the corners of the resulting polygon (Fig. 3). Practically, only the corners of the polygon have to be examined for the highest growth rate at each simulation step. Respective model results are given in the subsequent paragraphs.

Of highest interest for process modeling is the partitioning of the carbon flux to different functional macromolecules. This leads to changes in biomass composition with respect to the major classes namely proteins,

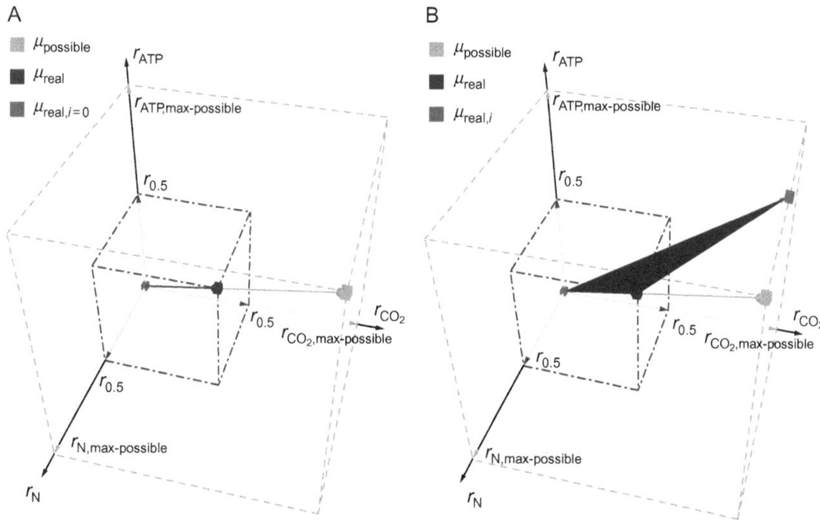

Figure 3 The principle of linear programming. The parameter space here consisting of three metabolic fluxes represents the space of possible metabolic states. The cube represents the space, where no stoichiometry applies, in case all fluxes are interconnected it shrinks to the green line, of which the length represents the specific growth rate. The cell reaches its maximum specific growth rate at the edge of the *cube* (A), green (light gray in the print version) without, blue (dark gray in the print version) with nitrogen limitation. In cases where a stoichiometry is missing, here the nitrogen balance during lipid formation (B), the *triangle* represents possible states. Assuming the maximum principle, specific growth rate is again at the edge of the *green* (*light gray* in the print version) *cube*.

carbohydrates, nucleic acids, and fatty acids. But also a more specific view has to applied, eg, carbon partitioning to antenna proteins or RuBisCo. Here no direct stoichiometry applies, leaving several degrees of freedom for intracellular control. Nevertheless, different pathways need different amounts of ATP and NADH; the respective balances are possible constraints. Nitrogen and phosphate availability are another candidate to formulate constrains for carbon partitioning into different products. In contrast to the fast reactions on the flux level, here slower reactions on the epigenetic level are involved leading to observable growth dynamics with timescales from hours to days. To find appropriate equations for modeling, some ideas and correlations are given in literature:

- Under low light conditions chlorophyll content is increased.
- Under nitrogen starvation starch and oil content are increased. This is considered in model linking carbon and nitrogen flux.
- Under suboptimum temperature conditions starch can accumulate.

- Removal of extracellular polysaccharides from the medium leads to increased polysaccharide formation.

2.4 Modeling the Reactor Level

The reactor level is mainly characterized by transport phenomena. Similarly as in heterotrophic cultivation material balances for dissolved compounds arc written as:

$$\frac{d\mathbf{c}}{dt} = D \cdot (\mathbf{c}_{in} - \mathbf{c}) - \mathbf{r} \cdot c_X, \tag{20}$$

This equation represents accumulation, net transport as influx minus outflux, and uptake/reaction by the biomass. For batch cultivations, the dilution rate $q_{feed}/V_R = 0$. For dissolved gases a transport term via the gas phase has to be amended. For the carbon dioxide transfer rate (CTR) this reads

$$\mathrm{CTR_{gas}} = k_L a_{CO_2} \cdot \left(c^*_{CO_2} - c_{CO_2}\right) \tag{21}$$

Details can be found in bioprocess engineering textbooks (Doran, 2013; Nikolaou et al, 2015). For the total CO_2 transport input by a presaturated medium or losses via outflow have to be considered due to the high solubility. For oxygen in heterotrophic cultures that is usually neglected.

2.5 Simulation Example

The equations derived above form a complete representation of a simple generic phototrophic organism. Parameter values, which are typical for such processes, are chosen as given in Tables 1 and 2.

The model represents the most basic features of a phototrophic batch process. In the first phase, there are only small light gradients leading to a uniform illumination of all cells (Fig. 4). Light intensity is strong enough

Table 1 Set of Parameter for the Physiological Model

$r_{X,max}\left(\frac{g}{g \cdot h}\right)$	$k_I\left(\frac{\mu E}{m^2 \cdot s}\right)$	$r_{X,resp}\left(\frac{g}{g \cdot h}\right)$	$r_N\left(\frac{g}{g \cdot h}\right)$	$k_N\left(\frac{g}{L}\right)$
1.6	180	0.25	0.3	0.4

Table 2 Parameter Set for the Reactor Model

$I_0\left(\frac{\mu E}{m^2 \cdot s}\right)$	$\varepsilon\left(\frac{L}{m \cdot g}\right)$	$c_{X,0}\left(\frac{g}{L}\right)$	$c_{N,0}\left(\frac{g}{L}\right)$	$D_R(m)$
500	200	0.027	0.5	0.01

to allow for light saturation and maximum growth rate, so exponential growth can evolve. In the transition phase between phase I and II transmitted light decreases due to higher biomass concentration, in the dark part of the reactor remote from the light source cells cannot longer achieve $r_X = r_{X,max}$, so the mean specific growth rate is reduced. Finally in phase II all light is absorbed. This is in analogy with a fed-batch cultivation with a linear substrate supply. The reactor equations reduce to

$$\frac{dc_X}{dt} = y_{X,I} \cdot I_0 \tag{22}$$

predicting a linearly increasing biomass concentration. Phase II is therefore called the linear phase. Note that with increasing biomass and constant energy supply the specific growth rate diminishes. The transition from phase II to phase III is characterized by the onset of ammonia limitation. A similar behavior of the cultivation can be observed, when with increasing biomass concentration the specific growth rate comes closer to the compensation point and energy (light) consumption is used more and more for maintenance purposes. Growth is further reduced reaching finally zero in phase III due to complete nutrient limitation.

3. BUILDING BLOCKS FOR PHOTOTROPHIC PROCESS MODELS

Modeling of photobioprocesses on the cellular level underlies the same principles as for heterotrophic ones. However, there are specific features of photosynthetic cells which have to be included into modeling. Some of them are pointed out in this paragraph. These include:

- The most important point is of course photosynthesis in its two steps of light-dependent water splitting and carbon fixation. Photosynthesis has been quantitatively studied for decades from stoichiometry down to the molecular basis with respect to spatial organization and details of electron transfer. To include light into modeling is one of the challenges.
- Carbon uptake is an important issue as well, as carbon dioxide partial pressure is limiting in nature and in many cultivations, while gas transfer is a cost issue. CO_2 uptake is a unique feature in phototrophs and is determined by activity of the enzyme ribulose-1,5-bisphosphate carboxylase/oxygenase (RuBisCo). The influence of oxygen on this step is a reason for reduced productivity.

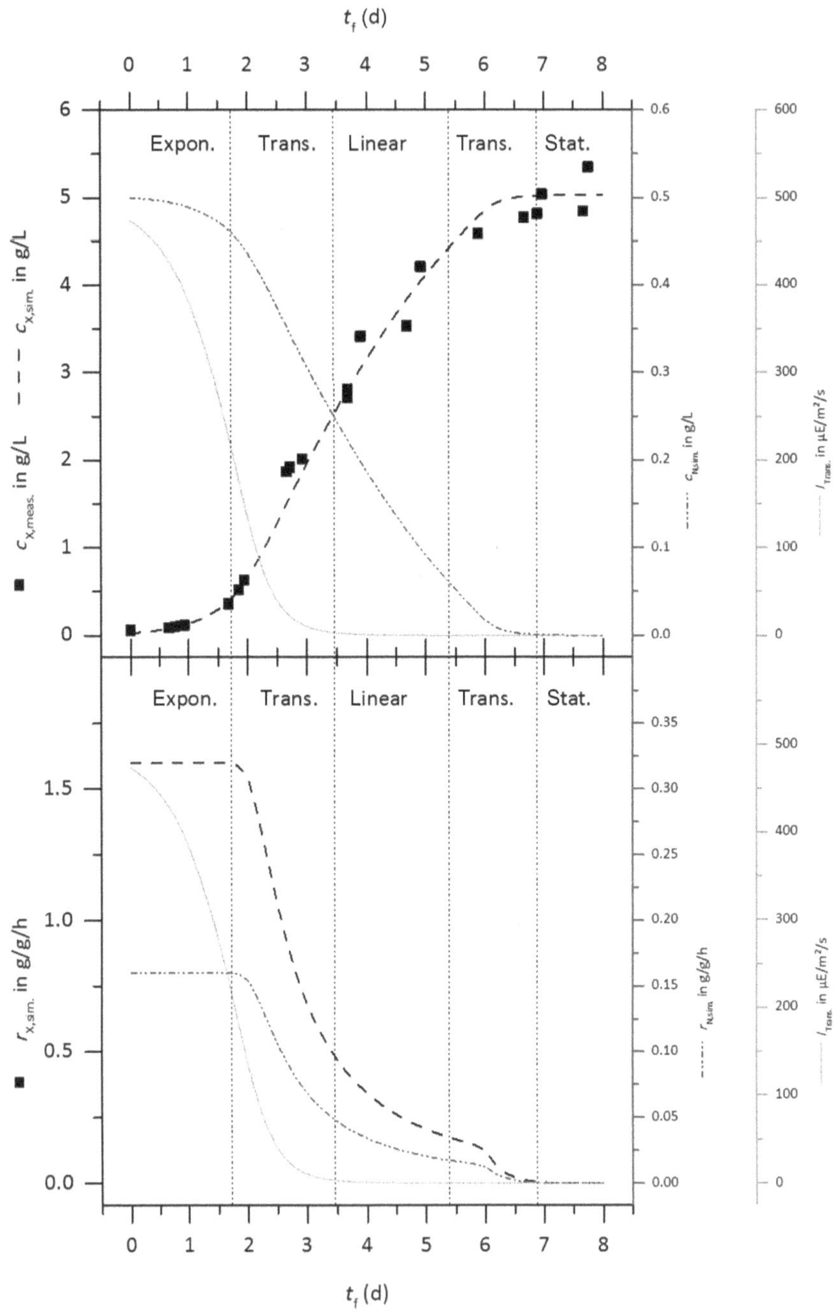

Figure 4 Simulation of the model described above. *Solid lines* are simulation results and *symbols* are measurements from a real process.

- In natural habitats but also in bioreactors nutrient availability is often the limiting factor, where precise measurement of kinetics is necessary for process optimization.
- As microalgae are cultivated for their ability to produce high-value products and accumulate them to high intracellular concentrations, carbon partition into these different cell compounds depending on the environmental conditions is introduced as well.

For background knowledge on phototrophic microorganisms, a well-elaborated and understandable treatise is given, eg, in Falkowski and Raven (2007).

3.1 Photosynthesis and PI Curve

Photosynthetic activity as a function of light intensity is widely measured in photobiology and represented as light response curve. In the so-called PI curve (photosynthesis vs irradiance curve) oxygen production or carbon dioxide uptake is measured for increasing light intensity. Maintenance is visible for low light intensities, a linear relation for medium light intensities until for higher intensities a plateau is reached, representing an intracellular limitation. This bottleneck could be located inside the light reaction itself, coupled to the RuBisCo activity, or caused by another rate-limiting bottleneck in the metabolism.

The PI curves are usually spectroscopically performed in cuvette scale and in minute timescale. To measure the specific growth rate as a function of light intensity in analogy to the Monod equation, cultivations have to be performed allowing appropriate acclimation of the cells. Aiba (1982) formulated the observable growth rate as substrate inhibition:

$$\mu = \mu_{\max} \cdot \frac{I}{k_I + I + \frac{I^2}{k_{I,i}}} \tag{23}$$

It reflects a part with increasing growth, saturation, and inhibition similarly as the classical PI curve. The Monod equations are nowadays motivated by an enzymatic substrate uptake (Michaelis and Menten, 1913 (2013)) step followed by a linear relationship between substrate uptake and growth (Pirt, 1965). In contrast, photosynthetic activity is typically understood as a sequence of a transport step, being light absorbance in the antennas, followed by a limiting bottleneck in the late enzymatic steps of the photosystems or even later in the metabolism. Unlike enzyme kinetics with a first reversible reaction between substrate and enzyme, light absorbance and transport are not reversible.

To look closer to light response kinetics specific growth rates have been determined experimentally (Lehr et al, 2012). In Fig. 5 a typical growth vs light response curve is shown.

Such curves cannot be represented by usual enzymatic kinetics. The linear part for low light intensities justifies the description of the growth curve as a linear relationship between growth and light intensity reflecting the irreversible transport step. The slope of this curve, usually denoted as α, is called the photon efficiency, here named $\gamma_{X,I}$ for consistency. Here the parameter is based on biomass, while chlorophyll as a reference is also a choice depending on the detailing of the model.

For higher light intensities saturation occurs, here again a linear curve with a small slope. Data from other authors and for other organisms could show a constant saturation part. To find out what it is in a specific case is one task of modeling. This part of the curve can be subjected to acclimation or

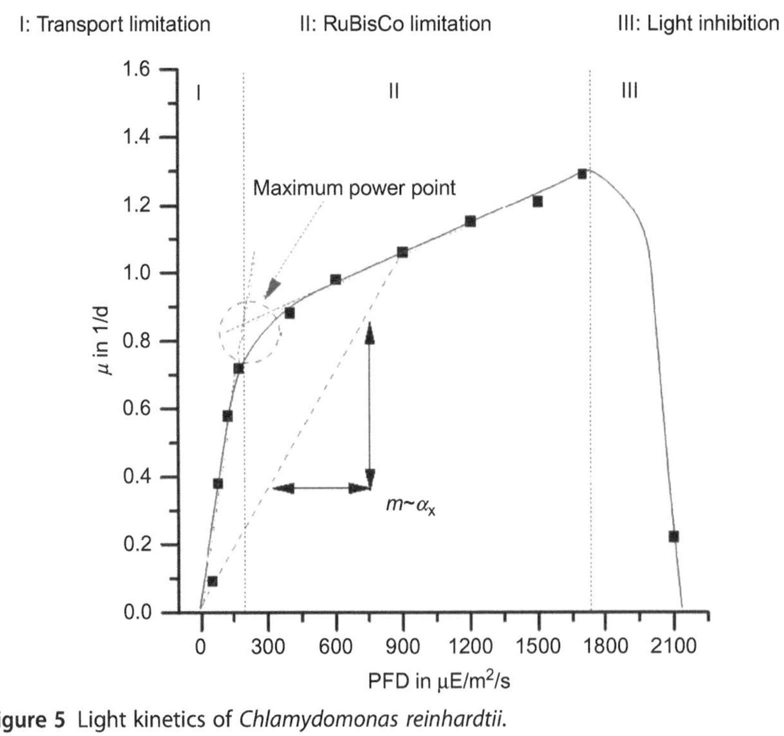

Figure 5 Light kinetics of *Chlamydomonas reinhardtii*.

depends on nutrient availability. Respective model equations will represent this part of the curve in the framework of the whole model.

Such light response curves can be used as a first lumped approach for data assessment, calculating μ-integration, reactor layout, or control. Nevertheless, a more detailed mechanistic model needs more implementation of quantitative elements. Especially light absorbance and usage of the photosynthetic products ATP and NADPH have to be separated.

To follow further the way of light energy, the absorption process is considered for calculation of photons contributing to energy supply of the cells. This is done by the absorption cross section σ_X of the biomass. Finally, the photon balance inside the antennas can be written as

$$q_{h\nu,\mathrm{abs}} = \sigma_X \cdot I_{h\nu} - q_{h\nu,\mathrm{diss}} \tag{24}$$

While in regular enzyme kinetics a limiting step at the end of a cascade feeds back to the beginning via metabolite accumulation and back-reactions, in photosynthesis the energy balance in the antennas dissipates energy, which cannot be further processed, as heat and fluorescent light, here referred to as $q_{h\nu,\mathrm{diss}}$. The fluorescence measurement signal is a powerful tool to estimate photosynthetic efficiency and allows to distinguish between light and nutrient limitation. Further introduction to this methodology and application for determination of kinetics can be found at Gargano et al (2015).

The following paragraphs refer to Fig. 6, where the basic fluxes are given.

The stoichiometric gross equation for photosynthesis

$$2 \cdot \mathrm{H_2O} + \mathrm{CO_2} + 8 \cdot h\nu \rightarrow (\mathrm{CH_2O}) + \mathrm{H_2O} + \mathrm{O_2} \tag{25}$$

is the basis of almost all photochemical processes and can serve as the first step into stoichiometric equations for a lumped model. This leads to three formal stoichiometric equations for the four unknown fluxes

$$q_{\mathrm{CH_2O,PS}} - q_{\mathrm{CO_2,PS}} = 0 \tag{26}$$

$$q_{\mathrm{CH_2O,PS}} - q_{\mathrm{O_2,PS}} = 0 \tag{27}$$

$$q_{\mathrm{CH_2O,PS}} - q_{\mathrm{H_2O,PS}} = 0. \tag{28}$$

To set up stoichiometric equations for other relations between the fluxes is possible as well, but not additionally, as it would lead to linear dependent equations. The photosynthetic quotient turns out be

$$\mathrm{PQ} = \frac{q_{\mathrm{O_2,PS}}}{q_{\mathrm{CO_2,PS}}} = 1 \tag{29}$$

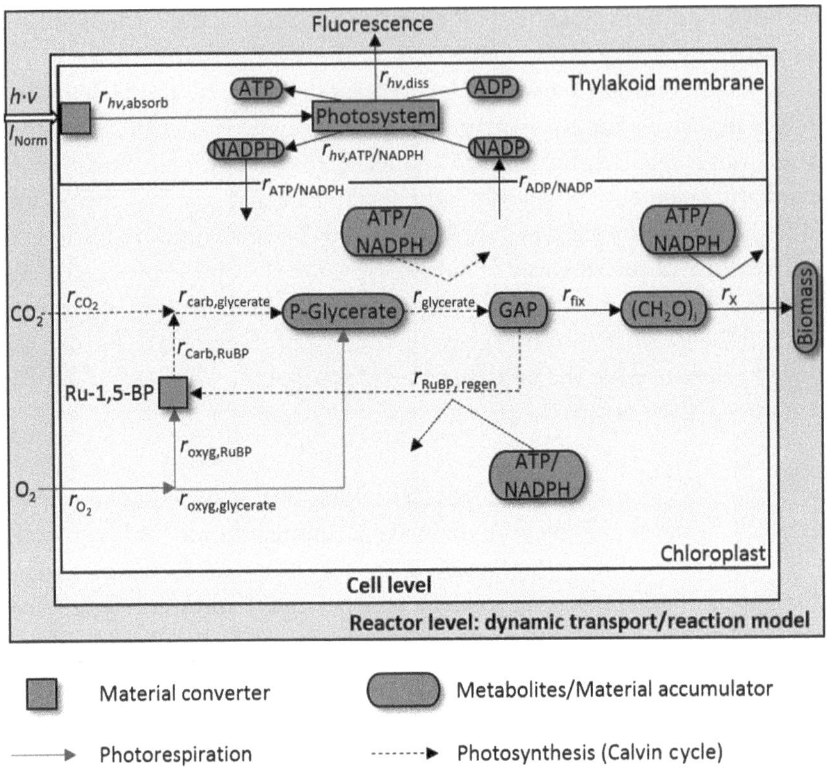

Figure 6 Flux distribution in a photosynthetic cell, especially the light reaction in the thylakoid membrane (*upper part*) and the Calvin cycle are shown.

The fourth equation has to come from the light as the energy source into the system. In the kinetic equation for the energetic activity of the photons

$$q_{CH_2O,\,PS} - y_{CH_2O,\,h\nu} \cdot q_{h\nu,\,abs} = 0 \tag{30}$$

the yield factor is theoretically determined to be 8 but is practically counted to be 10–12 because different steps like photorespiration or efficiency of the transmembrane ATPase are not strictly stoichiometrically controlled. So we have three stoichiometric relations for four fluxes. The fourth one is assumed to be known from the light absorbance part of the model.

To have a closer look at photosynthesis, we consider the two steps, of which it is organized, the light reaction

Table 3 Parameters of the Photosynthesis Submodel

	$y_{X,I}\left(\frac{g/(g\cdot d)}{\mu mol/(m^2\cdot s)}\right)$	$y_{X,h\nu}\left(\frac{g}{mol}\right)$	$y_{ATP,h\nu}\left(\frac{mol}{mol}\right)$	$y_{NADP,h\nu}\left(\frac{mol}{mol}\right)$	$y_{CH_2O,h\nu}\left(\frac{mol}{mol}\right)$
Theoretical[a]			3	2	8
Measured[b]	<0.02	~0.8	~4.13	~2.75	10–12

Theoretical values from molecular mechanisms and lumped values from measurements.
[a]Falkowski and Raven (2007).
[b]Dillschneider et al (2013).

$$2\cdot H_2O+2\cdot NADP^+ +3\cdot ADP+3\cdot P_i \rightarrow O_2+2\cdot NADPH/H^+ +3\cdot ATP+3\cdot H_2O \quad (31)$$

and the dark reaction in the Calvin cycle

$$3\cdot CO_2+9\cdot ATP+6\cdot NADPH+6\cdot H^+ \rightarrow C_3H_6O_3-P_i(GAP)+9\cdot ADP+8\cdot P_i+6\cdot NADP^+ +3\cdot H_2O \quad (32)$$

Phosphate in GAP is lumped into the ATP pool. The exact stoichiometry is not really known, but the energy and the reductant flux from the light reaction match exactly the demand for glucose formation. So this submodel for the linear photophosphorylation can be lumped and decoupled from the anabolic part of the model.

For an educated guess as starting point for data evaluation and parameter estimation on experimental data sets, Table 3 gives parameter values from references.

3.2 CO_2 Uptake Kinetics and Light Respiration

Besides light supply, carbon dioxide uptake is most important to be considered. CO_2 transfer to photobioreactors has been shown to be one of the reasons for energy dissipation. Furthermore, it influences pH value what on the other hand can be used for control purposes. CO_2 kinetics is therefore an important issue for optimizing cultivation conditions.

Measured data for growth at different p_{CO_2} are shown in Fig. 7 (own data).

Optimum growth rates are possible for partial pressures above 0.5%. The shape of the curve cannot be fitted exactly by Michaelis–Menten kinetics; the maximum growth rate is rather given by another intracellular limiting step. The apparent inhibition is probably induced by light respiration as p_{O_2} in the reactor increases with increasing growth rate. Again there is the necessity during modeling to refine the formal μ vs p_{CO_2} kinetics by a substrate uptake step and a submodel for maximum growth rate.

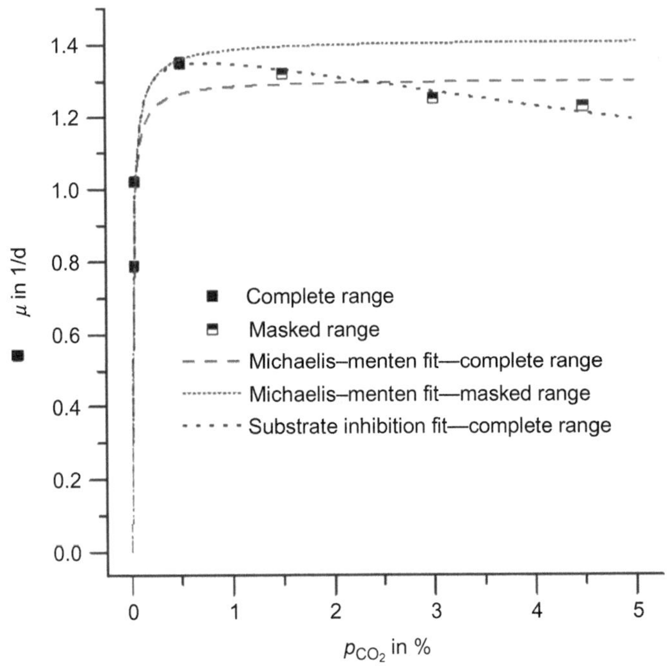

Figure 7 Measured CO_2 growth kinetics under otherwise ideal conditions; different model assumptions have been tested.

Carbon dioxide can diffuse easily through the cell membrane, but CO_2 uptake is determined by activity of the RuBisCo, involved in the major step of carbon fixation forming 2 glycerate-3P from ribulose 1,5-bP and CO_2 (Fig. 6). Oxygen is converted by RuBisCo as well forming glycolate-2P. This metabolite can be refixed to glycerate-3P but with loss of carbon. This has to be refixed in the Calvin cycle but on cost of ATP and NADPH. This cycle is referred to as photorespiration. Oxygen competes with carbon dioxide at the active site of the RuBisCo/ribulose complex. Application of mass action law and quasi-stationary conditions as well as assuming intracellular ribulose concentration being in excess (formally infinity) leads to a double Michaelis–Menten-type kinetics with competitive inhibition for the other gas (Jordan and Ogren, 1981):

$$q_{CO_2,\,carb} = q_{CO_2,\,carb,\,max} \cdot \frac{c_{CO_2}}{c_{CO_2} + k_{CO_2,\,carb} \cdot \left(1 + \frac{c_{O_2}}{k_{O_2,\,oxyg}}\right)} \tag{33}$$

for carboxylation, and

$$q_{O_2,\,oxyg} = q_{O_2,\,oxyg,\,max} \cdot \frac{c_{O_2}}{c_{O_2} + k_{O_2,\,oxyg} \cdot \left(1 + \frac{c_{CO_2}}{k_{CO_2,\,carb}}\right)} \quad (34)$$

for oxygenation.

These equations are valid for carbon limitation and contain four unknown parameters.

Both reaction rates are coupled by the concentrations of the dissolved gases:

$$\frac{q_{CO_2,\,carb}}{q_{O_2,\,oxyg}} = s_{C,O} \cdot \frac{c_{O_2}}{c_{CO_2}} \quad (35)$$

with the selectivity factor $s_{C,O}$. This equation holds for light limitation as well, while for Eqs. (33) and (34) "less than" applies.

Despite the relevance of this step, not many experimental investigations are available measuring the related RuBisCo parameters in vivo. This is necessary, as carboxylation and oxygenation are subject to different intracellular activation and deactivation mechanisms. Furthermore, intracellular concentration of the dissolved gases is not necessarily the same as in the medium, although both gases can easily diffuse through the cell membrane. Many algae possess also carbon concentration mechanisms leading to higher concentrations at the reaction site of RuBisCo. Some algae groups can take up hydrogen carbonate as well. Biologically speaking RuBisCo has a low affinity to its substrate CO_2 as it developed in evolution in eras of high carbon dioxide concentration in the atmosphere. It is present in the cell in large amounts and is the most abundant protein on earth. Oxygen is not inhibiting in the strict sense, but influences carboxylation under typical conditions in phototrophic cultivations. Kliphuis et al (2011) measured a PCE reduction of about 30% for typical cultivations against an ideal situation with high p_{CO_2} and low p_{O_2}. For modeling purposes this double kinetics has to be considered setting up the ATP balance.

Data for the respective kinetic parameters of the isolated enzyme are provided in references. For in vitro cultivations it is not easy to distinguish between CO_2 and O_2 turnover from carboxylation and (light) respiration. Nevertheless data for the respective kinetic parameters are provided in references as well. Some mutants of Chlamydomonas are lacking the refixation and excrete glycolate (Wilhelm et al, 2006) allowing calculation of the oxygenation reaction.

For the parameter values the following data have been given in literature (Table 4).

Table 4 Parameters for Carbon Dioxide Uptake Kinetics

	$k_{CO_2,carb}\left(\frac{\mu M}{L}\right)$	$k_{O_2,oxyg}\left(\frac{\mu M}{L}\right)$	$s_{C,O}=\frac{r_{CO_2}}{r_{O_2}}\left(\frac{\mu M}{L}\right)$
In vitro[a]	29	480	3.7
In vivo typical[b]	30–100		3–5

[a]Isolated enzyme, values converted from μmol (Jordan and Ogren, 1981).
[b]After Raven and Falkowski (2007).

Note that k_{CO_2} is significantly higher than typical concentrations in natural habitats, eg, in the ocean at 10 μM. Carboxylation rate is higher than the macroscopically measurable CO_2 turnover due to respiration (see below).

The five metabolic fluxes handled by RuBisCo are therefore fully described for carbon limitation by the two kinetic equations and the 1:1 stoichiometry between CO_2/O_2 and glycerate conversion and RuBP, respectively. For the carbon split at GAP no stoichiometry but mass balance applies.

3.3 Kinetics for Nutrient Uptake

Even less is known about kinetics of nutrients. Usually they are given for batch experiments in excess but are not measured in high enough sample frequency to resolve the limiting concentrations. The final biomass concentration reflects the stoichiometry but does not allow for determining limitation constants. These are necessary to calculate more advanced process strategies. If during turbidostat operation the nutrient concentration was too high, it would be a loss of nutrients, microfeeding is a means to induce limitation without adverse secondary effects.

3.4 Stoichiometry and Carbon Partitioning

Photosynthesis is mainly determined by stoichiometry and delivers finally glucose for anabolic reactions. In the concept ofbalanced growth it is assumed that carbon flux is maintained so that the cell machinery remains in an optimal constant state with constant cell composition. This idea does no longer hold for changing environmental conditions like nutrient availability or changing light supply. Microalgae respond with metabolic changes in the macromolecular composition. The cell can direct the carbon flux to the different cellular macromolecules like proteins, carbohydrates, nucleic acids, or lipids. This capability, referred to as carbon allocation or carbon partitioning, is of highest technical interest as these macromolecules are often the final products. While only some balance constraints are applicable, the cells remain with several degrees of freedom for carbon partitioning. This is where biological knowledge comes in to formulate additional model equations.

Carbon flux is the main skeleton of flux distribution. It shuttles from the mitochondria to the cytosol as GAP and is referred here as carbohydrate equivalent (r_{CH_2O}). This part of the metabolism is comparable to heterotrophic growth and the structure of related equations can be translated. For process modeling it is convenient to lump the different steps in the TCA and the PPP and relate all pooled carbon, energy, and reductant flows to the carbohydrate equivalent.

The flux is first split into a part for anabolic biomass formation and a part for respiration

$$r_{CH_2O} - r_{CH_2O,\,ana} - r_{CH_2O,\,resp} = 0 \quad \text{(carbon split)}. \tag{36}$$

For this knot no specific stoichiometry applies, but the ratio is adjusted by the cell to gain the balance between carbon, reductant, and energy for biomass formation.

For respiration the usual net stoichiometry

$$CH_2O + O_2 \rightarrow CO_2 + H_2O + 6 \cdot ATP \tag{37}$$

applies, assuming that redox equivalents for the respiratory chain are produced in different parts of the metabolism under net release of CO_2.

The anabolic branch then is targeted to the different macromolecules proteins, carbohydrates, nucleic acids, and lipids and possibly further detailing to cell wall carbohydrates and starch as storage material. These are formed with rates r_{Macro} of which the cell consists so that the mass balance can be set up as

$$r_{CH_2O,\,ana} - \sum r_{Macro,\,i} = 0 \tag{38}$$

and the appropriate balance equations as well. The different macromolecules and different amounts of ATP and NADH are necessary. For details a list in the supplementary material of Kliphuis et al (2012) is recommended. This gives a frame for the possible space in which the cell can act. Depending on the number of compounds included in the model, some degrees of freedom remain. These can be considered in modeling by the optimization concept shown above. Specific growth rate r_X and PCE have been proposed in literature. For the biological kernel of the model related physiological state-dependent stoichiometries or intracellular control laws have to be found. However, this approach is not always feasible to set up process models.

For simplification of the above outlined model approach the concept of "active biomass" is introduced, where the functional macromolecules are combined to form the active part $r_{X,act}$ and accumulated lipids or carbohydrates contribute to the apparent biomass formation $r_{X,app}$. Lipids themselves are part of the cell dry mass but do not contribute to light capture and do not support the cell machinery directly. This leads to the mass balance, neglecting hydration or de/carboxylation reactions

$$r_{CH_2O,\,ana} - r_{X,\,app} - r_{X,\,act} - r_{Lipid} = 0. \qquad (39)$$

A first attempt to handle this condensed approach can be made using elemental balances. Nitrogen is mainly used for building up proteins and nucleic acids. The N-balance is a powerful means to model the distribution of nitrogen-free compounds like lipids on the one hand and other macromolecules on the other

$$e_{N,\,NH_3} \cdot r_{NH_3} - e_{N,\,Prot} \cdot r_{Prot} - e_{N,\,NucAc} \cdot r_{NucAc} = 0 \quad \text{(N-balance)}. \qquad (40)$$

Ammonia has been set here as an example. Other macroelements (S, P) can be handled the same way. While phosphate can accumulate as polyphosphate or phytin, nitrogen can be fixed in cyanophycin. For such storage compounds additional terms have to be added in this equation if necessary. Minor phosphate pools are phospholipids, NADP, and ATP.

A simulation is shown in Fig. 8. The full set of equations including ATP and NADH balance with different needs for lipids and active biomass and employing the optimization approach is given in Dillschneider et al (2013).

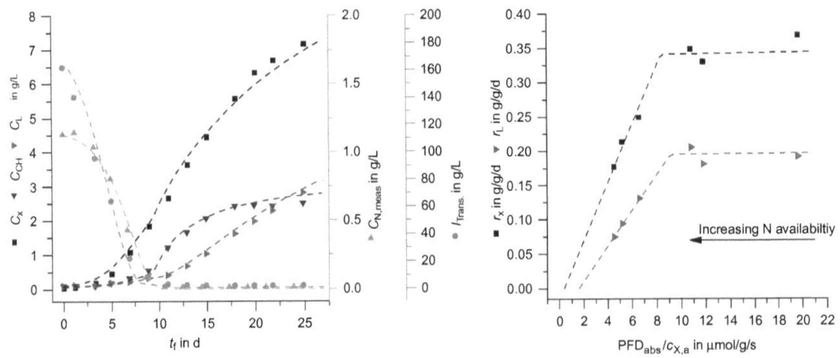

Figure 8 Simulation of the full set of equations (including ATP and NADH balance).

Table 5 Theoretical and Lumped Parameters for Carbon Allocation

	$y_{ATP,CH_2O,resp}\left(\frac{mol}{mol}\right)$	$y_{X,CH_2O}\left(\frac{g}{g}\right)$	$e_{N,X}\left(\frac{g}{g}\right)$
Theoretical	36		
Educated guess, experimental		0.5	0.1

It is remarkable that the overall efficiency of the process (PCE = 4.5%) does not depend on the growth phase. During N-saturation at the beginning and N-limitation with lipid formation in the second phase, photosynthesis and metabolism work with constant energetic yields. Stress phenomena are not explicitly formulated in the model. The apparent growth rate based on biodrymass is reduced due to the higher heat of combustion of lipids in comparison to carbohydrates and proteins. ATP consumption is accordingly higher. Other authors report on stress-induced reduction of metabolic activities due to lower light absorption, high turnover rates of lipids, or damages of the cell machinery due to lacking protein repair. The latter issue could be addressed by N-micro-feeding, what is a useful application of the model. Table 5 gives some hints for parameter values from literature to model carbon allocation to respiration and biomass fractions.

The use of energy for anabolic processes cannot be determined by counting ATP-dependent steps, as protein turnover, repair mechanisms, or spatial ordering of the cells require additional energy. Appropriate parameters have therefore been experimentally determined.

Modeling of an exact stoichiometric distribution of different intracellular fluxes with the aim to understand and optimize intracellular product formation is indeed a field where much more work could be done. Stoichiometry can be shifted in one direction as the lipid formation example shows. In most cases it is not known how far the shift can go and to what extent it affects the costs for growth and productivity.

3.5 Dynamics on Cell Level and Acclimation

Microalgae experience in nature and in photobioreactors many different changes in the environmental conditions over time. They react to these changes with changes in the macromolecular composition, measurable as dynamic time responses. Metabolic flux distribution is changing accordingly. Modeling of this so-called acclimation is the next challenge for modeling. A basic approach to model the role of intracellular nutrients by the concept of active biomass and storage compounds is given by Droop (Bernard, 2011). Access to intracellular measurements in reasonable sample frequency is usually a problem as well as insight into the biological effect, clear enough to formulate an intracellular control law above the level of numerous single mechanisms. Some acclimation processes are of specific importance for photobioprocess modeling:

- Under outdoor conditions microalgae are subjected to day–night cycles and can show autonomous diurnal growth cycles. Light acclimation leads inter alia to changing pigmentation and changing kinetics depending on current solar irradiation. High light conditions usually lead to lower pigment concentrations, eg, Fleck-Schneider et al (2007).
- Besides light, changing temperature in outdoor photobioreactors over time is a main reason for reduced productivity. A better model-based understanding is an important issue. An example is given below.
- In photobioreactors microalgae experience fast dark/light cycles induced by turbulences and strong light gradients. The reaction to this so-called flashing light or intermittent light is important to understand growth behavior. Vejrazka and Wijffels give a model to represent respective growth effects (Vejrazka et al, 2012, 2013).
- Acclimation to light colors occurs in nature depending on the site, eg, shadowing by leaves lead to shift between linear and nonlinear electron transfer. In photobioreactors artificial illumination with differently colored LED light is an option for high-value products. Here it is worth to consider that red light is basically efficient and sufficient to run photosynthesis, but a small fraction of blue light is necessary to excite sensory pigments, which are present in microalgae like in plants besides the light-harvesting pigments (Wagner et al, 2015).
- A field more or less empty for the time being is modeling sensory reactions of microalgae. These include not only light sensing but also mechanical, turgor, and quorum sensing. An example for product sensing has been described, where extracellular polysaccharide concentration is kept constant by the algae what could be modeled as an intracellular Pi controller.
- Continuous cultivation under turbidostat conditions can be performed with the same productivity for low biomass concentration and high dilution rate or vice versa. Pruvost et al (2011) give a reactor model-based calculation of the best "window of operation." Furthermore, this results

in a different age distribution important for kinetics and product quality, where not much is known about.

Accumulation of macromolecules has to be modeled as a dynamic process with time constants in the range of minutes to hours or days. Furthermore, macromolecules act as feedback controller on the metabolic flux pattern. The following paragraphs give examples for some of these microalgal capabilities. Comprehensive models are often not available but some model ideas, bricks, or lumped approaches. Much more work can possibly be done in this field. An example for combined temperature/light kinetics is given below:

Suboptimal changing temperatures during outdoor cultivation lead to reduced productivities. Actually, microalgae have often a quite narrow optimal growth range. Temperature control is possible or economically feasible only to a limited extend. Modeling of temperature-depended growth behavior is important to optimize especially outdoor processes from strain selection to cooling measures. The usual approach is based on Arrhenius kinetics and given here according to Yan and Hunt (1999) as:

$$r_{X,\max,T}(T) = r_{X,\max} \cdot \left(\frac{T_{\max,r_X} - T}{T_{\max,r_X} - T_{\text{opt},\mu}}\right) \cdot \left(\frac{T}{T_{\max}}\right)^{\frac{T_{\text{opt}}}{T_{\max,r_X} - T_{\text{opt},r_X}}} \tag{41}$$

Maximum specific growth rate $r_{X,\max}$ is meant as a state, where no external substrate is limiting, but a—in most cases not known—intracellular step. The metabolism of the cell consists of physical and enzymatically catalyzed steps, which may differently react to different temperatures. Especially for microalgae, the photosynthetic apparatus down to proton transport through the thylakoid membrane as a transport step could react differently from the anabolic reactions. Explanations and first model approaches for combined temperature/light effects are given in Béchet et al (2013), Bernard and Rémond (2012), and Yun and Park (2003).

Data for the specific growth rate of Chlorella for different temperatures and light conditions (optimal data are extracted from the growth curves) and otherwise optimal conditions are given in Fig. 9.

The typical growth-irradiance curve is maintained for the two different conditions but with different slopes and maximum values. An evaluation for both growth states reveals that the data can indeed be represented by the

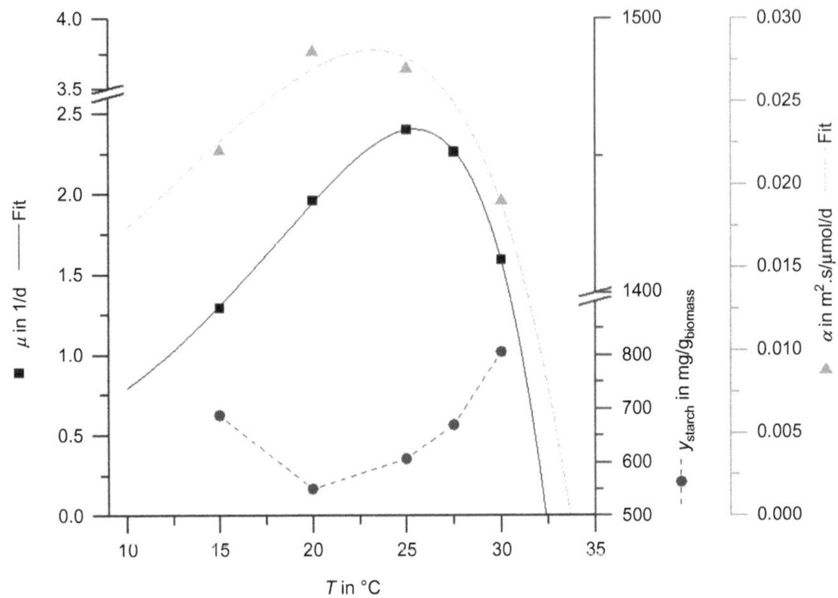

Figure 9 Specific growth rate of Chlorella for different temperature conditions.

Arrhenius equation (see Eq. 41) but with different parameter values. Growth efficiency under light limitation is less affected by suboptimal temperatures in comparison to growth under light saturation. This is underlined by the high starch content under suboptimal temperature conditions. This is interpreted in a way that photosynthesis is working faster as the anabolic steps, leading to accumulation of starch as the first photosynthetic product. This behavior can be modeled by two different temperature-related parameter sets for photosynthesis and growth accordingly.

These experiments have been performed under constant light and temperature conditions. Simulation with this temperature model of outdoor cultivations under real temperature changes showed nevertheless a smaller temperature impact underestimating productivity (Fig. 10).

The reason for the conversion of accumulated starch to active biomass during the night allows the cell to catch up losses in productivity during the day. A model has to include these processes by including day/night cycles. Furthermore, future modeling could look at different limitation conditions and the related light/temperature behavior.

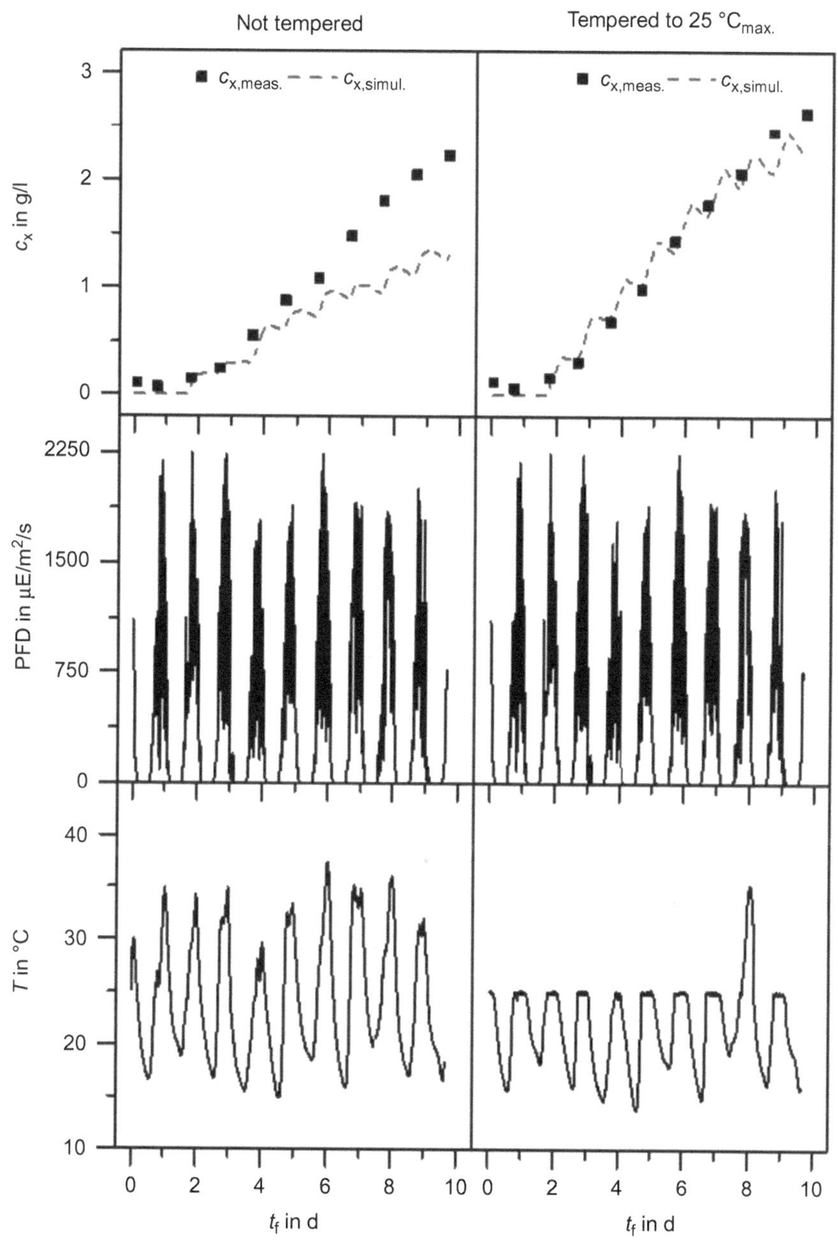

Figure 10 Model performance under constant light and temperature conditions.

REFERENCES

Aiba S: Growth kinetics of photosynthetic microorganisms. In *Microbial reactions,* vol 23, Berlin Heidelberg, 1982, Springer, pp 85–156.

Béchet Q, Shilton A, Guieysse B: Modeling the effects of light and temperature on algae growth: state of the art and critical assessment for productivity prediction during outdoor cultivation, *Biotechnol Adv* 31:1648–1663, 2013.

Bernard O: Hurdles and challenges for modelling and control of microalgae for CO2 mitigation and biofuel production, *J Process Control* 21:1378–1389, 2011.

Bernard O, Rémond B: Validation of a simple model accounting for light and temperature effect on microalgal growth, *Bioresource Technology (Special Issue: Biofuels - II: Algal Biofuels and Microbial Fuel Cell)* 123:520–527, 2012.

Dillschneider R, Steinweg C, Rosello-Sastre R, Posten C: Biofuels from microalgae: photoconversion efficiency during lipid accumulation, *Bioresour Technol* 142:647–654, 2013.

Doran PM: *Bioprocess engineering principles,* ed 2, Waltham, MA, 2013, Academic Press.

Falkowski PG, Raven JA: *Aquatic photosynthesis,* Princeton, 2007, Princeton University Press.

Fleck-Schneider P, Lehr F, Posten C: Modelling of growth and product formation of Porphyridium purpureum, *Adv Biochem Eng Sci* 132:134–141, 2007.

Gargano I, Olivieri G, Spasiano D, et al: Kinetic characterization of the photosynthetic reaction centres in microalgae by means of fluorescence methodology, *J Biotechnol* 212:1–10, 2015.

Jordan DB, Ogren WL: Species variation in the specificity of ribulose biphosphate carboxylase/oxygenase, *Nature* 291:513–515, 1981.

Kliphuis AM, Klok AJ, Martens DE, Lamers PP, Janssen M, Wijffels RH: Metabolic modeling of Chlamydomonas reinhardtii: energy requirements for photoautotrophic growth and maintenance, *J Appl Phycol* 24:253–266, 2012.

Kliphuis AM, Martens DE, Janssen M, Wijffels RH: Effect of O_2:CO_2 ratio on the primary metabolism of Chlamydomonas reinhardtii, *Biotechnol Bioeng* 108:2390–2402, 2011.

Lehr F, Morweiser M, Rosello Sastre R, Kruse O, Posten C: Process development for hydrogen production with Chlamydomonas reinhardtii based on growth and product formation kinetics, *J Biotechnol* 162:89–96, 2012.

Michaelis L, Menten ML: Die Kinetik der Invertinwirkung [The kinetics of invertin action], *Biochem Z* [*FEBS Lett*] 49(587):333–369 (2712–2720), 1913 (2013).

Nikolaou A, Bernardi A, Meneghesso A, Bezzo F, Morosinotto T, Chachuat B: A model of chlorophyll fluorescence in microalgae integrating photoproduction, photoinhibition and photoregulation, *J Biotechnol* 194:91–99, 2015.

Pirt SJ: The maintenance energy of bacteria in growing cultures, *Proc R Soc B Biol Sci* 163:224–231, 1965.

Pruvost J, van Vooren G, Le Gouic B, Couzinet-Mossion A, Legrand J: Systematic investigation of biomass and lipid productivity by microalgae in photobioreactors for biodiesel application, *Bioresour Technol* 102:150–158, 2011.

Raven JA, Falkowski PG: *Aquatic photosynthesis,* ed 2, Princeton, NJ, 2007, Princeton University Press.

Roels JA: *Energetics and kinetics in biotechnology. Relaxation times and their relevance to the construction of kinetic models,* Amsterdam, New York, 1983, Elsevier Biomedical Press.

Stephanopoulos G, Aristidou AA, Nielsen JH: *Metabolic engineering. Principles and methodologies,* San Diego, 1998, Academic Press.

Vejrazka C, Janssen M, Streefland M, Wijffels RH: Photosynthetic efficiency of *Chlamydomonas reinhardtii* in attenuated, flashing light, *Biotechnol Bioeng* 109:2567–2574, 2012.

Vejrazka C, Janssen M, Benvenuti G, Streefland M, Wijffels R: Photosynthetic efficiency and oxygen evolution of *Chlamydomonas reinhardtii* under continuous and flashing light, *Appl Microbiol Biotechnol* 97:1523–1532, 2013.

Wagner I, Braun M, Slenzka K, Posten C: Photobioreactors in life support systems, *Adv Biochem Eng Biotechnol* 153:143–184, 2015.
Wilhelm C, Büchel C, Fisahn J, et al: The regulation of carbon and nutrient assimilation in diatoms is significantly different from green algae, *Protist* 157:91–124, 2006.
Yan W, Hunt LA: An equation for modelling the temperature response of plants using only the cardinal temperatures, *Ann Bot* 84:607–614, 1999.
Yun Y-S, Park JM: Kinetic modeling of the light-dependent photosynthetic activity of the green microalga Chlorella vulgaris, *Biotechnol Bioeng* 83:303–311, 2003.

FURTHER READING

Aslan S, Kapdan IK: Batch kinetics of nitrogen and phosphorus removal from synthetic wastewater by algae, *Ecol Eng* 28:64–70, 2006.
Baroukh C, Muñoz-Tamayo R, Bernard O, Steyer J-P: Mathematical modeling of unicellular microalgae and cyanobacteria metabolism for biofuel production, *Curr Opin Biotechnol* 33:198–205, 2015a.
Baroukh C, Muñoz-Tamayo R, Steyer J-P, Bernard O: A state of the art of metabolic networks of unicellular microalgae and cyanobacteria for biofuel production, *Metab Eng* 30:49–60, 2015b.
Baumert H, Uhlmann D: Theory of the upper limit to phytoplankton production per unit area in natural waters, *Int Revue ges Hydrobiol Hydrogr* 68:753–783, 1983.
Bernardi A, Perin G, Sforza E, Galvanin F, Morosinotto T, Bezzo F: An identifiable state model to describe light intensity influence on microalgae growth, *Ind Eng Chem Res* 53:6738–6749, 2014.
Boyle NR, Morgan JA: Flux balance analysis of primary metabolism in Chlamydomonas reinhardtii, *BMC Syst Biol* 3:4, 2009.
Breuer G, Lamers PP, Martens DE, Draaisma RB, Wijffels RH: The impact of nitrogen starvation on the dynamics of triacylglycerol accumulation in nine microalgae strains, *Bioresour Technol* 124:217–226, 2012.
Breuer G, Lamers PP, Janssen M, Wijffels RH, Martens DE: Opportunities to improve the areal oil productivity of microalgae, *Bioresour Technol* 186:294–302, 2015a.
Breuer G, Martens DE, Draaisma RB, Wijffels RH, Lamers PP: Photosynthetic efficiency and carbon partitioning in nitrogen-starved Scenedesmus obliquus, *Algal Res* 9:254–262, 2015b.
Chang RL, Ghamsari L, Manichaikul A, et al: Metabolic network reconstruction of Chlamydomonas offers insight into light-driven algal metabolism, *Mol Syst Biol* 7:518, 2011.
Chen Y, Tang X, Kapoore RV, Xu C, Vaidyanathan S: Influence of nutrient status on the accumulation of biomass and lipid in Nannochloropsis salina and Dunaliella salina, *Energy Convers Manag* 106:61–72, 2015.
Cogne G, Gros J-B, Dussap C-G: Identification of a metabolic network structure representative of Arthrospira (spirulina) platensis metabolism, *Biotechnol Bioeng* 84:667–676, 2003.
Cogne G, Rügen M, Bockmayr A, et al: A model-based method for investigating bioenergetic processes in autotrophically growing eukaryotic microalgae: application to the green algae *Chlamydomonas reinhardtii*, *Biotechnol Prog* 27:631–640, 2011.
Doran PM: *Bioprocess engineering principles,* ed 2, Waltham, MA, 2013, Academic Press.
Fachet M, Flassig RJ, Rihko-Struckmann L, Sundmacher K: A dynamic growth model of Dunaliella salina: parameter identification and profile likelihood analysis, *Bioresour Technol* 173:21–31, 2014.
Gérin S, Mathy G, Franck F: Modeling the dependence of respiration and photosynthesis upon light, acetate, carbon dioxide, nitrate and ammonium in Chlamydomonas reinhardtii using design of experiments and multiple regression, *BMC Syst Biol* 8:96, 2014.

Hartmann P, Nikolaou A, Chachuat B, Bernard O: A dynamic model coupling photoacclimation and photoinhibition in microalgae. In *2013 European control conference, ECC 2013*, 2013, pp 4178–4183.

Janssen M, Janssen M, Winter M, et al: Efficiency of light utilization of Chlamydomonas reinhardtii under medium-duration light/dark cycles, *J Biotechnol* 78:123–137, 2000.

Kasiri S, Ulrich A, Prasad V: Kinetic modeling and optimization of carbon dioxide fixation using microalgae cultivated in oil-sands process water, *Chem Eng Sci* 137:697–711, 2015.

Kim HU, Kim TY, Lee SY: Metabolic flux analysis and metabolic engineering of microorganisms, *Mol BioSyst* 4:113–120, 2008.

Kliphuis AM, Janssen M, Martens DE, van den End EJ, Wijffels RH: Light respiration in Chlorella sorokiniana, *J Appl Phycol* 23:935–947, 2010a.

Kliphuis AM, Winter L, de Vejrazka C, Martens DE, Janssen M, Wijffels RH: Photosynthetic efficiency of Chlorella sorokiniana in a turbulently mixed short light-path photobioreactor, *Biotechnol Prog* 26:687–696, 2010b.

Maguer J-F, L'Helguen S, Caradec J, Klein C: Size-dependent uptake of nitrate and ammonium as a function of light in well-mixed temperate coastal waters, *Cont Shelf Res* 31:1620–1631, 2011.

Manichaikul A, Ghamsari L, Hom Erik FY, et al: Metabolic network analysis integrated with transcript verification for sequenced genomes, *Nat Methods* 6:589–592, 2009.

Markou G, Nerantzis E: Microalgae for high-value compounds and biofuels production: a review with focus on cultivation under stress conditions, *Biotechnol Adv* 31:1532–1542, 2013.

Markou G, Vandamme D, Muylaert K: Microalgal and cyanobacterial cultivation: the supply of nutrients, *Water Res* 65:186–202, 2014.

Nikolaou A, Bernardi A, Bezzo F, Morosinotto T, Chachuat B: A dynamic model of photoproduction, photoregulation and photoinhibition in microalgae using chlorophyll fluorescence. In *IFAC proceedings volumes (IFAC-PapersOnline)*, 2014, pp 4370–4375.

Palabhanvi B, Kumar V, Muthuraj M, Das D: Preferential utilization of intracellular nutrients supports microalgal growth under nutrient starvation: multi-nutrient mechanistic model and experimental validation, *Bioresour Technol* 173:245–255, 2014.

Procházková G, Brányiková I, Zachleder V, Brányik T: Effect of nutrient supply status on biomass composition of eukaryotic green microalgae, *J Appl Phycol* 26:1359–1377, 2014.

Sforza E, Gris B, De Farias Silva CE, Morosinotto T, Bertucco A: Effects of light on cultivation of scenedesmus obliquus in batch and continuous flat plate photobioreactor, *Chem Eng Trans* 38:211–216, 2014.

Sforza E, Calvaruso C, Meneghesso A, Morosinotto T, Bertucco A: Effect of specific light supply rate on photosynthetic efficiency of Nannochloropsis salina in a continuous flat plate photobioreactor, *Appl Microbiol Biotechnol* 99:8309–8318, 2015.

Simionato D, Basso S, Giacometti GM, Morosinotto T: Optimization of light use efficiency for biofuel production in algae, *Biophys Chem* 182:71–78, 2013.

Yang C, Hua Q, Shimizu K: Energetics and carbon metabolism during growth of microalgal cells under photoautotrophic, mixotrophic and cyclic light-autotrophic/dark-heterotrophic conditions, *Biochem Eng J* 6:87–102, 2000.

Zhang D, Dechatiwongse P, Del-Rio-Chanona EA, Hellgardt K, Maitland GC, Vassiliadis VS: Analysis of the cyanobacterial hydrogen photoproduction process via model identification and process simulation, *Chem Eng Sci* 128:130–146, 2015.

CHAPTER FOUR

Microalgal Photosynthesis and Growth in Mass Culture

Marcel Janssen[1]
AlgaePARC, Bioprocess Engineering, Wageningen University and Research Centre, Wageningen, The Netherlands
[1]Corresponding author: e-mail address: marcel.janssen@wur.nl

Contents

Abstract

The development of large-scale outdoor microalgae production requires a thorough understanding of microalgal growth which should be encompassed in a mathematical model. The model should be as simple as possible allowing use in outdoor practice by persons with varying backgrounds. This chapter provides a basis for such a model connecting microalgal growth and photobioreactor productivity to light exposure. Only light exposure is included as an environmental variable because sunlight irradiance will ultimately limit areal productivity and all other cultivation parameters must be balanced to that number.

Within microalgal mass cultures inside photobioreactors a light gradient will develop determining microalgal growth. This light gradient depends on microalgal

Advances in Chemical Engineering, Volume 48
ISSN 0065-2377
http://dx.doi.org/10.1016/bs.ache.2015.11.001

specific light absorption and biomass concentration. Based on the light gradient the local rates of photosynthesis are calculated and integrated over reactor volume. The model is connected to our current understanding of photosynthesis by adopting proven photosynthesis models developed by Blackman and Jassby & Platt, and employing efficiency parameters based on theoretical evaluations and practical experiments. The wavelength dependency of light absorption is included. Photosynthesis is then connected to microalgal growth adopting the model of Pirt and distinguishing between maintenance-related respiration and growth-related respiration.

The model is used to analyze productivity of simple photobioreactor geometry (1-dimensional light path) and calculate the limits of light-use efficiency. At the end of the chapter the assumptions and simplifications made are discussed in the light of possible effects of photoacclimation, mixing along the light gradient, day/night cycles, and the complexity of accurately modeling the light field.

LIST OF SYMBOLS

λ wavelength (nm)
α initial slope q_S^c–I_{ph} curve [$(\mathrm{mol_S\ mol_x^{-1}})(\mathrm{mol_{ph}\ m^{-2}})^{-1}$]
μ specific growth rate (s^{-1})
μ_m maximal specific growth rate (s^{-1})
μ_{PI} μ calculated from q_S^c–I_{ph} curve (s^{-1})
$\mu_{PI}(z)$ μ_{PI} at location z in photobioreactor (s^{-1})
$\bar{\mu}$ average specific growth rate microalgae in photobioreactor (s^{-1})
$A_{r,light}$ light-exposed surface area of reactor (m^2)
a_x spectrally averaged specific light absorption coefficient ($\mathrm{m^2\ mol_x^{-1}}$)
$a_{x,\lambda}$ wavelength-specific light absorption coefficient ($\mathrm{m^2\ mol_x^{-1}}$)
C_x biomass concentration in photobioreactor system ($\mathrm{mol_x\ m^{-3}}$)
$C_{x,opt}$ optimal biomass concentration in photobioreactor (at max r_x) ($\mathrm{mol_x\ m^{-3}}$)
d optical depth of photobioreactor (m)
D dilution rate of photobioreactor (s^{-1})
F liquid flow rate through photobioreactor ($\mathrm{m^3\ s^{-1}}$)
HRT hydraulic retention time (s)
I_{ph} PAR photon flux density ($\mathrm{mol_{ph}\ m^{-2}\ s^{-1}}$)
$I_{ph}(z)$ I_{ph} at location z in photobioreactor ($\mathrm{mol_{ph}\ m^{-2}\ s^{-1}}$)
$I_{ph}(0)$ I_{ph} at location $z=0$ (light-exposed surface) in photobioreactor ($\mathrm{mol_{ph}\ m^{-2}\ s^{-1}}$)
$I_{ph}(d)$ I_{ph} at location $z=d$ (dark surface) in photobioreactor ($\mathrm{mol_{ph}\ m^{-2}\ s^{-1}}$)
$I_{ph,c}$ compensation photon flux density for photoautotrophic growth ($\mathrm{mol_{ph}\ m^{-2}\ s^{-1}}$)
$I_{ph,s}$ saturation photon flux density for photosynthesis ($\mathrm{mol_{ph}\ m^{-2}\ s^{-1}}$)
m_S specific sugar consumption rate for maintenance ($\mathrm{mol_S\ mol_x^{-1}\ s^{-1}}$)
q_{ph} specific photon consumption rate ($\mathrm{mol_{ph}\ mol_x^{-1}\ s^{-1}}$)
q_s^c specific sugar (CH_2O) production rate in chloroplast ($\mathrm{mol_s\ mol_x^{-1}\ s^{-1}}$)
$q_{s,m}^c$ maximal specific sugar (CH_2O) production in the chloroplast ($\mathrm{mol_s\ mol_x^{-1}\ s^{-1}}$)
$q_s^c(z)$ q_s^c at location z in microalgal culture ($\mathrm{mol_s\ mol_x^{-1}\ s^{-1}}$)
q_s specific sugar (CH_2O) consumption rate in microalgal cell ($\mathrm{mol_s\ mol_x^{-1}\ s^{-1}}$)
r_x volumetric biomass production rate ($\mathrm{mol_x\ m^{-3}\ s^{-1}}$)
V_r liquid volume of reactor (m^3)

$Y^{c}_{s/ph}$ yield of sugar (CH_2O) produced on photons ($mol_s\ mol_{ph}^{-1}$)
$Y^{c}_{s/ph,m}$ maximal yield of sugar (CH_2O) produced on photons ($mol_s\ mol_{ph}^{-1}$)
$Y_{x/s}$ yield of biomass on sugar (CH_2O) ($mol_x\ mol_s^{-1}$)
$Y_{x/ph}$ yield of biomass on photons ($mol_x\ mol_{ph}^{-1}$)
$Y_{x/ph,m}$ maximal yield of biomass on photons ($m_s = 0$) ($mol_x\ mol_{ph}^{-1}$)
z shortest distance to light-exposed surface photobioreactor (m)
z_s depth of photobioreactor zone with saturating light, $I_{ph} = I_{ph,s}$ (m)

SUBSCRIPTS AND SUPERSCRIPTS

λ wavelength specific
c chloroplast or compensation
m maximal
ph photons
r reactor
s sugar (CH_2O) or saturating
x biomass

ABBREVIATIONS

PAR photosynthetic active radiation, 400–700 nm
PS photosystem

Large-scale production of microalgae and cyanobacteria carries a big promise to fill in our future needs for sustainable feedstock for food, feed, chemicals, and possibly fuels. Currently large-scale production is still in its infancy and the few larger production systems in the world are productions plants in the order of a few hectares (ha). A realistic outlook for future scale-up potential is relating microalgae production to greenhouse production of food crops. This comparison is based on the overlap in economics and technology of greenhouse horticulture and that of envisioned large-scale microalgae production. Greenhouse horticulture in The Netherlands, for example, has a total production area of 10,000 ha, and worldwide production is one to two orders of magnitude larger.

In order to facilitate the process of up-scaling a thorough understanding of the microalgal production process is required. This chapter provides a simple methodology to analyze microalgal production systems as related to light exposure. The focus on light is based on the fact that sunlight irradiance will ultimately limit areal productivity and that all other cultivation parameters must be balanced to that number. The close relation to sunlight exposure determines the shape and operation conditions of microalgae productions systems, which will be called photobioreactors. Also pond-based production systems will be considered to be photobioreactors. Other important aspects with respect to photobioreactor design and operation are the adequate supply of carbon dioxide, the removal of oxygen, and temperature control. Gas exchange in bioreactors is well described in numerous engineering books and scientific publications. Temperature control is predominantly limited by technological and environmental limitations.

The light-based modeling approach adopted in this chapter is simple allowing for a wide audience while still providing sufficient accuracy and process insight. Despite its simplicity the model is closely linked to the biological understanding of the process of photosynthesis and the fundamental limits of this process. In Sections 4.1–4.4 the assumptions and simplifications made will be discussed allowing for a reader evaluation of which topics must be explored further.

1. FUNDAMENTALS OF PHOTOAUTOTROPHIC GROWTH AND LIGHT

1.1 Photoautotrophic Growth

Microalgae grow based on photosynthesis. In Fig. 1 this process is schematically introduced. In the chloroplast light energy is absorbed by a large

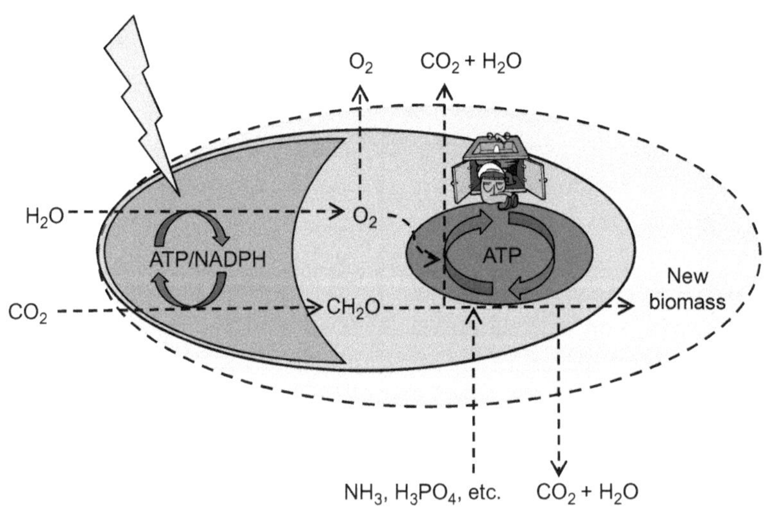

Figure 1 Schematic presentation of photoautotrophic growth of microalgae. The dark green (gray in the print version) organelle represents the chloroplast where photosynthesis takes place. The blue (dark gray in the print version) organelle represents the mitochondria where the respiratory machinery is located. The compound CH_2O represents sugar.

number of photosystems. The photosystems are located on the thylakoid membranes enclosing a space called the lumen. To facilitate light absorption each photosystem (PS) is equipped with an antenna complex composed of a large number of pigment molecules (chlorophylls and carotenoids). Only photons in the wavelength range of 400–700 nm are absorbed. This range is called photosynthetic active radiation (PAR). When talking about photons in the remainder of this chapter these always reflect PAR photons and the nature of light will be discussed in more detail in the next section. The energy of the absorbed photons is used within the photosystems to transfer electrons from water (H_2O) to oxidized nicotinamide adenine dinucleotide phosphate ($NADP^+$) to yield reduced NADPH and oxygen (O_2). At the same time adenosine triphosphate (ATP) is produced using a proton motive force which is created over the thylakoid membranes. This so-called linear photosynthetic electron transport requires the coordinated action of two types of photosystems, PSII followed by PSI. Two photosystems are needed in order to provide sufficient driving force to move electrons from water to $NADP^+$.

The NADPH and ATP generated are used within the chloroplast to fix carbon dioxide (CO_2) and reduce it to sugar. This sequence of reactions is called the Calvin–Benson–Bassham cycle and this cycle results in the production of the phosphorylated 3-carbon sugar glyceraldehyde-3-phosphate, also called triose. This triose sugar is exported out of the chloroplast and in this course we will refer to it as CH_2O which is the chemical composition of sugar when normalized to 1 carbon atom.

Sugar (CH_2O) is the building block for new microalgal biomass in the growth reactions. But not all triose ends up in the microalgal biomass. A significant part is broken down and oxidized (respired) in the mitochondria to generate energy in the form of ATP. This ATP is needed to drive, or "push," the growth reactions, as well as the maintenance reactions. Maintenance is defined as the collection of cellular processes needed to survive not including growth-related processes.

All the sugar (CH_2O) needs to be formed by photosynthesis in the chloroplast. In the Calvin–Benson–Bassham cycle CO_2 is reduced to the level of sugar (CH_2O) according to the following reaction and stoichiometry:

$$CO_2 + 3 \cdot ATP + 2 \cdot NADPH + 2 \cdot H^+ \rightarrow CH_2O \\ + 3 \cdot ADP + 3 \cdot P_i + 2 \cdot NADP^+ + H_2O$$

The ATP and NADPH required to reduce CO_2 are produced in the light reactions of photosynthesis

$$2 \cdot H_2O + 2 \cdot NADP^+ + 3 \cdot ADP + 3 \cdot P_i + \text{photons} \rightarrow O_2 + 2 \cdot NADPH + 2 \cdot H^+ + 3 \cdot ATP.$$

According to this stoichiometry, which is often reported, the amount of ATP and NADPH needed to fix 1 (mol) of CO_2 is equal to the amount which is generated in the light reactions when 2 (mol) H_2O is split generating 1 (mol) O_2. The theoretical minimal amount of light (=photons) needed to generate 2 NADPH and thus 1 O_2 is 8 photons. This number is based on the functioning of linear photosynthetic electron transport requiring the coordinated action of two types of photosystems (PSII and PSI) to extract 4 electrons from two water molecules leading to two molecules of NADPH.

In the stoichiometry of the light reactions described earlier it is implicitly assumed that 8 protons (H^+) are transported over the thylakoid membranes during linear electron transport assuming a fully functional Q-cycle (Cape et al, 2006). Together with the 4 protons (H^+) released when oxidizing water the lumen is thus enriched with 12 H^+. It is assumed that these 12 H^+ drive the production of 3 ATP in the thylakoid ATPase enzyme complex. The real stoichiometry between protons (H^+) translocated and ATP produced by the ATPase enzyme complex of microalgae probably is different from this 12:3 (H^+:ATP) ratio. For spinach chloroplasts the ratio was determined to be 14:3 (Seelert et al, 2000; Vollmar et al, 2009), and also for *Chlamydomonas* and cyanobacteria ratios of 13:3, 14:3, and 15:3 have been reported (Meyer zu Tittingdorf et al, 2004; Pogoryelov et al, 2007). This results in less ATP produced then required in the Calvin–Benson cycle since we cannot allow for the accumulation of reduced NADPH. The resulting lack of ATP then is suggested to be compensated by cyclic photosynthetic electron transport around photosystem I (PSI) yielding only ATP (Allen, 2003) by generating a proton motive force across the thylakoid membrane. Cyclic electron transport driven by PSI allows for the transport of 2 protons (H^+) over the thylakoid membranes per photon absorbed. Consequently, minimally 9 photons are used to reduce one CO_2 to the level of sugar

$$2 \cdot H_2O + 2 \cdot NADP^+ + 3 \cdot ADP + 3 \cdot P_i + 9 \cdot h\nu \rightarrow O_2 + 2 \cdot NADPH + 2 \cdot H^+ + 3 \cdot ATP.$$

By combining the Calvin–Benson–Bassham cycle with the light reactions the following general reaction of photosynthesis is then obtained

$$CO_2 + H_2O + 9 \cdot h\nu \rightarrow CH_2O + O_2.$$

Under optimal conditions in the laboratory at low photon flux density a photon requirement (or quantum requirement) of 9 for O_2 production or CO_2 fixation, however, is rarely measured. The most realistic assumption of the minimal quantum requirement seems to be a value of 10. This is based on an analysis of dedicated research in this field: oxygen exchange measurements with leaf–disc electrodes for higher plants (Björkman and Demmig, 1987; Evans, 1987); laser-induced oxygen flash yield measurements (Dubinsky et al, 1986; Ley and Mauzerall, 1982); and photoacoustic and photothermal calorimetry, as reflected upon by Malkin and Fork (1996).

A quantum requirement of 10 thus means that the maximal yield of sugar (CH_2O) on photons is equal to 0.10. This number will be called the maximal yield of sugar on photons $Y^c_{s/ph,m}$ with unit $mol_S\ mol_{ph}^{-1}$. This parameter will be very important later when describing photoautotrophic growth in a mathematical model.

In Fig. 2 the most important production and consumption rates in photoautotrophic growth are presented. Here "q" rates are used which represent biomass-specific production and consumption rates with unit mol $mol_x^{-1}\ s^{-1}$ (symbol "x" reflects biomass). Later when moving from a cellular approach to a reactor approach volumetric production and consumption rates will be introduced designated with symbol "r" and unit mol $m^{-3}\ s^{-1}$. Volumetric

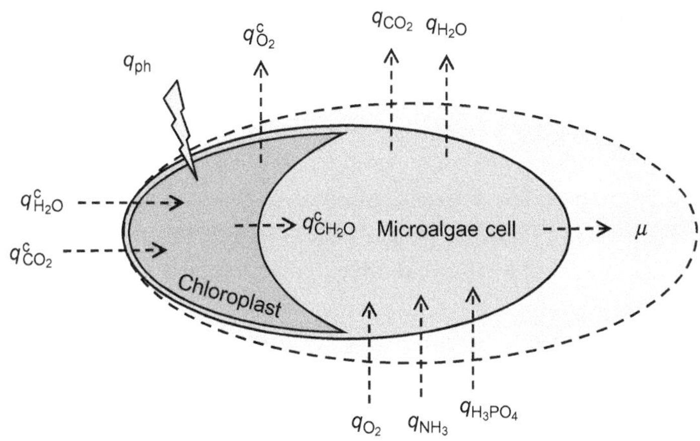

Figure 2 Representation of most important specific consumption and production rates for the quantitative analysis of photoautotrophic growth. The superscript "c" reflects the chloroplast.

rates are the products of specific rates "q" and the concentration of biomass in the bioreactor C_x.

Based on the reaction stoichiometry the following can be said about the corresponding specific consumption and production rates in the chloroplast:

$$-q^c_{CO_2} = -q^c_{H_2O} = q^c_s = q^c_{O_2}\ (\text{mol s}^{-1})$$

The specific rate of sugar production in the chloroplast q^c_s thus is a direct measure of the rate of photosynthesis and is equivalent to the specific rate of oxygen (O_2) production or carbon dioxide (CO_2) fixation, which are also used as measures of photosynthesis. The superscript "c" is used to denote that these reactions take place in the chloroplast.

The sugar (CH_2O) produced in the chloroplast is the building block of real biomass. Below a generalized reaction is shown without any stoichiometry where the compound $CH_{xH}O_{xO}N_{xN}P_{xP}$ represents biomass. So, 1 mol of this compound $CH_{xH}O_{xO}N_{xN}P_{xP}$ will be called 1 mol$_x$. All other elements present in biomass (S, K, Mg, Ca, Fe, and others) are not included because they do not contribute a lot on a mass basis

$$CH_2O + O_2 + NH_3 + H_3PO_4 \rightarrow CH_{XH}O_{XO}N_{XN}P_{XP} + CO_2 + H_2O.$$

Not all sugar CH_2O produced by photosynthesis ends up in the biomass. A considerable part of the sugar (CH_2O) needs to be broken down and oxidized to provide energy (ATP) to support the growth reactions. In addition, additional energy (ATP) is needed for maintenance. Maintenance refers to the collection of cellular processes needed to survive not including growth-related processes.

The growth-related sugar consumption is a very important aspect for an accurate quantitative description of microalgal growth. In the mitochondria of microalgae part of the triose sugar is thus degraded in the citric acid cycle and the generated reductant is oxidized (i.e., oxidative phosphorylation) generating considerable ATP necessary to drive all growth reactions. In several published analyses of microalgal growth a constant respiration is assumed. In reality respiration is thus coupled to growth, and this has been well documented (Geider and Osborne, 1989; Kliphuis et al, 2011a). In fact, the chloroplast can be seen as a sugar factory. Accordingly, considering the general process occurring outside of the chloroplast this process can be well described by aerobic chemoheterotrophic growth.

A simple substrate balance according to Pirt (1965) can be used to couple the specific consumption rate of sugar outside of the chloroplast q_s to the specific growth rate μ of the microalgae

$$q_s = -\frac{\mu}{Y_{x/s}} - m_s.$$

Please note that this specific sugar consumption rate q_s is equal to the specific sugar production rate in the chloroplast q_s^c but of opposite sign. The parameter μ represents the specific growth rate of the microalgae and has unit s^{-1}. The parameter m_s is the specific sugar consumption rate to provide energy (ATP) for maintenance with unit $mol_s\ mol_x^{-1}\ s^{-1}$. Its value is difficult to measure and also depends on the microalgal species and growth conditions (for example, temperature). Typical values for microalgae found in the laboratory are in the range of 1×10^{-6} to 5×10^{-6} $mol_s\ mol_x^{-1}\ s^{-1}$.

The parameter $Y_{x/s}$ represents the yield of microalgal biomass on sugar with units $mol_x\ mol_s^{-1}$. Studies under heterotrophic and mixotrophic conditions with glucose as a carbon source show a maximal yield of about 0.5 $g\ g^{-1}$ translating in 0.625 $mol_x\ mol_s^{-1}$ (Chen and Johns, 1991; Lee et al, 1996; Li et al, 2014; Shi et al, 1997). It has to be stressed that these numbers reflect nutrient replete growth with "normal" protein content. Biomass accumulating starch will result in somewhat higher yields. Microalgae accumulating lipids probably will result in somewhat lower yields. In this chapter we will not discuss such product accumulation and we will only consider nutrient replete growth. Also the nitrogen source will affect the biomass yield on sugar since nitrate is much more oxidized than ammonia or urea. The higher yields of 0.625 $mol_x\ mol_s^{-1}$ are usually reported with ammonium and urea as a nitrogen source. For example, a recalculation of the data of Kliphuis et al (2012) results in a biomass yield of 0.52 $mol_x\ mol_s^{-1}$ for growth on nitrate of the green microalga *Chlamydomonas*.

The separation of photosynthetic sugar production in the chloroplast from actual microalgal biomass production from triose sugars allows for a comparison with aerobic chemoheterotrophic growth. The maximal biomass yield on sugar for aerobic chemoheterotrophic growth (bacteria and yeasts) is reported to range between 0.4 and 0.7 $mol_x\ mol_s^{-1}$, and it is bound by thermodynamic constraints (Heijnen, 1994; Heijnen and Van Dijken, 1992; von Stockar and Liu, 1999). Based on this comparison with yeast and bacteria, it will be assumed that the microalgal biomass yield on photosynthetically derived sugar probably will be somewhere between 0.5 and 0.65 $mol_x\ mol_s^{-1}$, and that it will not be higher than 0.7 $mol_x\ mol_s^{-1}$.

The analysis of microalgal growth presented in this chapter in principle could be extended to include growth of photoautotrophic cyanobacteria. Cyanobacteria do not have separate organelles for photosynthesis and respiration (i.e., chloroplasts and mitochondria). The photosynthetic and respiratory chains are located on the same thylakoid membranes which form stacks of parallel sheets close to the cytoplasmic membrane. The photosynthetic and respiratory electron transport chains share a number of components leading to alternative electron transport routes (Mullineaux, 2014; Vermaas, 2001). Without going into detail in the light cyanobacteria can generate additional ATP via either increased cyclic photosynthetic electron transport around PSI, or redirecting reducing equivalents generated by PSII to respiratory components on the thylakoids (Mullineaux, 2014). Similar to microalgae this additional ATP is required to support the growth reactions.

The amount of ATP required to support growth starting from sugar will be comparable between cyanobacteria, microalgae, and other aerobic microorganisms, with a similar macromolecular composition. But one could ask the question what is the most efficient way to produce ATP considering light as a limiting substrate. Cyclic electron transport around PSI seems a simple process but it requires 2.35 photons to generate 1 ATP (Allen, 2003). An analysis of Kliphuis (Kliphuis et al, 2012) showed that photosynthetic production of triose sugar in the chloroplast, followed by its breakdown in the mitochondria, requires 1.5 photons to generate 1 ATP. Of course the latter pathway involves more components (proteins) representing a bigger burden for the cell. Only complex metabolic modeling including the costs of the biosynthetic machinery required will allow for the identification of the most efficient pathways (Kramer and Evans, 2011; Nogales et al, 2012). At this stage it is important to note there is no proof that either microalgae or cyanobacteria are more efficient in employing light for biomass growth.

Under short periods of darkness also cyanobacteria will rely on the breakdown of sugar-based metabolites to generate reductant to generate ATP via oxidative phosphorylation. Under long periods of darkness both microalgae and cyanobacteria mobilize sugar from starch or glycogen reserves to support growth and maintenance processes.

To summarize, three important constants necessary for a quantitative model description of microalgal growth have been introduced:

- $Y^{c}_{s/ph,m}$—maximal yield of sugar (CH_2O) on photons in the chloroplast ($mol_s\ mol_{ph}^{-1}$)
- $Y_{x/s}$—yield of biomass on sugar (CH_2O) ($mol_x\ mol_s^{-1}$)

- m_s—specific sugar consumption rate for maintenance ($mol_s\ mol_x^{-1}\ s^{-1}$).

In addition, two important variables have been introduced which will appear in a growth model:

- q_s^c—specific sugar (CH_2O) production rate in chloroplast ($mol_s\ mol_x^{-1}\ s^{-1}$)
- μ—specific growth rate of microalgae ($mol_x\ mol_x^{-1}\ s^{-1} = s^{-1}$).

Please note that: $q_s^c = -q_s$, with q_s the specific sugar consumption rate of the microalgae.

1.2 Light and Photosynthesis

In a scientific context light is electromagnetic radiation of any wavelength. On the other hand, it is also defined as that part of the electromagnetic spectrum which can be sensed by the human eye. For this part of the spectrum, located between 400 and 700 nm, it is therefore best to talk about visible light. In Fig. 3 the relative sensitivity of the human eye to visible light of different wavelengths is presented. It is clear that we are most sensitive for green light (555 nm).

Microalgae absorb and use light from the whole visible wavelength range, 400–700 nm. But, as will be shown later, the sensitivity of microalgae for visible light of a specific wavelength differs dramatically from our sensitivity. This is also the reason why the lumen, the unit for the luminous flux, cannot, and should not, be used in photosynthesis. The lumen, derived from the candela (the base SI unit), is weighed by the sensitivity of the human eye

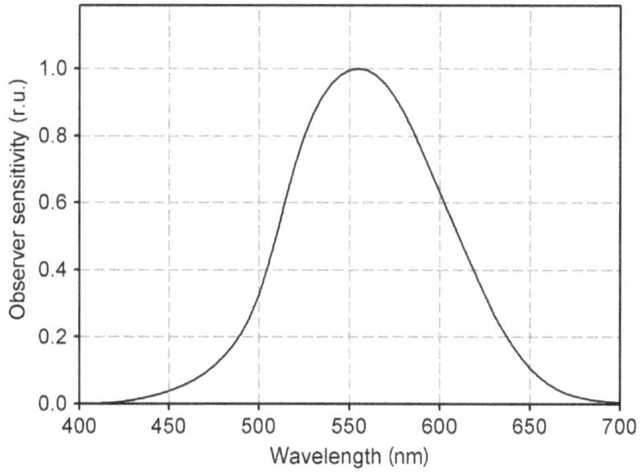

Figure 3 CIE 1931 Standard Colorimetric Observer curve. *CIE*, Commission Internationale de l'Eclairage.

(Fig. 3). For this reason, always radiometric units must be used in photosynthesis instead of photometric units. The radiometric analogue of the lumen (luminous flux) is the watt (radiant flux). At each wavelength both are coupled according to the standard observer curve: at 555 nm 1 W of radiant flux corresponds to a luminous flux of 683 lumens by definition. As a comparison, at 650 nm, 1 W corresponds to 73 lumens only.

Photosynthesis is a quantum process. The actual absorption of quanta or photons by the microalgal photosystems sets in motion the cascade of reactions ultimately resulting in growth. As such, the radiant flux in watts, W ($=\mathrm{J\ s^{-1}}$), is usually converted into a photon flux in $\mathrm{mol_{ph}\ s^{-1}}$ employing the Planck–Einstein relation ($E = h \cdot \nu$). Light meters used in photosynthesis research are adapted to immediately give the photon flux density in $\mathrm{mol_{ph}\ m^{-2}\ s^{-1}}$ within the PAR wavelength range of 400–700 nm. Please note that in this chapter the symbol I_{ph} is used for the photon flux density in the PAR range and that the photon flux density I_{ph} sometimes is referred to by the term "light intensity."

In Fig. 4 the spectral response of a typical light meter used in photosynthesis research is demonstrated. Using specific filters the meter is made equally sensitive for any photon within the range of PAR which corresponds to the 400–700 nm range.

Microalgae, and all other photosynthetic organisms, need light energy (photons) to grow. Outdoors it is the sun supplying all light energy. On the other hand, also artificial light can be used to grow algae. For the production of very high value compounds microalgal cultivation on artificial light can be an attractive alternative. But, even when taking into account the continual development of LED lighting in combination with very efficient photosynthesis the combined costs for investment in lamps and consumed power will add about $16 of production costs per kg of dry biomass produced (Blanken et al, 2013). The impact of employing artificial light for microalgae production is not only a cost factor but also includes an energy factor. Converting electricity to light (photons) leads to large energy losses (65% or more) depending on the light source used. Moreover, all electricity needed must be generated.

A wide variety of lamps exist and the spectral distribution of a number of lamps used in microalgae research is shown in Fig. 5. Only the distribution within the PAR range is shown and all spectra are normalized such that the integral photon flux density over the PAR range is unity. The spectral distribution of sunlight is shown in Fig. 6. Between 42% and 43% of the sun's radiant flux reaching Earth's surface falls within the PAR between 400 and

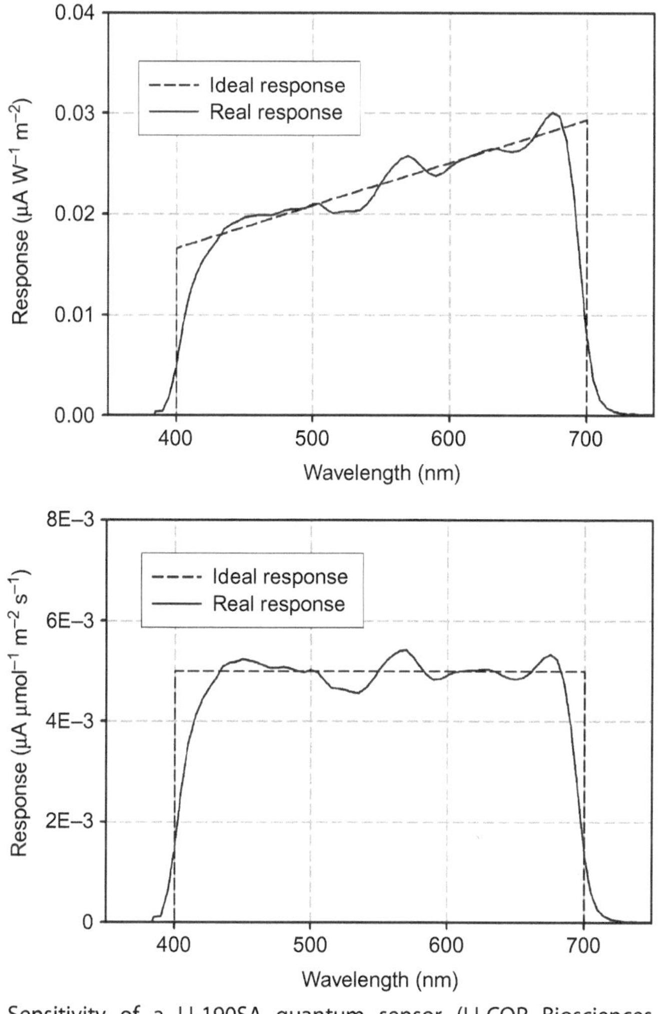

Figure 4 Sensitivity of a LI-190SA quantum sensor (LI-COR Biosciences, Lincoln, Nebraska, United States) for photosynthetically active radiation (PAR). The graph was redrawn from the datasheet delivered with the sensor.

700 nm. Comparing the spectra of all these different lamps with those of the sun it is evident that all are different from the sun. That should not be a problem because photosynthesis can be fueled with any photon within the PAR range although certain spectral effects cannot be excluded. Timing of cell division, for example, has been shown to depend on light color (Oldenhof et al, 2006).

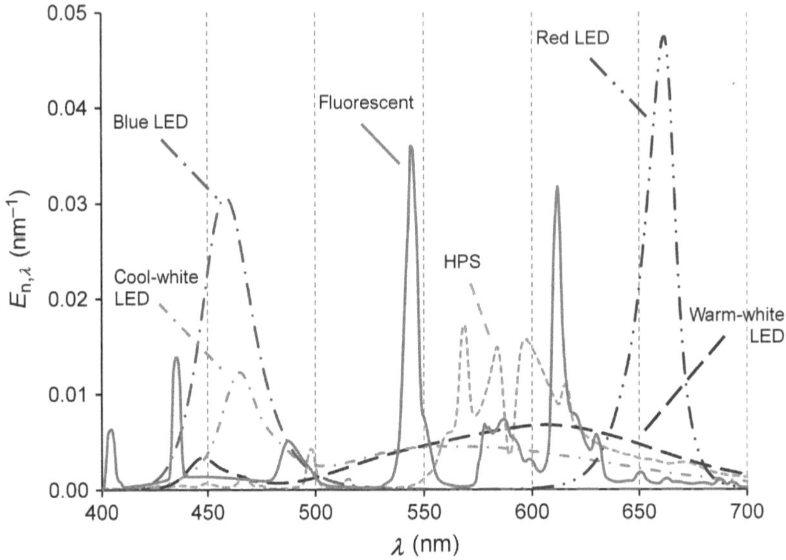

Figure 5 Spectral distribution of artificial lamps employed in microalgae research. The spectra are normalized to PAR light, $E_{n,\lambda}$ with units nm^{-1} (i.e., the area under all curves is unity). The tungsten-halogen lamps were placed behind a 1-cm water filter.

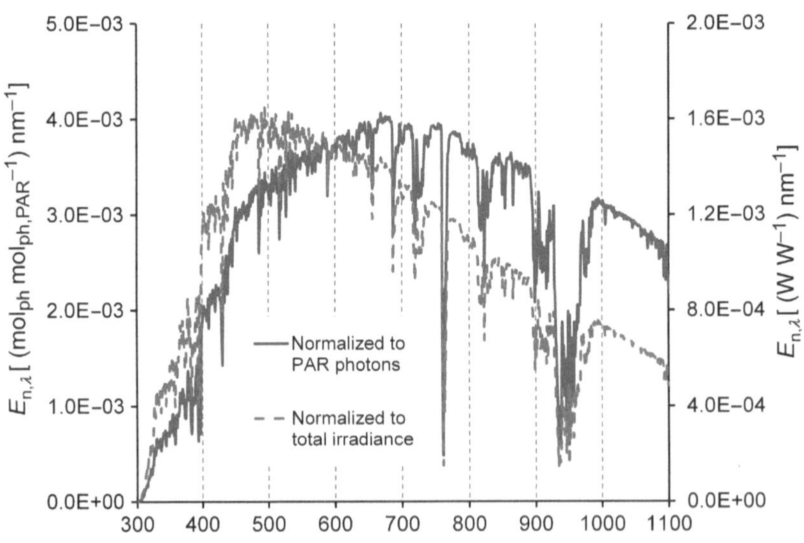

Figure 6 Spectral distribution of sunlight at ground level (ASTM G173-03). The spectrum is normalized to PAR light as well as total irradiance.

Irradiance from the sun depends on the geographical position, the season, and the hour. As an example the average hourly irradiance in the month of July in Athens (Greece) and Amsterdam (The Netherlands) is presented in Fig. 7. The maximal irradiance in Athens is 860 W m^{-2}, which corresponds to more than 1.65×10^{-3} mol_{ph} m^{-2} s^{-1} in the PAR range (400–700 nm). As will be explained later this is a very high light intensity and it is more than sufficient to saturate photosynthesis. At higher latitudes such as Amsterdam

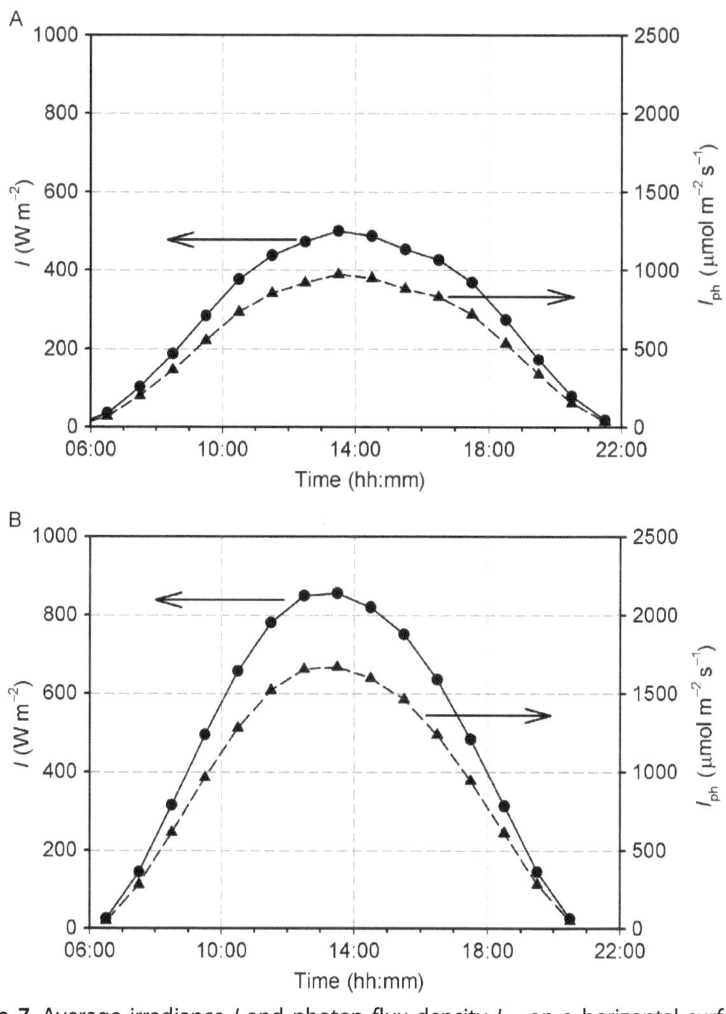

Figure 7 Average irradiance *I* and photon flux density I_{ph} on a horizontal surface in Athens (A), Greece, Lat: 37°58′N, and Amsterdam (B), The Netherlands, Lat: 52°21′N, in the month of July. Data are based on values collected from 1996 to 2000 made available by S@tel-Light (The European database of daylight and solar radiation).

the average irradiance is less as can be seen in Fig. 7. In July average peak irradiance is 516 W m^{-2}, corresponding to 1.00×10^{-3} mol_{ph} m^{-2} s^{-1} of PAR. This value is lower than the irradiance in Athens predominantly because of increased cloud cover. Also in Amsterdam on a clear-sky day the photon flux density will increase to values of 1.50×10^{-3} mol_{ph} m^{-2} s^{-1} around solar noon.

As can be seen in Figs. 5 and 6 the light emission of artificial lamps and sunlight varies over the PAR wavelength range (400–700 nm). Using PAR quantum sensors we do not see how light is distributed over the PAR wavelength range. As will be discussed elsewhere in this chapter microalgal light absorption also varies over the PAR wavelength range. This implies that when calculating light absorption we have to take this distribution into account, and for this reason we will distinguish between the photon flux density I_{ph} and the wavelength-dependent photon flux density $I_{ph,\lambda}$. The I_{ph} is the total photon flux density in the PAR range expressed in mol_{ph} m^{-2} s^{-1}. $I_{ph,\lambda}$, on the other hand, gives the photon flux density at wavelength λ in a 1-nm interval and consequently has units mol_{ph} m^{-2} s^{-1} nm^{-1}.

The I_{ph} can be calculated from $I_{ph,\lambda}$ by integrating $I_{ph,\lambda}$ over the PAR range according to:

$$I_{ph} = \int_{400}^{700} I_{ph,\lambda} \cdot d\lambda \left(mol_{ph} m^{-2} s^{-1}\right)$$

This integral is equivalent to the area under the curve in Fig. 8 between 400 and 700 nm. The integral suggest that $I_{ph,\lambda}$ can be accurately described by a mathematical equation. In reality this is not always evident and this "integral" therefore is commonly solved numerically based on the measured $I_{ph,\lambda}$ over small wavelength intervals ($\Delta\lambda$), preferably as small as 1 nm. Mathematically this can be described by using the summation symbol Σ according to:

$$I_{ph} = \sum_{\lambda=400}^{\lambda=700} I_{ph,\lambda} \cdot \Delta\lambda \left(mol_{ph} m^{-2} s^{-1}\right)$$

Based on I_{ph} and $I_{ph,\lambda}$ we can now define a normalized spectral distribution of the PAR photons, $E_{n,\lambda}$, according to:

$$E_{n,\lambda} = \frac{I_{ph,\lambda}}{I_{ph}} \left(nm^{-1}\right)$$

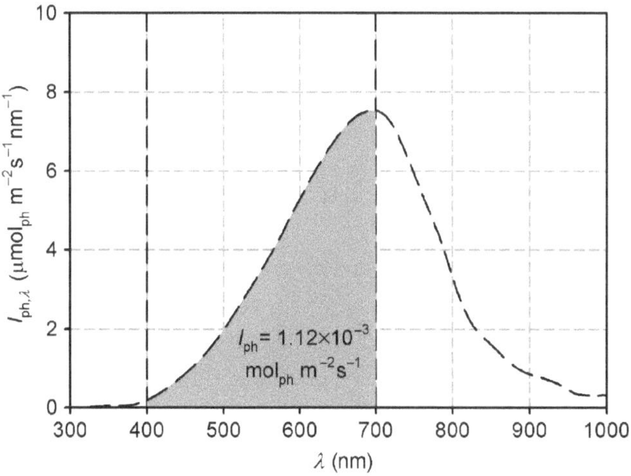

Figure 8 Wavelength-dependent photon flux density ($I_{ph,\lambda}$ in $\mu mol_{ph}\ m^{-2}\ s^{-1}\ nm^{-1}$) measured inside a bench-scale photobioreactor illuminated with tungsten-halogen lamps. The lamps were placed behind a 1-cm water filter. The photon flux density over the whole PAR range (I_{ph} in $mol_{ph} \cdot m^{-2} \cdot s^{-1}$) is also given and is equivalent to the area below the $I_{ph,\lambda}$–curve between 400 and 700 nm.

The parameter $E_{n,\lambda}$ has units nm^{-1} and it gives the fraction of PAR photons present in a 1-nm wavelength interval. Sunlight has its own characteristic distribution $E_{n,\lambda}$, just like any artificial light source. The parameter $E_{n,\lambda}$ is very useful in calculations as will be shown elsewhere in this chapter and this parameter was also used in Fig. 8 to illustrate the spectral distribution of different light sources.

2. QUANTIFYING LIGHT-LIMITED MICROALGAL GROWTH

2.1 Light Absorption

Light is absorbed by the pigments in the antennae of the photosystems, which are located on the thylakoid membranes in the chloroplast. The amount of pigments and their packing determine which fraction of the light falling on a microalgal cell will be absorbed. This fraction can be directly calculated from the specific light absorption coefficient which is also called the optical cross section of microalgae.

When a light beam hits a microalgal cell a fraction of the light is absorbed by the pigments, number 1 in Fig. 9. A large fraction of the light, however, will simply pass through the cells unabsorbed. This fraction usually is

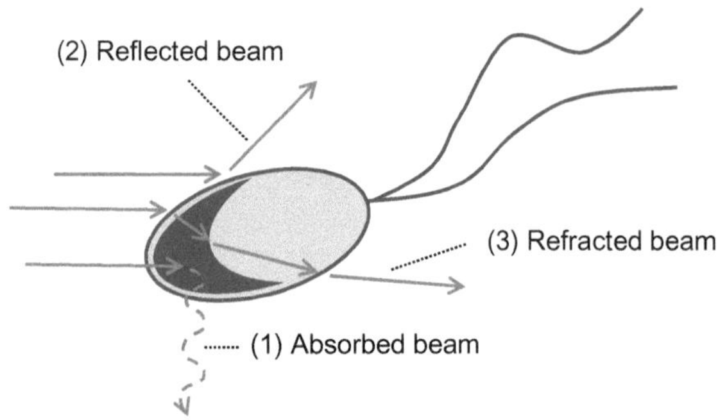

Figure 9 Light absorption and scattering by microalgal cells. Scattering is the sum of reflection and refraction events of light rays by the (sub)cellular structures with different indices of refraction than that of the surrounding water.

refracted over a small angle due to the difference of the indices of refraction of the different cellular compartments/structures with the index of refraction of the surrounding water, beam number 3 in Fig. 9. Light can also be reflected right at the cellular surface and change direction, number 2 in Fig. 9. The reflected and refracted light is still available for other microalgal cells and is not lost. The sum of reflection and refraction is called scattering. Besides using geometrical optics scattering can also be explained and described based on electromagnetic theory employing the MIE theory (Kirk, 1994).

In general scattering occurs over small angles depending on the size and shape of the microalgal cells as well as their intracellular composition and compartmentalization (Kirk, 1994). In the remainder of this chapter we will neglect the influence of scattering and assume that light is either absorbed, or passes through the cell without changing direction. At the end of the chapter the impact of this assumption will be qualitatively addressed. Light absorption by microalgal cells can be calculated based on the size of the light-absorbing surface perpendicular to the light beam/rays. This optical cross section of the cell is also called the specific light absorption coefficient, $a_{x,\lambda}$ (Fig. 10). This coefficient will be different for light of different wavelengths and it depends on the actual pigment composition of the microalgal cells. For green microalgae, for example, the optical cross section for green light is much smaller than for blue light and red light (see Fig. 11). The optical cross section can only be measured in specialized spectrophotometers which only measure light absorption and thus correct for the effect of light scattering (Davies-Colley et al, 1986; Merzlyak and Razi Naqvi, 2000).

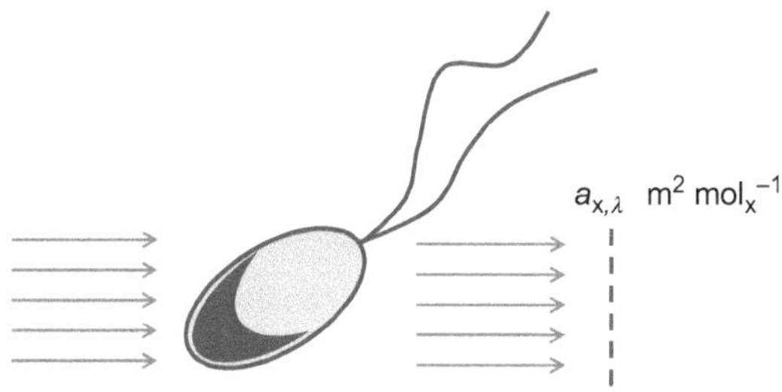

Figure 10 Optical cross section of microalgal cells: the specific absorption coefficient $a_{x,\lambda}$.

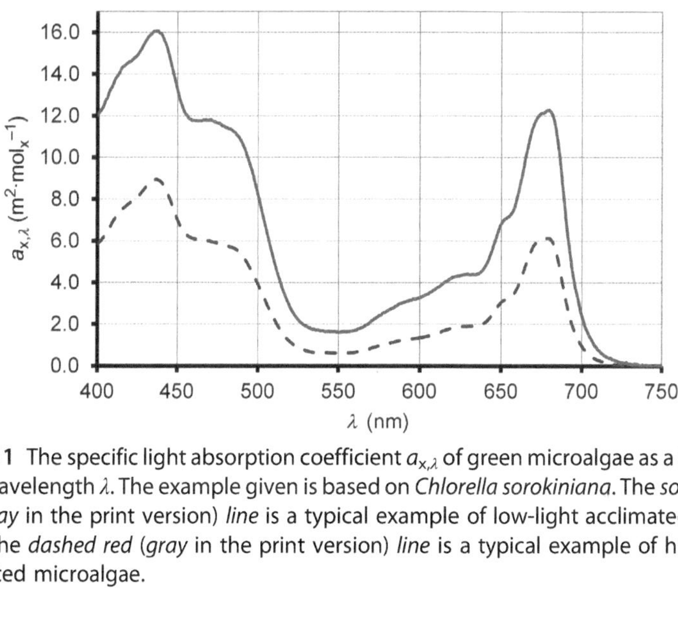

Figure 11 The specific light absorption coefficient $a_{x,\lambda}$ of green microalgae as a function of the wavelength λ. The example given is based on *Chlorella sorokiniana*. The *solid green* (*dark gray* in the print version) *line* is a typical example of low-light acclimated microalgae. The *dashed red* (*gray* in the print version) *line* is a typical example of high-light acclimated microalgae.

Microalgae will acclimate to the light regime they experience. In case the algae are light limited they will respond by increasing their pigmentation and, as such, their specific absorption coefficient $a_{x,\lambda}$. This process is called photoacclimation (Dubinsky and Stambler, 2009). High-light acclimated cells have smaller absorption coefficient than low-light acclimated cells and this is illustrated in Fig. 11. Results of the laboratory of Bioprocess Engineering (Wageningen University, The Netherlands) with microalgae of the species *Chlorella*, *Chlamydomonas*, *Scenedesmus*, *Neochloris* show that under nutrient-replete conditions low-light acclimated microalgae express a specific absorption coefficient which is not more than two times larger than those of high-light acclimated microalgae.

The specific light absorption coefficient is a crucial parameter when analyzing microalgae production processes. Later it will be discussed how this parameter can be used to estimate light penetration in microalgal cultures inside photobioreactors while neglecting the effect of light scattering. Here it will first be explained that this parameter can also be used to calculate the specific photon absorption rate q_{ph} of microalgae.

When taking into account the spectral distribution of light the specific photon absorption rate q_{ph} can be calculated as follows:

$$-q_{ph} = \sum_{\lambda=400}^{\lambda=700} a_{x,\lambda} \cdot I_{ph,\lambda} \cdot \Delta\lambda \left(\mathrm{mol_{ph}\,mol_x^{-1}\,s^{-1}}\right)$$

Please note that a minus "−" sign is added to q_{ph} because its real value should be negative since it represents a consumption term. This expression can be simplified by assuming a constant specific light absorption coefficient over the PAR range a_x:

$$-q_{ph} = a_x \cdot I_{ph} \left(\mathrm{mol_{ph}\,mol_x^{-1}\,s^{-1}}\right)$$

Both relations above are in agreement when a_x is defined as follows:

$$a_x = \sum_{\lambda=400}^{\lambda=700} a_{x,\lambda} \cdot E_{n,\lambda} \cdot \Delta\lambda \left(\mathrm{m^2\,mol_x^{-1}}\right)$$

$$E_{n,\lambda} = \frac{I_{ph,\lambda}}{I_{ph}} \left(\mathrm{nm^{-1}}\right)$$

In other words, the parameter a_x is a spectrally averaged specific absorption coefficient. Typical values for a_x for green microalgae averaged for sunlight are:

Low-light acclimated cells: $a_x \approx 5 - 7 \left(\mathrm{m^2\,mol_x^{-1}}\right)$

High-light acclimated cells: $a_x \approx 2.5 - 3.5 \left(\mathrm{m^2\,mol_x^{-1}}\right)$

2.2 Photosynthesis

In well-designed photobioreactors light is usually the growth-limiting "substrate." So, how does microalgal growth depend on light? First models will be introduced describing photosynthetic sugar production rate in the chloroplast as a function of photon flux density. The triose sugar produced in the chloroplast is consumed again in the rest of the cell. From the specific sugar consumption rate the specific growth rate of the microalgae can be calculated as explained hereafter.

The specific sugar production rate in the chloroplast q_s^c thus depends on the photon flux density I_{ph}. This relation can be described by the following model based on the hyperbolic tangent function, the model of Jassby and Platt (1976)

$$q_s^c = q_{s,m}^c \cdot \tanh\left(\frac{\alpha \cdot I_{ph}}{q_{s,m}^c}\right) \left(\mathrm{mol_s\, mol_x^{-1}\, s^{-1}}\right).$$

Sugar is defined as the 1-carbon equivalent of any sugar molecule (e.g., triose or hexose) thus having elemental composition CH_2O.

The hyperbolic tangent function is composed of exponential components and can also be written differently:

$$\tanh x = \frac{\sinh x}{\cosh x} = \frac{e^x - e^{-x}}{e^x + e^{-x}} = \frac{e^{2x} - 1}{e^{2x} + 1}$$

This relation between the specific rate of photosynthesis and photon flux density is shown in Fig. 12. The term photosynthesis thus reflects only sugar production in the chloroplast and not biomass growth as a whole which will be covered hereafter. The specific rate of photosynthesis first increases rapidly with increasing photon flux density. At higher photon flux density the increase slows down and eventually the rate of photosynthesis reaches a

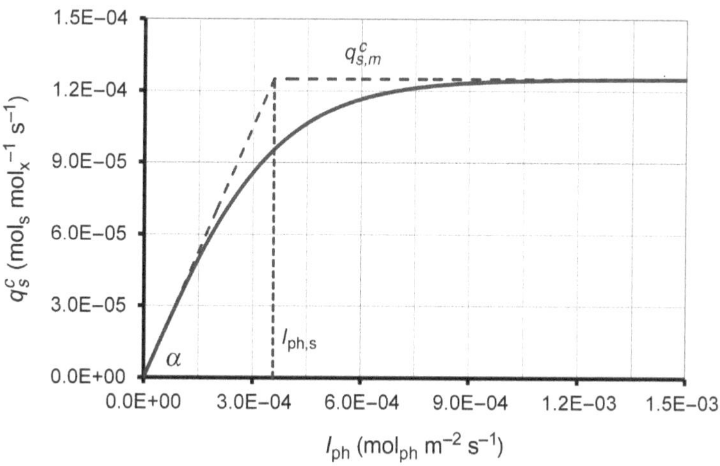

Figure 12 The specific rate of photosynthesis as a function of photon flux density I_{ph} according to the model of Jassby and Platt. Photosynthesis is represented by the specific sugar production rate in the chloroplast q_s^c. Parameter values based on high-light acclimated *Chlorella sorokiniana*: $a_x = 3.5$ m^2 mol$_x^{-1}$; $Y_{s/ph,m}^c = 0.10$ mol$_s$ mol$_{ph}^{-1}$; $q_{s,m}^c = 1.25 \times 10^{-4}$ mol$_s$ mol$_x^{-1}$ s^{-1}; $M_x = 24$ g mol$_x^{-1}$.

maximum. As can be seen in Fig. 12 the specific rate of photosynthesis q_s^c will approach $q_{s,m}^c$ at high photon flux density. The parameter $q_{s,m}^c$ thus represents the maximal photosynthetic capacity of the cell and it depends on microalgal species as well as cultivation temperature. The parameter α represents the initial slope of the curve when the photons flux density equals $0\ mol_{ph}\ m^{-2}\ s^{-1}$.

Thc parameter α also carries a biological meaning. It can be described as the product of the maximal yield of sugar on photons (in the chloroplast), $Y_{s/ph,m}^c$, and the spectrally averaged absorption coefficient a_x:

$$\alpha = Y_{s/ph,m}^c \cdot a_x$$

A good estimate of $Y_{s/ph,m}^c$ is 0.10 and this was already discussed before. The parameter α can thus be replaced with the product of two parameters with a biological meaning and then the following expression is obtained:

$$q_s^c = q_{s,m}^c \cdot \tanh\left(\frac{Y_{s/ph,m}^c \cdot a_x \cdot I_{ph,\lambda}}{q_{s,m}^c}\right) \left(mol_s\, mol_x^{-1}\, s^{-1}\right)$$

The parameter a_x can also be multiplied with variable I_{ph} giving the specific photon absorption rate. Please note that the following relation for the specific rate of photon absorption was already introduced:

$$-q_{ph} = a_x \cdot I_{ph} \left(mol_{ph}\, mol_x^{-1}\, s^{-1}\right)$$

And thus also the following relation for the specific rate of sugar production can be obtained:

$$q_s^c = q_{s,m}^c \cdot \tanh\left(\frac{Y_{s/ph,m}^c \cdot -q_{ph}}{q_{s,m}^c}\right) \left(mol_s\, mol_x^{-1}\, s^{-1}\right)$$

The last form is particularly useful when calculating photobioreactor productivity as will be discussed later.

For the sake of simplicity an alternative model will be introduced which will prove to be very useful when analyzing photobioreactor productivity. This one is based on Blackman kinetics (Blackman, 1905), and it follows the dashed lines in Fig. 12. According to this model the rate of photosynthesis increases linearly with increasing photon flux density with the proportionality constant being α. When a saturating photon flux density of $I_{ph,s}$ is reached, photosynthesis reaches its maximal level and remains constant at $q_{s,m}^c$.

Mathematically this model is described as follows:

$$I_{ph} \leq I_{ph,s} \quad q_s^c = \alpha \cdot I_{ph}$$
$$I_{ph} > I_{ph,s} \quad q_s^c = q_{s,m}^c$$

As explained before the parameter α carries a biological meaning and it can be described as follows:

$$\alpha = Y_{s/ph,m}^c \cdot a_x$$

So, the saturating photon flux density $I_{ph,s}$ can be calculated as follows:

$$I_{ph,s} = \frac{q_{s,m}^c}{\alpha} = \frac{q_{s,m}^c}{a_x \cdot Y_{s/ph,m}^c}$$

In Fig. 13 the photosynthesis response according to both models (Blackman and Jassby & Platt) is shown. As discussed before microalgae will

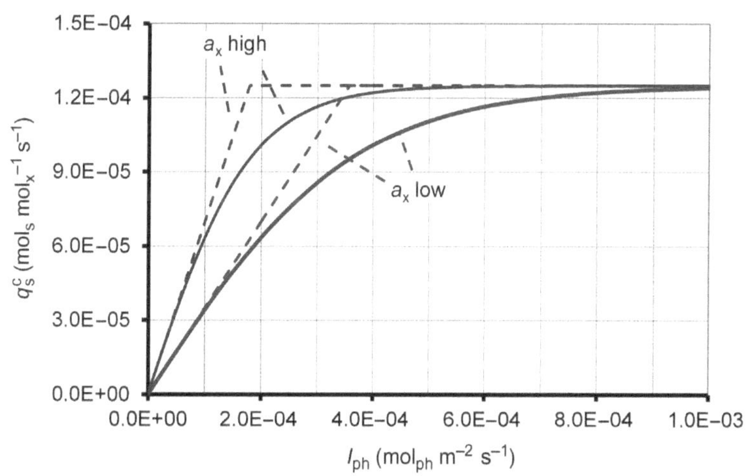

Figure 13 The specific rate of photosynthesis as a function of photon flux density I_{ph}. Photosynthesis is represented by the specific sugar production rate in the chloroplast q_s^c. Two models are shown: Jassby and Platt (*solid lines*) and Blackman (*dashed lines*). The *curves* in *red* (*dark gray* in the print version) reflect the behavior of microalgal cells acclimated to low photon flux density having high pigment content, and thus high a_x. The *curves* in *blue* (*dark gray* in the print version) reflect the behavior of microalgal cells acclimated to high photon flux density having low pigment content, and thus low a_x. Parameter values based on *Chlorella sorokiniana*: $a_x = 3.5$ m^2 mol$_x^{-1}$ (high light); $a_x = 7.0$ m^2 mol$_x^{-1}$ (low light); $Y_{s/ph,m}^c = 0.10$ mol$_s$ mol$_{ph}^{-1}$; $q_{s,m}^c = 1.25 \times 10^{-4}$ mol$_s$ mol$_x^{-1}$ s^{-1}; $M_x = 24$ g mol$_x^{-1}$.

acclimate to the light conditions experienced. In case the algae are light limited they will respond by increasing their pigmentation and, as such, their specific light absorption coefficient a_x. This process is called photoacclimation. It has implications for the photosynthetic rates at low photon flux density. Cells with a large a_x will absorb more light and will thus express a higher photosynthetic rate under light-limited conditions. Light limitation is defined as a photon flux density under which the maximal rate of photosynthesis is not reached yet, and no other nutrients are limiting growth.

The photosynthetic response shown in Figs. 12 and 13 represents instantaneous rates of microalgae acclimated to a certain photon flux density and it is thus assumed that a_x is constant. In the situation microalgal cells acclimated to light-limited (low light) conditions would be transferred to a high photon flux density; they would slowly decrease their pigment content and thus a_x. This acclimation process takes many hours.

2.2.1 Comparison of Photosynthesis Models

The model of Jassby and Platt is reported to best describe of the photosynthetic response to light (Chalker, 1980; Jassby and Platt, 1976). In this chapter the model of Jassby and Platt will be used to evaluate microalgal growth in photobioreactors, but also the simpler Blackman model. The Blackman model can be easily solved analytically and is well suited to highlight and discuss the specific aspects of phototrophic production processes. Besides these two models there are alternative models. Two models which are frequently used are the model of Webb and a Monod model:

$$\text{The model of Webb}: \quad q_s^c = q_{s,m}^c \cdot \left(1 - e^{\left(\frac{-\alpha \cdot I_{ph}}{q_{s,m}^c}\right)}\right)$$

$$\text{A Monod model}: \quad q_s^c = q_{s,m}^c \cdot \frac{I_{ph}}{K_s + I_{ph}}, \quad K_s = \frac{q_{s,m}^c}{\alpha}$$

$$\text{And thus}: \quad q_s^c = q_{s,m}^c \cdot \frac{I_{ph}}{\frac{q_{s,m}^c}{\alpha} + I_{ph}}$$

$$\text{Also in these models}: \quad \alpha = a_x \cdot Y_{s/ph,m}^c$$

In Fig. 14 all four models introduced are compared. The parameters α and $q_{s,m}^c$ are chosen equal for all four models because these represent inherent

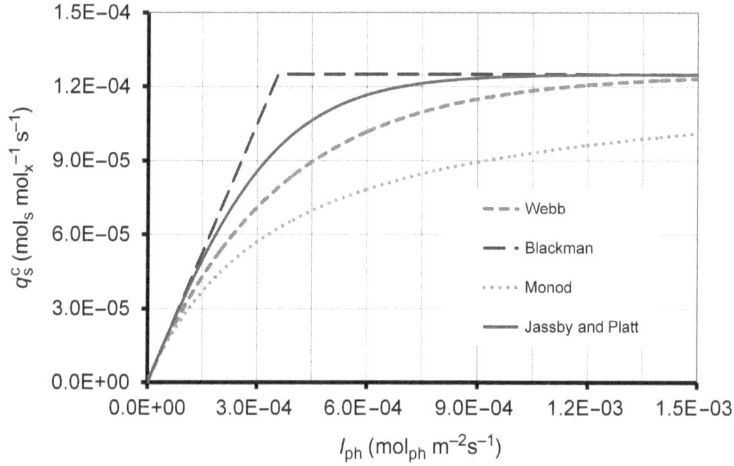

Figure 14 Comparison of photosynthesis models. Parameter values based on high-light acclimated *Chlorella sorokiniana*: $a_x = 3.5\ m^2\ mol_x^{-1}$; $Y_{s/ph,m}^c = 0.10\ mol_s\ mol_{ph}^{-1}$; $q_{s,m}^c = 1.25 \times 10^{-4}\ mol_s\ mol_x^{-1}\ s^{-1}$; $M_x = 24\ g\ mol_x^{-1}$.

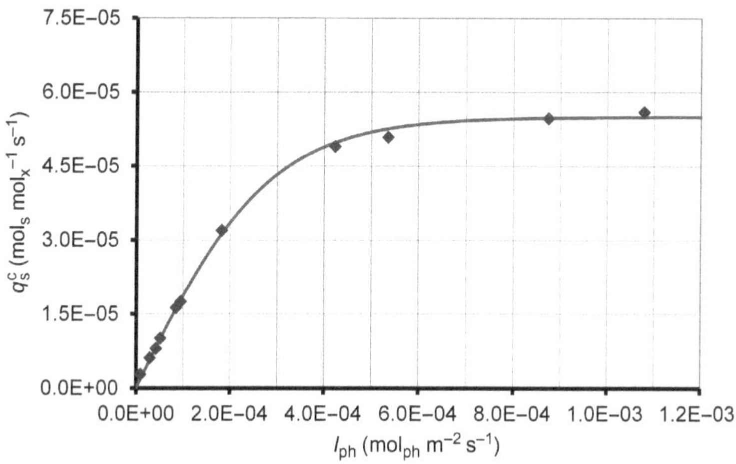

Figure 15 Curve-fit of the model of Jassby and Platt to the photosynthesis response on increasing photon flux density as measured for the green microalga *Chlamydomonas reinhardtii. Data derived from measurements by Vejrazka C, Janssen M, Benvenuti G, et al: Photosynthetic efficiency and oxygen evolution of* Chlamydomonas reinhardtii *under continuous and flashing light,* Appl Microbiol Biotechnol *97(4):1523–1532, 2013.*

biological characteristics. Only the model of Jassby and Plat is able to accurately describe the real (measured) photosynthesis response on light (as demonstrated in Fig. 15). The model of Webb is second best and is used frequently in microalgal research. When comparing the model of Webb with the model of Jassby and Platt it can be seen that the rate of photosynthesis levels off too rapidly with increasing photon flux density. The Monod model is used in modeling of other bioprocesses but it levels off even more rapidly.

2.3 Microalgal Growth

Microalgal growth can be coupled to photosynthetic sugar production employing a simple substrate balance according to Pirt (1965). The specific consumption rate of sugar outside of the chloroplast q_s (unit $mol_S\, mol_x^{-1}\, s^{-1}$) then can be coupled to the specific growth rate μ (unit s^{-1}) of the microalgae:

$$q_s = -\frac{\mu}{Y_{x/s}} - m_s \left(mol_s\, mol_x^{-1} s^{-1}\right)$$

This relation can be rewritten to obtain a function to calculate the specific growth rate:

$$\mu = (-q_s - m_s) \cdot Y_{x/s} \left(s^{-1}\right)$$

Moreover, it was already discussed that the specific consumption rate of sugar (CH_2O) is directly coupled to its production in the chloroplast q_s^c:

$$-q_s = q_s^c \left(mol_s\, mol_x^{-1} s^{-1}\right)$$

This results in a final expression for the specific growth rate:

$$\mu = \left(q_s^c - m_s\right) \cdot Y_{x/s} \left(s^{-1}\right).$$

Please note that the maximal specific growth rate μ_m of a microalgal species can be described as a function of the maximal rate of photosynthesis $q_{s,m}^c$:

$$\mu_m = \left(q_{s,m}^c - m_s\right) \cdot Y_{x/s}$$

The variable q_s^c depends on the photon flux density I_{ph} as discussed in the previous section where the photosynthesis models of Jassby & Platt and Blackman were introduced. Consequently, the specific growth rate μ can be written as a function of the photon flux density as illustrated in Fig. 16:

$$\mu = f\left(I_{ph}\right) = \left(q_s^c\left(I_{ph}\right) - m_s\right) \cdot Y_{x/s}$$

Not surprisingly, the relation between specific growth rate and photon flux density has an almost identical shape as the relation between the specific

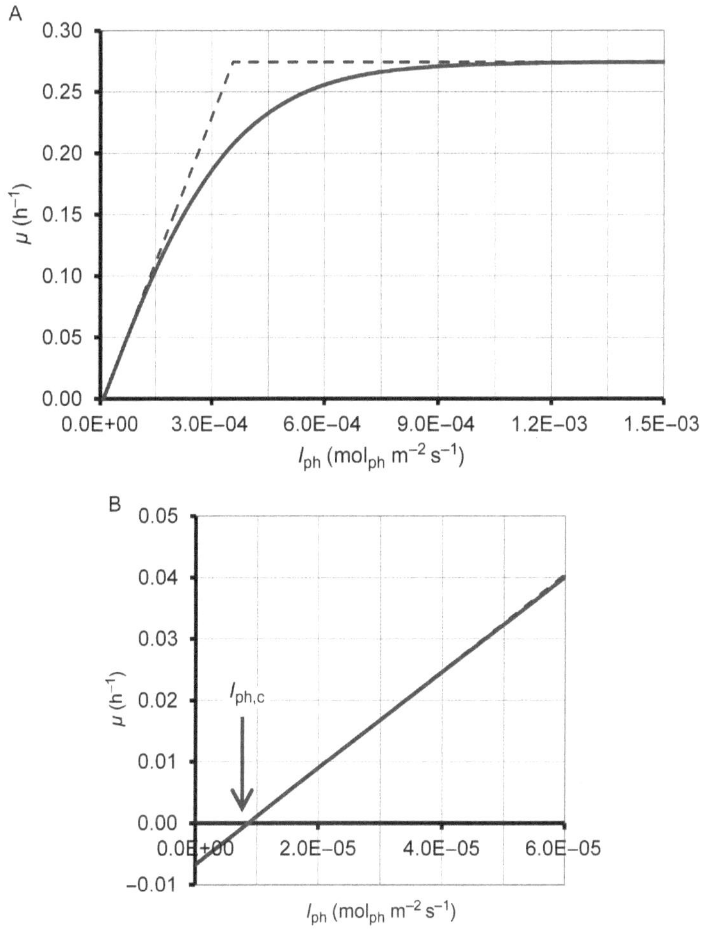

Figure 16 Specific growth rate μ of a microalga as a function of photon flux density I_{ph}. The specific growth rate is derived from a substrate balance according to Pirt in combination with the photosynthesis models of Jassby and Platt (*solid line*) or Blackman (*dashed line*). The insert illustrates the compensation point $I_{ph,c}$ where photosynthesis q_s^c is compensated by maintenance-associated respiration m_s. Parameter values based on high-light acclimated *Chlorella sorokiniana*: $a_x = 3.5\ m^2\ mol_x^{-1}$; $Y_{s/ph,m}^c = 0.10\ mol_s\ mol_{ph}^{-1}$; $q_{s,m}^c = 1.25 \times 10^{-4}\ mol_s\ mol_x^{-1}\ s^{-1}$; $Y_{x/s} = 0.625\ mol_x\ mol_s^{-1}$; $m_s = 3.0 \times 10^{-6}\ mol_s\ mol_x^{-1}\ s^{-1}$; $M_x = 24\ g\ mol_x^{-1}$.

rate of photosynthesis and photon flux density. But, because of the maintenance requirement for sugar m_s there will be "negative growth" in darkness meaning that sugar reserves are slowly depleted. In order to achieve positive growth the photon flux density I_{ph} needs to be higher than the so-called compensation point of photoautotrophic growth $I_{ph,c}$. At this photon flux density $I_{ph,c}$ just enough sugar is produced in the chloroplast to compensate for the amount lost due to cellular maintenance. The existence of the compensation point is illustrated in the inset of Fig. 16.

Based on the Blackman model the compensation point $I_{ph,c}$ can be calculated.

When $I_{ph} = I_{ph,c}$, then $\mu = 0$, and thus:

$$0 = \left(q_s^c\left(I_{ph,c}\right) - m_s\right) \cdot Y_{x/s} \Leftrightarrow q_s^c\left(I_{ph,c}\right) = m_s\left(\mathrm{mol_s\,mol_x^{-1}\,s^{-1}}\right)$$

Please note that $I_{ph,c} < I_{ph,s}$ and according to the Blackman model:

$$q_s^c\left(I_{ph,c}\right) = \alpha \cdot I_{ph,c} = Y^c_{s/ph,m} \cdot a_x \cdot I_{ph,c}\left(\mathrm{mol_s\,mol_x^{-1}\,s^{-1}}\right)$$

Combining both relations for $q_s^c(I_{ph,c})$ gives:

$$m_s = Y^c_{s/ph,m} \cdot a_x \cdot I_{ph,c}\left(\mathrm{mol_s\,mol_x^{-1}\,s^{-1}}\right)$$

Further rewriting gives the final expression for $I_{ph,c}$:

$$I_{ph,c} = \frac{m_s}{Y^c_{s/ph,m} \cdot a_x}\left(\mathrm{mol_{ph}\,m^{-2}\,s^{-1}}\right)$$

2.3.1 Definition of Instantaneous *Specific Growth Rate* μ_{PI}

On a timescale of several hours microalgae will acclimate to prevailing light conditions by modifying their pigment content and their specific light absorption coefficient (i.e., optical cross section). This process is called photoacclimation. Light-limited microalgae will respond by increasing their pigmentation and, as such, their specific absorption coefficient a_x. Cells with a large a_x will absorb more light and will thus express a higher photosynthetic rate q_s^c under light-limited conditions. High-light acclimated cells will give the opposite response.

Because the specific growth rate μ of microalgae is directly related to the rate of photosynthesis q_s^c, also the specific growth rate is affected by the acclimation state of the microalgae. It might seem very strange that there are multiple relations between specific growth rate and photon flux density: a different one for every possible value of the specific absorption coefficient a_x (see Fig. 17). But, this is the situation which needs to be considered.

As will be discussed elsewhere in this chapter microalgae are usually grown in dense cultures inside photobioreactors or raceway ponds. Consequently, microalgae will shade each other. Microalgae will experience maximal light levels at the light-exposed surface of the culture and they will

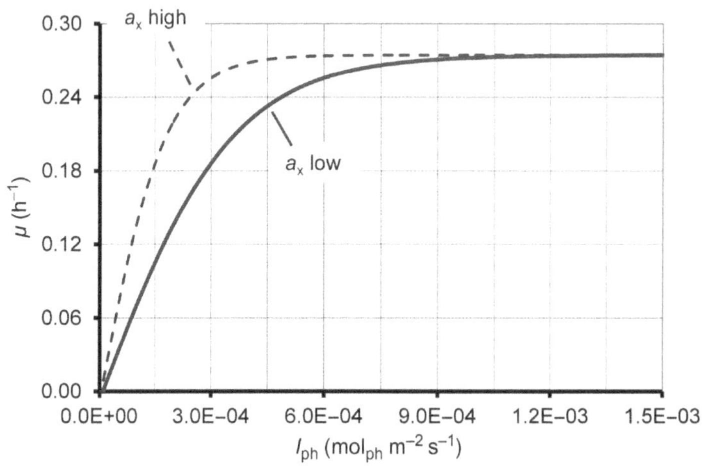

Figure 17 Instantaneous specific growth rate μ_{PI} of a microalga as a function of photon flux density I_{ph} and the influence of photoacclimation. The specific growth rate is derived from a substrate balance according to Pirt in combination with the photosynthesis models of Jassby and Platt. Parameter values based on *Chlorella sorokiniana*: $a_x = 3.5$ m^2 mol$_x^{-1}$ (high light); $a_x = 7.0$ m^2 mol$_x^{-1}$ (low light); $Y^c_{s/ph,m} = 0.10$ mol$_s$ mol$_{ph}^{-1}$; $q^c_{s,m} = 1.25 \times 10^{-4}$ mol$_s$ mol$_x^{-1}$ s^{-1}; $Y_{x/s} = 0.625$ mol$_x$ mol$_s^{-1}$; $m_s = 3.0 \times 10^{-6}$ mol$_s$ mol$_x^{-1}$ s^{-1}; $M_x = 24$ g mol$_x^{-1}$.

experience low light levels, or even darkness, at the bottom. In other words, they will experience a different photon flux density depending on their position within the culture. Due to mixing of the culture microalgae will then be temporarily exposed to different light levels. Depending on the cultivation system mixing times over this light gradient may vary from a few seconds, to multiple seconds, to minutes. The relation between the specific rate of photosynthesis and photon flux density proposed is very well suited to describe this temporal response within a microalgal culture

$$\mu = \left(q^c_s\left(I_{ph}\right) - m_s\right) \cdot Y_{x/s}.$$

This relation thus gives the specific growth rate of microalgae when the cells are temporarily exposed to a certain photon flux density at a certain location within a microalgal culture. The microalgae are acclimated to the "average" light regime in the culture. On the long term (hours to days) they could change their acclimation state if, for example, the biomass density changes, or if the photon flux density at the surface changes.

For all these reasons the specific growth rate calculated from the specific rate of photosynthesis q^c_s will be called the instantaneous specific growth rate.

The symbol μ_{PI} will be adopted for it. The subscript "PI" reflects the relation between photosynthesis and photon flux density (irradiance) derived for q_s^c.

2.4 Energetic Analysis of Microalgal Photosynthesis and Growth

In Fig. 18 the specific sugar production rate in the chloroplast q_s^c is plotted against the photon flux density I_{ph} according to the model of Jassby and Platt (solid line) as well as the model of Blackman (dashed line). In the same figure also the specific photon absorption rate q_{ph} is plotted. The specific photon absorption increases linearly with the photon flux density according to the following relationship:

$$-q_{\mathrm{ph}} = a_x \cdot I_{\mathrm{ph}}$$

Please note that a minus sign is added to q_{ph} because actual absorption (consumption) has a negative value. In contrast to the linear increase of

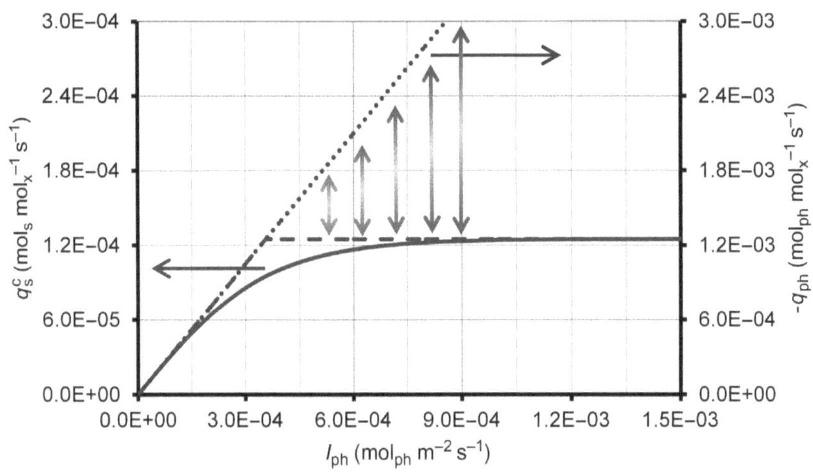

Figure 18 Specific rate of sugar production q_s^c (*blue* (*dark gray* in the print version) *lines*) and light absorption $-q_{\mathrm{ph}}$ (*dotted red* (*dark gray* in the print version) *line*) as a function of photon flux density I_{ph}. The *solid blue* (*dark gray* in the print version) *line* follows the model of Jassby and Platt; the *dashed blue* (*dark gray* in the print version) *line* follows the Blackman model. The *double-sided arrows* exemplify the mismatch between light use and light absorption at increasing photon flux density. Parameter values based on high-light acclimated *Chlorella sorokiniana*: $a_x = 3.5$ m^2 mol$_x^{-1}$; $Y_{s/ph,m}^c = 0.10$ mol$_s$ mol$_{ph}^{-1}$; $q_{s,m}^c = 1.25 \times 10^{-4}$ mol$_s$ mol$_x^{-1}$ s^{-1}; $M_x = 24$ g mol$_x^{-1}$. Please note that the scale of the q_{ph}-axis is exactly 10 times higher and that the maximal yield of sugar on photons $Y_{s/ph,m}^c$ is 0.10 mol$_s$ mol$_{ph}^{-1}$. Both facts combined explain the fact that at low photon flux density both curves have the same slope.

$-q_{ph}$, the rate of photosynthesis q_s^c levels off and reaches a maximal level $q_{s,m}^c$. It is evident from Fig. 24 that at increasing photon flux density a discrepancy develops between photosynthesis q_s^c and light absorption q_{ph}.

Solely focusing on what is happening in photosynthesis in the chloroplast the observable yield of sugar on photons absorbed $Y_{s/ph}^c$ can be calculated with the following relation:

$$Y_{s/ph}^c = \frac{q_s^c}{-q_{ph}} \left(\text{mol}_s\,\text{mol}_{ph}^{-1}\right)$$

This observable yield is different from the maximal yield of sugar on photons $Y_{s/ph,m}$ which was defined earlier and which is one of the parameters in the photosynthesis models. This observable yield is a measure of the actual efficiency at which light energy is used in photosynthesis. In Fig. 19 it can be seen that when the photon flux density I_{ph} approaches 0 $\text{mol}_{ph}\,\text{m}^{-2}\,\text{s}^{-1}$ the maximal yield of 0.10 $\text{mol}_s\,\text{mol}_{ph}^{-1}$ is reached.

On the contrary, at increasing photon flux density the yield drops and only part of the photons absorbed are really used in photosynthesis. The surplus of light absorbed by the pigments in the photosystems is dissipated as heat by the same photosystems. Dedicated biochemical and biophysical

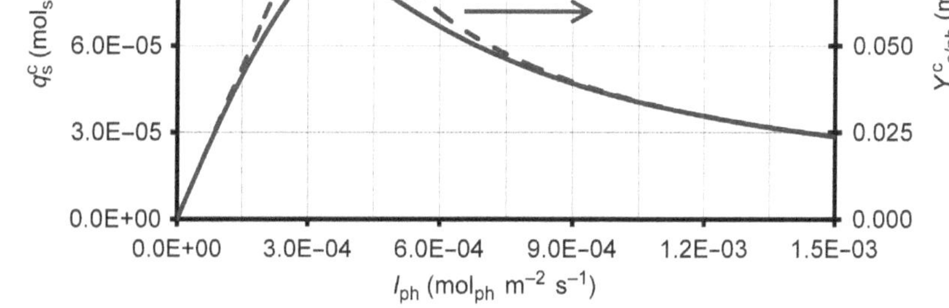

Figure 19 Specific rate of sugar production q_s^c (*blue* (*dark gray* in the print version) *lines*) and the observable yield of sugar on photons absorbed $Y_{s/ph}^c$ (*red* (*dark gray* in the print version) *lines*) as a function of photon flux density I_{ph}. The *solid lines* follow the model of Jassby and Platt; the *dashed lines* follow the Blackman model. Parameter values based on high-light acclimated *Chlorella sorokiniana*: $a_x = 3.5\ \text{m}^2\,\text{mol}_x^{-1}$; $Y_{s/ph,m}^c = 0.10\ \text{mol}_s\,\text{mol}_{ph}^{-1}$; $q_{s,m}^c = 1.25 \times 10^{-4}\ \text{mol}_s\,\text{mol}_x^{-1}\,\text{s}^{-1}$; $M_x = 24\ \text{g}\,\text{mol}_x^{-1}$.

processes inside the photosystems and their pigments make this possible (Foyer et al, 2012; Müller et al, 2001). In this way the surplus of light can be safely disposed of. This decrease in the efficiency of photosynthesis at increasing photon flux density is called photosaturation or light saturation.

Similarly, also the observable yield of biomass on photons absorbed can be calculated:

$$Y_{x/ph} = \frac{\mu_{PI}}{-q_{ph}} \left(mol_x\, mol_{ph}^{-1}\right)$$

$$\Leftrightarrow Y_{x/ph} = \frac{\left(q_s^c - m_s\right) \cdot Y_{x/s}}{-q_{ph}} \left(mol_x\, mol_{ph}^{-1}\right)$$

The symbol μ_{PI} thus reflects the instantaneous specific growth rate as calculated from the specific rate of photosynthesis q_s^c.

The relation between observable biomass yield on photons and photon flux density is shown in Fig. 20. Similar to the sugar yield also the biomass yield on photons absorbed $Y_{x/ph}$ decreases at increasing photon flux density because of the photosaturation effect. At low photon flux density a different

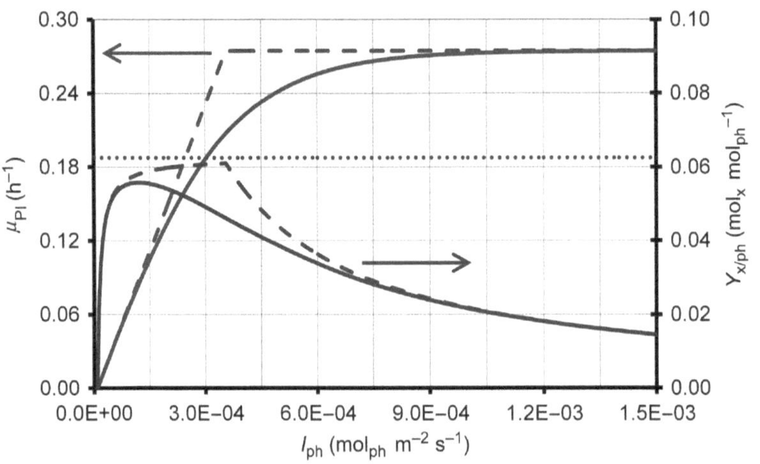

Figure 20 Instantaneous specific growth rate μ_{PI} (*blue* (*dark gray* in the print version) lines) and the observable yield of biomass on photons $Y_{x/ph}$ (*red* (*dark gray* in the print version) lines) as a function of photon flux density I_{ph}. The *solid lines* follow the model of Jassby and Platt; the *dashed lines* follow Blackman. The *dotted red* (*dark gray* in the print version) *line* represents the hypothetical maximal biomass yield on photons $Y_{x/ph,m}$ when maintenance requirement is neglected ($m_s = 0$). Parameter values based on high-light acclimated *Chlorella sorokiniana*: $a_x = 3.5$ m^2 mol$_x^{-1}$; $Y_{s/ph,m}^c = 0.10$ mol$_s$ mol$_{ph}^{-1}$; $q_{s,m}^c = 1.25 \times 10^{-4}$ mol$_s$ mol$_x^{-1}$ s^{-1}; $Y_{x/s} = 0.625$ mol$_x$ mol$_s^{-1}$; $m_s = 3.0 \times 10^{-6}$ mol$_s$ mol$_x^{-1}$ s^{-1}; $M_x = 24$ g mol$_x^{-1}$.

trend is observed. The biomass yield does not increase to a maximal value when I_{ph} approaches 0 $mol_{ph}\ m^{-2}\ s^{-1}$. On the contrary, $Y_{x/ph}$ drops dramatically at very low light levels. This is related to the fact that the larger part of the little light absorbed must be used for cellular maintenance. Sugars are still produced at high efficiency by photosynthesis in the chloroplast, but almost all sugar is broken down again to provide energy for maintenance. Consequently, little or no energy is left for biomass growth.

In a hypothetical situation where the cells have no maintenance requirements ($m_s = 0$) the maximal biomass yield can be calculated as follows:

$$Y_{x/ph,m} = Y^{c}_{s/ph,m} \cdot Y_{x/s}$$

The dotted horizontal line in Fig. 20 represents this hypothetical maximum which cannot be reached in practice. In the example presented in Fig. 20 it was assumed that the biomass yield on sugar was 0.625 $mol_x\ mol_s^{-1}$ and that the sugar yield on photons was 0.10 $mol_s\ mol_{ph}^{-1}$, and thus:

$$Y_{x/ph,m} = 0.10 \times 0.625 = 0.0625 \left(mol_x\, mol_{ph}^{-1}\right)$$

Knowing that the weight of 1 C-mol of biomass (i.e., 1 mol_x) equals about 24 g this means that 1 mol of PAR photons can be converted in 1.50 g of biomass in the hypothetical situation maintenance can be neglected. In practice maintenance will reduce the yield on photons as illustrated by the model description presented in Fig. 20 where the maximal yield $Y_{x/ph,m}$ is not reached.

It must be noted that this value for $Y_{x/ph,m}$ is an estimate because the yield of biomass on sugar $Y_{x/s}$ is usually not well known and this was already discussed before. Also the value of the maximal yield of sugar on photons $Y_{s/ph,m}$ was discussed before and it was shown that its value is close to 0.10 mol $mol_s\ mol_{ph}^{-1}$. The spectral dependence of the yield of photosynthesis $Y^{c}_{s/ph,m}$ was not discussed yet. In the old work of Emerson and Lewis (1943) and in the work of Tanada (1951) the spectral dependency of the yield of photosynthesis is described and some of their data are redrawn and presented in Fig. 21. Blue light (400–500 nm) appears to be used with a reduced efficiency. In the green part of the spectrum (500–600 nm) the efficiency increases up to a maximal level observed in the red region (600–700 nm). It is important to emphasize that the efficiency in the blue and green region is still more than 70% of the level in the red region. Within the framework of a simple model description for engineering purposes we

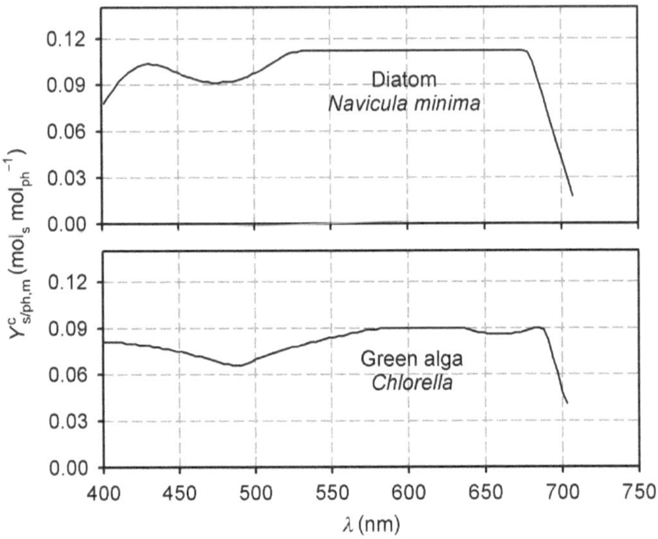

Figure 21 Maximum yield of sugar on photons in the chloroplast $Y^c_{s/ph,m}$ as a function of wavelength λ. $Y^c_{s/ph,m}$ is shown for a green alga, *Chlorella* (Emerson and Lewis, 1943); and a diatom, *Navicula minima* (Tanada, 1951).

propose we can assume a flat spectral response in the PAR range and thus assume a fixed sugar yield on photons irrespective of the wavelength.

The studies of Emerson and Lewis and Tanada already date from the middle of last century. Almost no attempts have been made to measure the spectral dependence of the efficiency of photosynthesis with latest technology or with other microalgal species. Only recently a new approach was presented by Tamburic et al (2014) for the marine eustigmatophyte microalga *Nannochloropsis*. This study confirms that all wavelengths across the PAR spectrum can be utilized with similar efficiency to drive photosynthesis. Although Tamburic reported differences between certain wavelengths the lowest values measured were still 70% of the maximal values and no consistent differences between the blue, green, and red spectral regions were reported.

Within the research arena of the higher plants (crop plants) much more work has been done on resolving the spectral dependency of the efficiency of light use in photosynthesis (Evans, 1987; McCree, 1971). The most recent study is the one of Hogewoning et al (2012) who measured the maximal yield of carbon dioxide (CO_2) fixation in cucumber leaves as a function of wavelength. CO_2 fixation in the chloroplasts is equivalent to sugar production. Hogewoning measured a yield of CO_2 fixation on photons between 0.06 and 0.09 $mol_{CO_2}\,mol_{ph}^{-1}$. The lower value of 0.06 was measured in the blue range (400–500 nm) and the higher value of 0.09 in the red range (600–700 nm). In the green part of the spectrum the yield increased from 0.06 $mol_{CO_2}\,mol_{ph}^{-1}$ at 500 nm to 0.09 $mol_{CO_2}\,mol_{ph}^{-1}$ at 600 nm. The lower yield in the blue range can be largely explained by the absorption of nonphotosynthetic pigments which attributed to about 20% of the total light absorption between 400 and 500 nm. The work of Hogewoning thus confirms the old work Emerson and Lewis on *Chlorella* (Fig. 21).

2.4.1 From Biomass Yield on Photons to Photosynthetic Efficiency

Commonly the potential of large-scale microalgae production is supported by referring to their high "photosynthetic efficiency" (*PE*). This efficiency is based on the fraction of light energy (%) stored as the chemical energy within the biomass. The biomass yield on photons is expressed as a mole fraction ($mol_x\,mol_{ph}^{-1}$) but can be converted into energy units taking into account the energy content of the photons as well as the energy content of the biomass. The results of such recalculations are shown in Table 1 while assuming

Table 1 Maximal Biomass Yield on Photons, and Maximal Photosynthetic Efficiency (PE) of Photoautotrophic Microalgal Growth Under Nutrient-Replete Conditions

		$Y_{x/ph}$		PE/%		
$Y_{s/ph,m}$ ($mol_s\,mol_{ph}^{-1}$)	$Y_{x/s}$ ($mol_x\,mol_s^{-1}$)	$mol_x\,mol_{ph}^{-1}$	$g\,mol_{ph}^{-1}$	680 nm	Sunlight PAR	Sunlight complete
0.100	0.5	0.050	1.20	15.9%	12.9%	5.5%
	0.6	0.060	1.44	19.1%	15.5%	6.6%
	0.7	0.070	1.68	22.2%	18.1%	7.7%
	Sugar (CH_2O) only →		*3.00*	*26.2%*	*21.3%*	*9.0%*
0.125	0.5	0.0625	1.50	19.8%	16.2%	6.8%
	0.6	0.075	1.80	23.8%	19.4%	8.2%
	0.7	0.0875	2.10	27.8%	22.6%	9.6%
	Sugar (CH_2O) only →		*3.75*	*32.7%*	*26.6%*	*11.3%*

Numbers used: PAR fraction sunlight: 42.3% (based on ASTM g173 spectrum); energy content PAR photons sunlight: 216 kJ mol_{ph}^{-1} (based on ASTM g173 spectrum); molar weight of biomass (1-carbon equivalent): $M_x = 24$ g mol_x^{-1}; enthalpy of combustion biomass: $\Delta H_c^0(s) = 559$ kJ mol_x^{-1}; degree of reduction biomass: $\gamma = 4.86$; molar weight of sugar (CH_2O): $M_x = 30$ g mol_x^{-1}; enthalpy of combustion sugar: $\Delta H_c^0(s) = 460$ kJ mol_x^{-1}; degree of reduction sugar: $\gamma = 4$. References: Cordier et al, 1987; Kliphuis et al, 2010, 2011b, 2012.

different values for the maximal sugar yield on photons $Y_{s/ph,m}$, and the biomass yield on sugar $Y_{x/s}$. The calculations are based on microalgae cultivated under nutrient-replete and thus light-limited conditions.

Based on the assumption that 10 photons are required for the production of 1-carbon equivalent of sugar ($Y_{s/ph,m}=0.10$ mol$_s$ mol$_{ph}^{-1}$) the PE of microalgal growth ranges between 5.5% and 7.7% depending on the assumed biomass yield on sugar ($Y_{x/s}$). The higher value of 7.7% corresponds to a biomass yield on sugar of 0.7 mol$_x$ mol$_s^{-1}$, which would be among the highest reported for aerobic chemoheterotrophic growth. When normalizing to the PAR range of 400–700 nm the PE ranges between 12.9% and 18.1% depending on $Y_{x/s}$. When narrowing down to the minimal bandgap required by photosystem II (i.e., 680 nm) the efficiency ranges between 15.9% and 22.2% depending on the value of $Y_{x/s}$.

The higher values of the biomass yield on sugar $Y_{x/s}$ probably can only be reached with ammonium or urea as a nitrogen source. In that situation the actual energy stored in the overall reaction representing microalgal growth is less than the enthalpy of combustion of biomass. This is related to the fact that the enthalpy of combustion of these reduced nitrogen compounds is considerable. Taking this into account maximal PE will drop from 7.7% to 7.1% (assuming $Y_{s/ph,m}=0.1$ mol$_s$ mol$_{ph}^{-1}$ and $Y_{x/s}=0.7$ mol$_x$ mol$_s^{-1}$). Moreover, in real microalgae production systems standing biomass will continuously consume photosynthetically derived sugar for maintenance purposes bringing down maximal PE further. This effect will be exemplified in the following section on photobioreactor productivity.

In the hypothetical situation that only 8 photons are required for the production of 1-carbon equivalent of sugar ($Y_{s/ph,m}=0.125$ mol$_s$ mol$_{ph}^{-1}$) the PE of microalgal growth ranges between 6.8% and 9.6% depending on the assumed biomass yield on sugar ($Y_{x/s}$). Statements of *PE* for microalgae production beyond 9.6% should thus be treated with great caution as it is very unlikely that the $Y_{x/s}$ is higher than 0.7. Moreover, as discussed before, the yield of sugar on photons achieved under ideal conditions is closer to 0.10 than 0.125 mol$_s$ mol$_{ph}^{-1}$. Possibly statements of PE of 10%, or more, could be based on the first steps of photosynthesis leading to the production of sugar. In an extra calculation referred to as "sugar only" in Table 1 is referred to a hypothetical situation microalgal cells are only producing and accumulating sugar (CH_2O). In that case a maximal PE on sunlight of 11.3% can be reached.

3. ESTIMATING PHOTOBIOREACTOR PRODUCTIVITY

There is a wide variety of photobioreactors ranging from closed photobioreactors of variable designs to open microalgal raceway ponds. The productivity of such photobioreactors will be analyzed in a general way under the assumption that they are operated under light-limited conditions. In other words it is assumed that all nutrients, including carbon dioxide, are available in excess. In addition, it is assumed that all culture parameters are maintained at their optimal range, e.g., pH, temperature, salinity, dissolved oxygen concentration.

3.1 Light Penetration in Microalgal Cultures

Microalgae absorb light and, consequently, the photon flux density will decrease along the path light takes through a microalgal culture. In order to quantify microalgal growth in photobioreactors the local photon flux density within the microalgal culture must be known. The specific light absorption coefficient $a_{x,\lambda}$ of the microalgae can be used to calculate light penetration. Its value is wavelength dependent meaning that light of certain wavelengths are absorbed more strongly than others.

For simplicity it is assumed that the effect of light scattering can be neglected. Later this simplification will be revisited and the effect of light scattering will be discussed in general terms. In the analysis to follow a unidirectional light field is assumed which is perpendicular to a flat light-exposed surface, e.g., raceway ponds or panel-type photobioreactors. In that case the law of Lambert–Beer can be used to calculate the photon flux density $I_{ph,\lambda}(z)$ at each location z within the microalgal culture. Lambert–Beer states that the attenuation of light over an infinitesimal small distance is proportional to the photon flux density itself

$$\frac{dI_{ph,\lambda}}{dz} = -a_{x,\lambda} \cdot C_x \cdot I_{ph,\lambda}.$$

The proportionality constant is the product of the biomass concentration C_x and the specific light absorption coefficient $a_{x,\lambda}$ (i.e., the optical cross section). Integrating from the light-exposed surface ($z=0$) to any location z then delivers an exponential decrease of the photon flux density over z:

$$I_{ph,\lambda}(z) = I_{ph,\lambda}(0) \cdot e^{-a_{x,\lambda} \cdot C_x \cdot z} \left(\text{mol}_{ph}\,\text{m}^{-2}\,\text{s}^{-1}\,\text{nm}^{-1}\right)$$

The photon flux density $I_{ph,\lambda}$ was introduced before and it represents the wavelength-dependent photon flux density with units $mol_{ph}\ m^{-2}\ s^{-1}\ nm^{-1}$. Its value at the light-exposed surface can be calculated as follows:

$$I_{ph,\lambda}(0) = I_{ph}(0) \cdot E_{n,\lambda} \left(mol_{ph} m^{-2} s^{-1} nm^{-1}\right)$$

The parameter $I_{ph}(0)$ represents the integrated photon flux density over the complete PAR range at the light-exposed surface of the photobioreactor. This parameter must be measured in order to estimate productivity of microalgal cultivation systems. It should be remembered that $E_{n,\lambda}$ (unit nm^{-1}) is the PAR-normalized spectrum of the light source used. More specifically, at each wavelength it gives the fraction of the total amount of PAR photons in a 1-nm interval explaining the unit nm^{-1}.

In the example shown in Fig. 22 only a small part of green light is able to penetrate to the "bottom" of the culture which is at 3 cm distance from the light-exposed surface. Of course this depends on the actual biomass concentration C_x, which is 100 $mol_x\ m^{-3}$ in this example, and this is equivalent to 2.4 kg m^{-3} ($M_x \approx 24$ g mol_x^{-1}). It is apparent from the results presented in Fig. 22 that also green light (500–600 nm) is largely absorbed in the culture and is thus far from useless. In fact, when referring to Fig. 21 and the

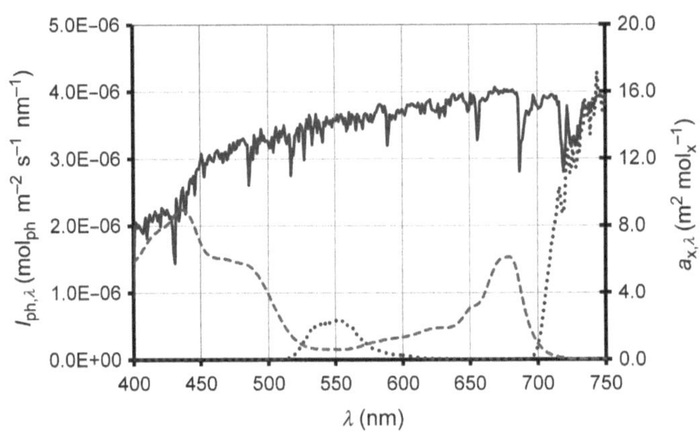

Figure 22 Light penetration in a microalgal culture according to Lambert–Beer as a function of the wavelength λ. The *dashed green* (*gray* in the print version) *line* is the specific light absorption coefficient $a_{x,\lambda}$. The *solid red* (*dark gray* in the print version) *line* represents the intensity of sunlight $I_{ph,\lambda}(0)$ at the light-exposed surface. The *dotted red* (*dark gray* in the print version) *line* represents the photon flux density $I_{ph,\lambda}$ in the microalgal culture at 0.03 m distance from the light-exposed surface. Parameter values based on high-light acclimated *Chlorella sorokiniana*: $C_x = 100\ mol_x\ m^{-3}$; $I_{ph}(0) = 1 \times 10^{-3}\ mol_{ph}\ m^{-2}\ s^{-1}$; $a_x = 3.33\ m^2\ mol_x^{-1}$; $E_{n,\lambda}$ for sunlight.

concomitant discussion on the spectral dependence of the maximal yield of sugar on photons $Y^{c}_{s/ph,m}$ it is evident that green light is still used efficiently in microalgal photosynthesis after it has been absorbed.

In order to calculate the integrated photon flux density over the complete PAR range at a depth z the contribution of all individual wavelengths from 400 to 700 nm has to be summed:

$$I_{ph}(z) = \sum_{400\,nm}^{700\,nm} I_{ph,\lambda}(0) \cdot e^{-a_{x,\lambda} \cdot C_x \cdot z} \cdot \Delta\lambda \left(mol_{ph}\, m^{-2} s^{-1}\right)$$

This approach can be simplified by adopting the spectrally averaged specific absorption coefficient a_x:

$$I_{ph}(z) = I_{ph}(0) \cdot e^{-a_x \cdot C_x \cdot z} \left(mol_{ph}\, m^{-2} s^{-1}\right)$$

with

$$a_x = \sum_{\lambda=400}^{\lambda=700} a_{x,\lambda} \cdot E_{n,\lambda} \cdot \Delta\lambda \left(m^2 mol_x^{-1}\right)$$

In theory an error is introduced when assuming a constant specific absorption coefficient a_x. This is related to the fact that the spectral distribution of light $E_{n,\lambda}$ changes when moving deeper inside a microalgal culture. This is also exemplified in Fig. 22 where the spectrum changes from white (sunlight) at the reactor surface to green at the bottom. But, only by adopting this simplification Lambert-Beer can be used later in combination with the Blackman model to set up a complete model description for photobioreactor productivity which can be readily solved. The result of this simplification is an exponential decrease of $I_{ph}(z)$ as a function of depth z as illustrated in Fig. 23.

3.2 Microalgae Cultivation in Photobioreactors: Calculating Productivity

Photobioreactors usually are operated according to a chemostat mode (constant dilution) or turbidostat mode (constant turbidity) of operation. Details will be explained later in this chapter but below the biomass balance over a microalgal cultivation system is already presented:

$$V_r \cdot \frac{dC_x}{dt} = F_{in} \cdot C_{x,in} - F_{out} \cdot C_{x,out} + r_x \cdot V_r$$

with

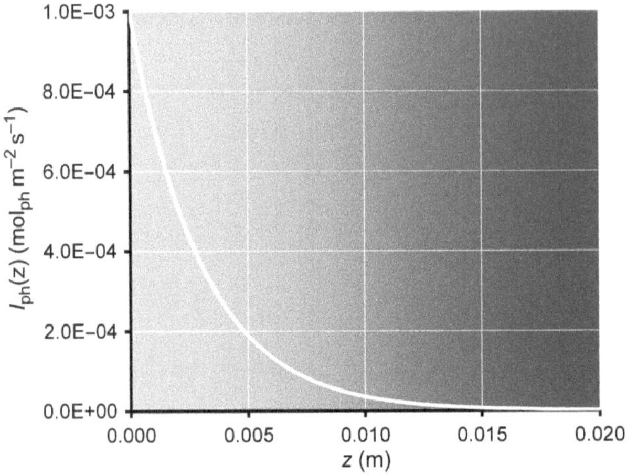

Figure 23 Light penetration in a microalgal culture neglecting wavelength dependency of light absorption, and thus assuming a constant a_x: I_{ph} as a function of depth z. The calculation of I_{ph} is based on Lambert–Beer with a constant specific light absorption coefficient a_x. Parameter values based on high-light acclimated *Chlorella sorokiniana*: $a_x = 3.33\ m^2\ mol_x^{-1}$; $C_x = 100\ mol_x\ m^{-3}$; $I_{ph}(0) = 1 \times 10^{-3}\ mol_{ph}\ m^{-2}\ s^{-1}$.

V_r = liquid volume in photobioreactor (m^3)
F_{in}/F_{out} = liquid flow rate in and out of photobioreactor ($m^3\ s^{-1}$)
C_x = biomass concentration in photobioreactor ($mol_x\ m^{-3}$)
$C_{x,in}/C_{x,out}$ = biomass concentration in inflow or outflow ($mol_x\ m^{-3}$)
r_x = volumetric biomass production rate ($mol_x\ m^{-3}\ s^{-1}$)

With respect to the design and operation of photobioreactors the production term r_x is of major importance. Parameter r_x represents the volumetric biomass production rate of a cultivation system and it is the product of the biomass concentration in the system and the average specific growth rate of the microalgae:

$$r_x = \bar{\mu} \cdot C_x \left(mol_x m^{-3} s^{-1}\right)$$

In order to assess the volumetric productivity of the system the microalgal biomass concentration and average specific growth rate need to be known. For the moment it will be assumed that the biomass concentration can be set at any desired level by means of a turbidostat mode of operation. Consequently, only the average specific growth rate of the microalgae in the photobioreactor needs to be calculated.

All theory necessary for this calculation is already introduced and is summarized in Fig. 24. The photon flux density at any location z in the microalgal culture can be described with Lambert–Beer. Based on this local

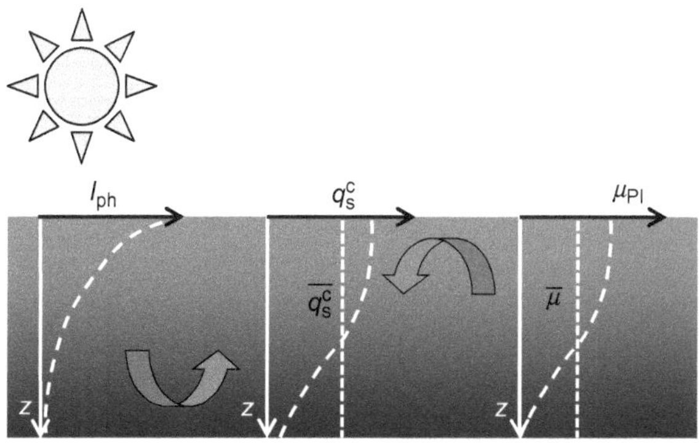

Figure 24 Schematic drawing of the light gradient inside microalgal cultures and consequent gradients in specific rates. In the drawing three different graphs are included describing the dependency of the following parameters on the location *z* within the microalgal culture: (*left*) photon flux density I_{ph}; (*middle*) specific rate of photosynthesis q_s^c; (*right*) instantaneous specific growth rate μ_{PI}.

photon flux density the local specific rate of sugar production in the chloroplast (photosynthesis) can be calculated using photosynthesis models such as those of Jassby & Platt or Blackman. The local specific sugar production rates then can be integrated over culture depth to obtain the average photosynthetic rate of the microalgal culture. Finally, based on the substrate balance for sugar (according to Pirt), the average specific growth rate of the microalgal culture can be calculated:

$$\overline{\mu} = \left(\overline{q_s^c} - m_s\right) \cdot Y_{x/s}$$

Below the calculation routines to calculate the average specific rate of sugar production in the chloroplast (photosynthesis) will be explained in more detail for the Blackman and Jassby & Platt models. In order to perform such calculations all necessary biological parameters must be known for the microalgal species to be cultivated: a_x; $Y^c_{s/ph,m}$; $q^c_{s,m}$; $Y_{x/s}$; m_s.

3.2.1 Integrating Photosynthesis Along Optical Path According to Blackman

When adopting the Blackman model a simple but useful expression for the volumetric productivity of photobioreactors can be derived. In the Blackman model a constant specific light absorption coefficient is used. In other words, the change of light spectrum by preferential absorption within a microalgal culture is not included.

As discussed earlier the average specific sugar production rate of the microalgal culture must be calculated in order to assess photobioreactor volumetric productivity. The following integral therefore must be calculated:

$$\overline{q_s^c} = \frac{\int_0^d q_s^c(z) \cdot dz}{d}$$

$$I_{ph}(z) \leq I_{ph,s}: \quad q_s^c = a_X \cdot Y_{s/ph,m} \cdot I_{ph}(z)$$

$$I_{ph}(z) \geq I_{ph,s}: \quad q_s^c = q_{s,m}^c$$

The saturating photon flux density $I_{ph,s}$ can be calculated as follows:

$$I_{ph,s} = \frac{q_{s,m}^c}{a_X \cdot Y_{s/ph,m}^c}$$

Based on the biomass concentration C_X and the value $I_{ph,s}$ Lambert–Beer can be used to calculate the depth z inside the photobioreactor where the local photon flux density $I_{ph}(z)$ is equal to $I_{ph,s}$. This depth will be called the saturation depth z_s

$$I_{ph}(z_s) = I_{ph,s} \text{ and } I_{ph}(z_s) = I_{ph}(0) \cdot e^{-a_X \cdot C_X \cdot z_s}.$$

Combining these relations gives:

$$z_s = \frac{\text{Ln}\left(\frac{I_{ph}(0)}{I_{ph,s}}\right)}{a_X \cdot C_X}$$

Consequently, we can split up the photobioreactor in two zones (Fig. 25):

1. $z = 0 \rightarrow z = z_s$ where $I_{ph}(z) \geq I_{ph,s}$, thus $q_s^c = q_{s,m}^c$

2. $z = z_s \rightarrow z = d$ where $I_{ph}(z) \leq I_{ph,s}$, thus $q_s^c = a_X \cdot Y_{s/ph,m} \cdot I_{ph}(z)$

Consequently the average specific sugar production rate of the microalgae can be described with the following integral:

$$\overline{q_s^c} = \frac{1}{d} \cdot \int_0^{z_s} q_{s,m}^c \cdot dz + \frac{1}{d} \cdot \int_{z_s}^d a_X \cdot Y_{s/ph,m}^c \cdot I_{ph}(z) \cdot dz$$

$$\text{with: } I_{ph}(z) = I_{ph}(0) \cdot e^{-a_X \cdot C_X \cdot z}$$

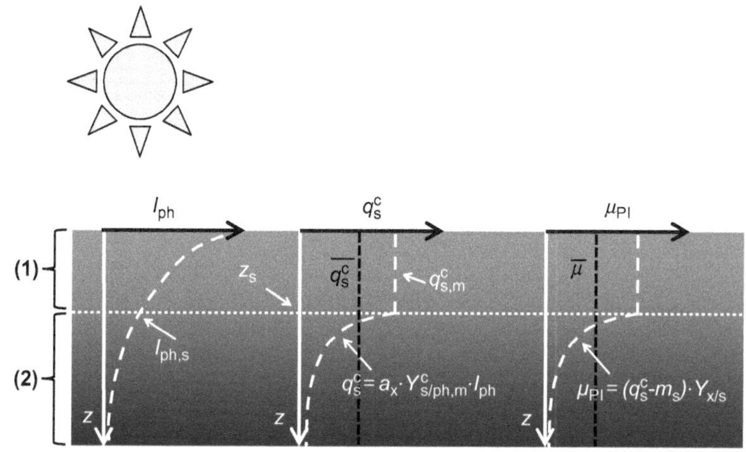

Figure 25 Schematic drawing of the light gradient inside microalgal cultures and the application of the Blackman model to calculate the average specific growth rate. The microalgal culture is subdivided into two zones: (1) $I_{ph}(z) \geq I_{ph,s}$ and (2) $I_{ph}(z) \leq I_{ph,s}$. See Fig. 24 for more details.

The analytical solution of this integral is:

$$\overline{q_s^c} = \frac{z_s}{d} \cdot q_{s,m}^c + \frac{Y_{s/ph,m}^c \cdot I_{ph}(0)}{d \cdot C_x} \left(e^{-a_x \cdot C_x \cdot z_s} - e^{-a_x \cdot C_x \cdot d} \right)$$

$$\text{with}: \quad z_s = \frac{\mathrm{Ln}\left(\dfrac{I_{ph}(0)}{I_{ph,s}} \right)}{a_x \cdot C_x} \quad \text{and} \quad I_{ph,s} = \frac{q_{s,m}^c}{a_x \cdot Y_{s/ph,m}^c}$$

The following two special situations can occur:

$$z_s \geq d, \quad \text{and thus } I_{ph}(z) \geq I_{ph,s} : \overline{q_s^c} = q_{s,m}^c$$

$$I_{ph}(0) \leq I_{ph,s}, \quad \text{and thus } z_s = 0 : \overline{q_s^c} = \frac{Y_{s/ph,m}^c \cdot I_{ph}(0)}{d \cdot C_x} \left(1 - e^{-a_x \cdot C_x \cdot d} \right)$$

Ultimately a single expression for the volumetric productivity of a microalgal culture in a photobioreactors can be derived:

$$r_x = \frac{Y_{x/s}}{d} \cdot \left\{ \frac{q_{s,m}^c}{a_x} \cdot \left(1 + \mathrm{Ln}\left(\frac{I_{ph}(0) \cdot a_x \cdot Y_{s/ph,m}^c}{q_{s,m}^c} \right) \right) - Y_{s/ph,m}^c \cdot I_{ph}(0) \cdot e^{-a_x \cdot C_x \cdot d} \right\} - Y_{x/s} \cdot m_s \cdot C_x$$

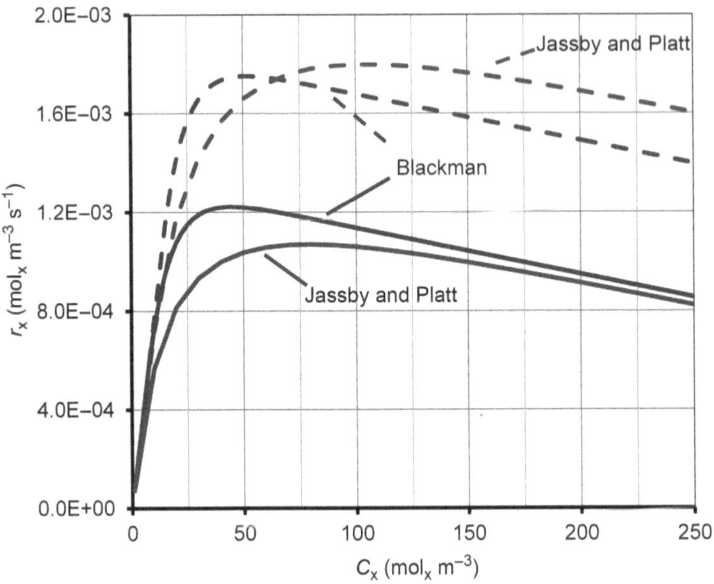

Figure 26 Volumetric productivity of a photobioreactor r_x as a function of biomass concentration C_x according to Blackman (*red* (*dark gray* in the print version) *lines*) and Jassby and Platt (*blue* (*dark gray* in the print version) *lines*). *Solid lines* represent results for $I_{ph}(0) = 0.75 \times 10^{-3}$ mol$_{ph}$ m^{-2} s^{-1}; *dashed lines* represent results for $I_{ph}(0) = 1.5 \times 10^{-3}$ mol$_{ph}$ m^{-2} s^{-1}. Photobioreactor: $d = 0.03$ m. Parameter values are based on high-light acclimated *Chlorella sorokiniana* and a sunlight spectrum: $a_x = 3.33$ m^2 mol$_x^{-1}$; $E_{n,\lambda}$ for sunlight; $Y^c_{s/ph,m} = 0.10$ mol$_s$ mol$_{ph}^{-1}$; $q^c_{s,m} = 1.25 \times 10^{-4}$ mol$_s$ mol$_x^{-1}$ s^{-1}; $Y_{x/s} = 0.625$ mol$_x$ mol$_s^{-1}$; $m_s = 3.0 \times 10^{-6}$ mol$_s$ mol$_x^{-1}$ s^{-1}.

This relation can be used, for example, to calculate volumetric productivity of a photobioreactor at different biomass concentrations C_x when assuming a constant incident photon flux density $I_{ph}(0)$. In Fig. 26 results are shown for a photobioreactor with an optical depth of 3 cm. The shape of this relation will be discussed later after having introduced a similar approach based on the model of Jassby and Platt.

3.2.2 Integrating Photosynthesis Along Optical Path According to Jassby and Platt

The model of Jassby and Platt results in a more accurate description of the relation between the specific rate of photosynthesis (sugar production) and photon flux density. Again the average specific sugar production rate of the microalgal culture must be calculated in order to assess the volumetric productivity of a photobioreactor. The following integral therefore must be calculated:

$$\overline{q_s^c} = \frac{\int_0^d q_s^c(z) \cdot dz}{d}$$

$$q_s^c(z) = q_{s,m}^c \cdot \tanh\left(\frac{Y_{s/ph,m}^c \cdot a_x \cdot I_{ph}(z)}{q_{s,m}^c}\right)$$

The integral in this calculation routine cannot be easily solved, especially in the situation when also spectral dependence of light absorption must be taken into account. In that case we also have to include the relation below:

$$I_{ph}(z) = \sum_{400\,nm}^{700\,nm} I_{ph,\lambda}(0) \cdot e^{-a_{x,\lambda} \cdot C_x \cdot z} \cdot \Delta\lambda$$

For this reason, such problems nowadays are solved numerically using computational power. Already in simple spreadsheet programs such problems can be accurately solved by subdividing the photobioreactor in a large but finite number of layers and following below routine:

1.

$$I_{ph,\lambda}(0) = I_{ph}(0) \cdot E_{n,\lambda}$$

2.

$$I_{ph}(z) = \sum_{400\,nm}^{700\,nm} I_{ph,\lambda}(0) \cdot e^{-a_{x,\lambda} \cdot C_x \cdot z} \cdot \Delta\lambda$$

3.

$$q_{ph}(z) = \frac{I_{ph}(z + {}^1/_2\Delta z) - I_{ph}(z - {}^1/_2\Delta z)}{C_x \cdot \Delta z}$$

4.

$$q_s^c(z) = q_{s,m}^c \cdot \tanh\left(\frac{Y_{s/ph,m}^c \cdot - q_{ph}(z)}{q_{s,m}^c}\right)$$

5.

$$\overline{q_s^c} = \frac{\sum_{n=1}^{n=N} q_s^c(z) \cdot \Delta z}{d}$$

In this approach it is assumed that q_{ph}, and $q_{\mathrm{s}}^{\mathrm{c}}$ are constant over the finite layer Δz. This assumption is only valid when the number of layers N is sufficiently large and Δz sufficiently small. In practice 100 layers within the photic zone proves to be accurate. The photic zone is defined as that part of the microalgal culture where the specific rate of sugar production in the chloroplast $q_{\mathrm{s}}^{\mathrm{c}}$ is higher than the maintenance requirement for sugar m_{s}. In this approach it is practical to express the specific rate of photosynthesis $q_{\mathrm{s}}^{\mathrm{c}}$ based on the local specific photon consumption rate q_{ph}, and not based on the local photon flux density I_{ph}. The local specific photon consumption rate q_{ph} can be calculated by setting up a light balance over the finite layer Δz. Having calculated the average specific rate of photosynthesis, the average specific growth rate can be calculated, and from that volumetric productivity of a photobioreactor.

3.2.3 Volumetric Productivity vs Biomass Concentration

In Fig. 26 the volumetric productivity r_{x} is presented as a function of the biomass concentration C_{x} for a photobioreactor with a 3-cm optical path. Results are shown both for the approach based on Blackman without spectral resolution and Jassby and Platt including spectral resolution.

Both models yield different numbers but the general trend is identical. Photobioreactor productivity is highest at a certain biomass concentration which will be called the optimal biomass concentration $C_{\mathrm{x,opt}}$. When the biomass concentration is lower than this optimal value light will pass the reactor unused. When the biomass concentration is higher than optimal the photon flux density in part of the reactor volume is below the compensation point of photosynthesis. In this part of the reactor intracellular sugar will be respired by the microalgae in order to fulfill maintenance requirements. This photosynthetically derived sugar is not available anymore for biomass growth and potential productivity is lost. This effect will be analyzed in more detail below based on the Blackman model.

The results of the calculations based on Blackman without spectral resolution and Jassby and Platt approach with spectral resolution are different. This is partly related to the hyperbolic tangent function which more accurately describes the change from light-limited to light-saturated photosynthesis. The Blackman model overestimates photosynthetic rate in the transition from light limitation to light saturation (200–750 $\mu\mathrm{mol_{ph}}\ \mathrm{m}^{-2}\ \mathrm{s}^{-1}$, see Fig. 19). This explains the fact that at 750 $\mu\mathrm{mol_{ph}}\ \mathrm{m}^{-2}\ \mathrm{s}^{-1}$ incident light productivity is higher according to Blackman (Fig. 26). In addition, spectral effects are important. At a higher photon flux density of 1.5×10^{-3} $\mathrm{mol_{ph}}\ \mathrm{m}^{-2}\ \mathrm{s}^{-1}$ the Jassby and Platt approach results in a higher productivity

(Fig. 26). This approach takes into account the change of spectrum when sunlight passes through the microalgal culture. The light loses the blue and red wavelengths faster than the green wavelengths. This results in an effective decrease of the optical cross section a_x of the microalgae and a reduced level of light saturation and increased volumetric productivity.

The model calculations presented in Fig. 26 were based on a high-light acclimated culture. In reality, at higher biomass concentration the microalgal culture as a whole will experience light limitation. Consequently, on a timescale of hours the microalgae will increase their pigmentation and their optical cross section (i.e., specific light absorption coefficient). The dynamics of photoacclimation processes and its effect on photobioreactor productivity will be discussed later in more detail, but it will alter the shape of the curves presented in Fig. 26 to some extent.

The existence of a point of optimal productivity can be analytically deduced employing the Blackman model without spectral resolution. The relation between volumetric productivity and biomass concentration plotted in Fig. 26 is given by the following relation already introduced:

$$r_x = \frac{Y_{x/s}}{d} \cdot \left\{ \frac{q^c_{s,m}}{a_x} \cdot \left(1 + \mathrm{Ln}\left(\frac{I_{ph}(0) \cdot a_x \cdot Y^c_{s/ph,m}}{q^c_{s,m}} \right) \right) - Y^c_{s/ph,m} \cdot I_{ph}(0) \cdot e^{-a_x \cdot C_x \cdot d} \right\} - Y_{x/s} \cdot m_s \cdot C_x$$

The function is concave meaning that the point of maximal productivity is characterized by the derivative of function $r_x = f(C_x)$ being equal to zero.

$$f'(C_x) = Y_{x/s} \cdot Y^c_{s/ph,m} \cdot a_x \cdot I_{ph}(0) \cdot e^{-a_x \cdot C_x \cdot d} - Y_{x/s} \cdot m_s$$

$$\Rightarrow 0 = Y_{x/s} \cdot Y^c_{s/ph,m} \cdot a_x \cdot I_{ph}(0) \cdot e^{-a_x \cdot C_x \cdot d} - Y_{x/s} \cdot m_s$$

$$\Leftrightarrow \frac{m_s}{Y^c_{s/ph,m} \cdot a_x} = I_{ph}(0) \cdot e^{-a_x \cdot C_x \cdot d}$$

This relation is equivalent to: $I_{ph,c} = I_{ph}(d)$

The outcome shows that at the point of maximal volumetric productivity the photon flux density at the "backside" of the photobioreactor $I_{ph}(d)$ is equal to the compensation point of photosynthesis $I_{ph,c}$ which was derived earlier:

$$I_{ph,c} = \frac{m_s}{Y^c_{s/ph,m} \cdot a_x}$$

Given a certain photon flux density $I_{ph}(0)$ at the light-exposed surface the optimal biomass concentration $C_{x,opt}$ can now be estimated based on the Blackman approach:

$$I_{ph}(d) = I_{ph,c} \quad \text{and} \quad I_{ph}(d) = I_{ph}(0) \cdot e^{-a_x \cdot C_{x,opt} \cdot d}$$

Combining these relations gives:

$$C_{x,opt} = \frac{1}{a_x \cdot d} \cdot \mathrm{Ln}\left(\frac{I_{ph}(0)}{I_{ph,c}}\right)$$

Also when employing more realistic and more complex models, such as Jassby and Platt with spectral resolution the same rule applies: "volumetric productivity of a photobioreactor is maximal when the biomass density is such that the photon flux density at the darkest zone of the reactor is equal to the compensation point of photoautotrophic growth." In practice it has also been confirmed that photobioreactors can be operated at maximal productivity when the biomass concentration is maintained at those levels that almost all light is absorbed within the microalgal culture but at the back still a minimal amount of light is left to compensate for maintenance purposes (Barbera et al, 2015; Takache et al, 2012; Zijffers et al, 2010).

3.2.4 Volumetric Productivity and Biomass Yield on Photons vs Photon Flux Density

Another interesting aspect about photobioreactor performance is the relation between volumetric productivity r_x and incident photon flux density $I_{ph}(0)$. Both the Blackman model without spectral resolution and the Jassby and Platt model with spectral resolution can be used to quantify this relation. The result is shown in Fig. 27. At all photon flux densities simulated the optimal biomass concentration was used, which needed to be calculated first. The volumetric productivity increases with increasing incident photon flux density and, consequently, light is limiting growth even at high photon flux density. The increase, however, is not linear and the rate of increase slowly levels off. This is a consequence of increased light saturation close to the reactor surface where light levels are far above the saturating photon flux density.

This trend of r_x vs $I_{ph}(0)$ can be better understood by calculating the efficiency of light use and thus calculating the biomass yield on photons $Y_{x/ph}$. For this we need the specific photon consumption rate q_{ph}. The q_{ph} can be estimated from the light balance over the photobioreactor if we assume that

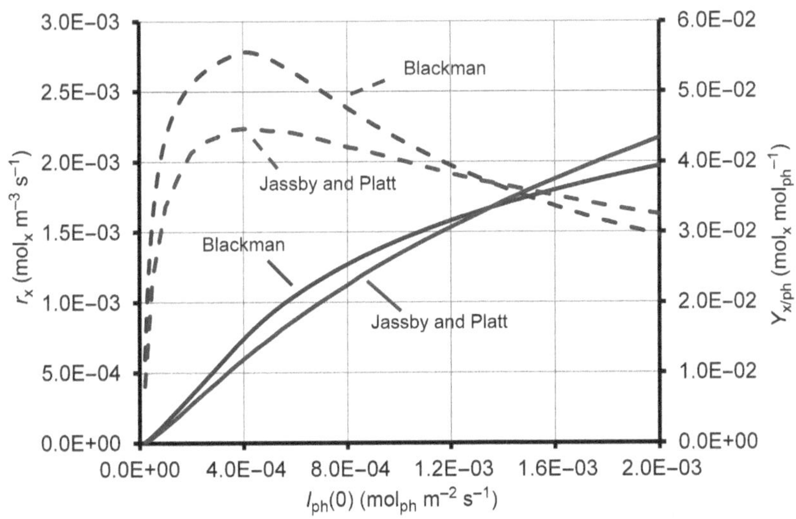

Figure 27 Volumetric productivity of a photobioreactor r_x as a function of photon flux density $I_{ph}(0)$ at the point of maximal productivity ($C_x = C_{x,opt}$) according to Blackman (*solid red* (*dark gray* in the print version) *line*) and Jassby and Platt (*solid blue* (*dark gray* in the print version) *line*). In addition, biomass yield on photons $Y_{x/ph}$ is shown as a function of $I_{ph}(0)$ according to Blackman (*dashed red* (*dark gray* in the print version) *line*) and Jassby and Platt (*dashed blue* (*dark gray* in the print version) line). Photobioreactor: $d = 0.03$ m. Parameter values are based on high-light acclimated *Chlorella sorokiniana* and a sunlight spectrum: $a_x = 3.33$ m^2 mol$_x^{-1}$; $E_{n,\lambda}$ for sunlight; $Y^c_{s/ph,m} = 0.10$ mol$_s$ mol$_{ph}^{-1}$; $q^c_{s,m} = 1.25 \times 10^{-4}$ mol$_s$ mol$_x^{-1}$ s^{-1}; $Y_{x/s} = 0.625$ mol$_x$ mol$_s^{-1}$; $m_s = 3.0 \times 10^{-6}$ mol$_s$ mol$_x^{-1}$ s^{-1}.

all the light incident on the reactor surface is absorbed by the microalgal biomass inside the reactor. This results in the following relation:

$$-q_{ph} = \frac{I_{ph}(0) \cdot A_{r,\text{light}}}{V_r \cdot C_{x,\text{opt}}} \left(\text{mol}_{ph}\,\text{mol}_x^{-1}\,\text{s}^{-1}\right)$$

The word "estimated" is used because at the back of the reactor a small amount of light will leave the system since $I_{ph}(d) = I_{ph,c} > 0$. This light in fact is not useful since it is less than needed to compensate for maintenance. The biomass yield on photons $Y_{x/ph}$ then can be calculated as follows:

$$Y_{x/ph} = \frac{\bar{\mu}}{-q_{ph}} \left(\text{mol}_x\,\text{mol}_{ph}^{-1}\right)$$

The result of this calculation is presented as well in Fig. 27. The biomass yield on photons is maximal at a photon flux density of about 400×10^{-6}

mol m^{-2} s^{-1} in this example. Above this light level the yield continuously decreases with increasing $I_{ph}(0)$ because of light saturation. At photon flux density lower than 400×10^{-6} mol m^{-2} s^{-1} the yield decreases because of the fact that an increasing part of the light absorbed is used for maintenance purposes and not for growth.

Clearly, light is used most efficiently when the photon flux density at the reactor surface is not too high, and not too low. A light level of 400×10^{-6} mol m^{-2} s^{-1} appears to be optimal for the example illustrated in Fig. 27 which is based on the green microalgae *Chlorella sorokiniana* with a high maximal specific growth rate. Moreover it is assumed that this microalga is high light acclimated. According to the more accurate Jassby and Platt approach a maximal biomass yield of 0.045 mol_x mol_{ph}^{-1} can be expected. This can be recalculated to 1.08 g mol_{ph}^{-1} assuming a biomass molar weight of 24 g mol_x^{-1}. In lab-scale experiments yields as high as 1.25 g mol_{ph}^{-1} have been measured (Cuaresma et al, 2011), which might be related to the relatively high estimate of the maintenance requirement in the model calculations. In case the maintenance is reduced from 3.0 to 1.0×10^{-6} mol_s mol_x s^{-1} a yield of 1.24 g mol_{ph}^{-1} can be simulated.

The two different model approaches, Blackman without spectral resolution and Jassby and Platt with spectral resolution, again lead to different results (Fig. 27). This aspect was already covered when discussing the results presented in Fig. 26. The Blackman model overestimates the photosynthetic rate in the transition from light limitation to light saturation leading to a higher volumetric productivity and biomass yield on photons. Only at photon flux density above 1.4×10^{-3} mol m^{-2} s^{-1} the Jassby and Platt approach results in higher productivity and biomass yield on photons. This is related to the incorporation of the spectral differences of specific light absorption. This results in an effective decrease of the optical cross section a_x of the microalgae and a reduced level of light saturation when light travels through the microalgal culture. This effect becomes dominant only at high photon flux density.

3.3 Microalgal Cultivation in Photobioreactors: Photobioreactor Operation

Microalgae can be grown batchwise in photobioreactors. In case biomass concentrations become too high the culture is diluted, and biomass is harvested, after which growth is allowed to continue again. Microalgae production systems can also be operated in a continuous manner based on a chemostat or turbidostat mode of operation. Here these three different operation modes will be shortly addressed. The biomass balance over the photobioreactor serves as a starting point of this evaluation:

$$V_r \cdot \frac{dC_x}{dt} = F_{in} \cdot C_{x,in} - F_{out} \cdot C_{x,out} + r_x \cdot V_r$$

3.3.1 Batch Operation

In a batch-wise cultivation there is no dilution with water and nutrients and the biomass balance can be simplified to the following relation:

$$\frac{dC_x}{dt} = \overline{\mu} \cdot C_x \Rightarrow \frac{1}{C_x} \cdot dC_x = \overline{\mu} \cdot dt$$

This relation can be integrated from time *0* to t provided the specific growth rate is constant and not dependent on the biomass concentration. The average specific growth rate in a microalgal culture in a photobioreactor, however, is strongly dependent on the biomass concentration. This dependency is expressed in below relation which was already derived from the Blackman model.

$$\overline{\mu} = Y_{x/CH_2O} \cdot \left\{ \frac{z_s}{d} \cdot q^c_{CH_2O,m} + \frac{Y^c_{CH_2O/ph,m} \cdot I_{ph}(0)}{d \cdot C_x} \left(e^{-a_x \cdot C_x \cdot z_s} - e^{-a_x \cdot C_x \cdot d} \right) \right\} - Y_{x/CH_2O} \cdot m_{CH_2O}$$

Clearly, no simple analytical solution of the differential equation derived from the biomass balance can be obtained. Adopting numerical methods based on this differential equation the increase in biomass in time in a batch culture can still be calculated. In Fig. 28 batch growth is simulated based on the model of Jassby and Platt including spectral resolution. This simulation is based on *C. sorokiniana* grown under a constant photon flux density of 1.5×10^{-3} mol_{ph} m^{-2} s^{-1}. The biomass concentration increases in time, but the rate of increase slows down. This is more apparent in the yield of biomass on photons which is maximal after about 24 h after which it decreases again. At 24 h the biomass concentration is optimal in the sense that at this point in time the volumetric productivity, as well as the biomass yield on light, is maximal.

The effect of photoacclimation is included in an arbitrary manner in this simulation (Fig. 28). After having reached the maximal productivity after 24 h it was assumed that the optical cross section a_x decreased linearly in time

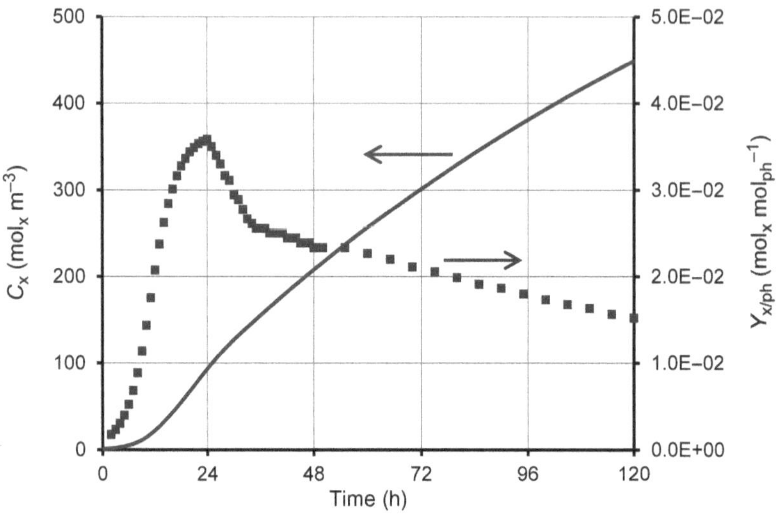

Figure 28 Simulated biomass concentration (*solid blue* (*dark gray* in the print version) *line*) and biomass yield on photons (*red* (*dark gray* in the print version) *square* markers) during batch growth of *Chlorella sorokiniana*. The biomass balance was solved based on a numerical routine taking fixed time steps of 1 min. Simulation was based on the Jassby and Platt model including spectral resolution. Photobioreactor: $d = 0.03$ m; $I_{ph}(0) = 1.5 \times 10^{-3}$ $mol_{ph}\ m^{-2}\ s^{-1}$. Parameter values: $a_x = 3.33$ $m^2\ mol_x^{-1}$; $E_{n,\lambda}$ for sunlight; $Y^c_{s/ph,m} = 0.10$ $mol_s\ mol_{ph}^{-1}$; $q^c_{s,m} = 1.25 \times 10^{-4}$ $mol_s\ mol_x^{-1}\ s^{-1}$; $Y_{x/s} = 0.625$ $mol_x\ mol_s^{-1}$; $m_s = 3.0 \times 10^{-6}$ $mol_s\ mol_x^{-1}\ s^{-1}$. Photoacclimation was simulated in an arbitrary manner: after having reached the maximal productivity after 24 h it was assumed that the specific light absorption coefficient a_x decreased linearly in time from the minimal level of 3.33 $m^2\ mol_x^{-1}$ to a maximal level of 6.66 $m^2\ mol_x^{-1}$ in a period of 10 h.

from its minimal level to its maximal level in a period of 10 h. Such time frame is in accordance with recent results obtained for *C. sorokiniana* in the Bioprocess Engineering group of Wageningen University and an example will be given later in this chapter. This simulated process of photoacclimation results in a rapid decline of the biomass yield on light and thus photobioreactor volumetric productivity in the 10-h period after having reached the maximal productivity. This trend is confirmed by published work of the same group (Kliphuis et al, 2010, 2011a) in which clear peaks in the oxygen and carbon dioxide exchange rates were observed during batch growth in a photobioreactor. This trend also confirms the expectation that low-pigmented microalgae yield higher photobioreactor productivity and higher biomass yield on light (Formighieri et al, 2012; Sukenik et al, 1987).

When operating photobioreactors or raceway ponds in batch mode a compromise must be found between achieving a high volumetric productivity (i.e., high biomass yield on light) or a high biomass concentration. A high biomass concentration results in reduced costs for processing of the microalgal suspension as less water needs to be removed and treated. A high volumetric productivity of the photobioreactor guarantees efficient use of money invested in the production system. Models describing microalgal growth as a function of microalgal biomass concentration as discussed here must be used to select the most economical and/or most sustainable operation process for photobioreactors.

3.3.2 Chemostat and Turbidostat Operation Under Continuous Light

Microalgae can also be grown continuously meaning that the microalgal culture is continuously harvested and the liquid volume removed is continuously replaced with fresh water with nutrients. In the case of chemostat operation the dilution rate is fixed. Assuming a constant photon flux density a steady state will be reached where the biomass concentration does not change anymore and is constant. The influent, water with nutrients, usually does not contain any microalgae. Furthermore, the liquid volume is usually maintained constant, so $F_{in} = F_{out} = F$. Finally, it will be assumed that the liquid inside the photobioreactor is perfectly mixed, so $C_{x,out} = C_x$. Then the biomass balance over the photobioreactor can be simplified as follows:

$$F \cdot C_x = \bar{\mu} \cdot C_x \cdot V_r$$
$$\Leftrightarrow F = \bar{\mu} \cdot V_r$$
$$\Leftrightarrow \bar{\mu} = \frac{F}{V_r} = D\left(s^{-1}\right)$$

The parameter D is the dilution rate with unit s^{-1}. The inverse of the dilution rate is the hydraulic retention time (HRT) of the liquid inside the photobioreactor with unit s. It is the average residence time of the liquid and microalgae inside the photobioreactor

$$\Leftrightarrow \frac{1}{D} = \frac{V_r}{F} = \text{HRT}\,(s).$$

When operating a photobioreactor as a chemostat the dilution rate is fixed and the microalgae are forced to grow with a specific growth rate equaling the dilution rate. The optimal dilution rate can be selected based on the relations derived before coupling the average specific growth and the biomass concentration in a photobioreactor. Please note that the product of biomass concentration and specific growth rate gives the volumetric productivity of the photobioreactor. Also in case of chemostat operation a balance needs to be found between achieving a high volumetric productivity of the photobioreactor and a high biomass concentration in the outflow.

A turbidostat mode of operation relies on the use of a sensor to measure the biomass concentration in the photobioreactor. Such a sensor can be based on the online measurement of culture turbidity, for example, but any other biomass sensor can be used as well. Based on such a continuous biomass measurement the flow rate through the system then is automatically adjusted such that the biomass concentration is maintained at a preselected value (Fig. 29).

The biomass balance over the photobioreactor then results in the same relation as the one derived for the chemostat:

$$\overline{\mu} = \frac{F}{V_r} = D$$

What happens is that the biomass concentration is fixed and determines the light gradient in the system. The local photon flux density then determines the local specific rates of photosynthesis resulting in an average specific growth rate when photosynthesis is integrated over the whole reactor volume. The biomass concentration thus determines the average specific growth rate and the turbidostat control forces the dilution rate D to become

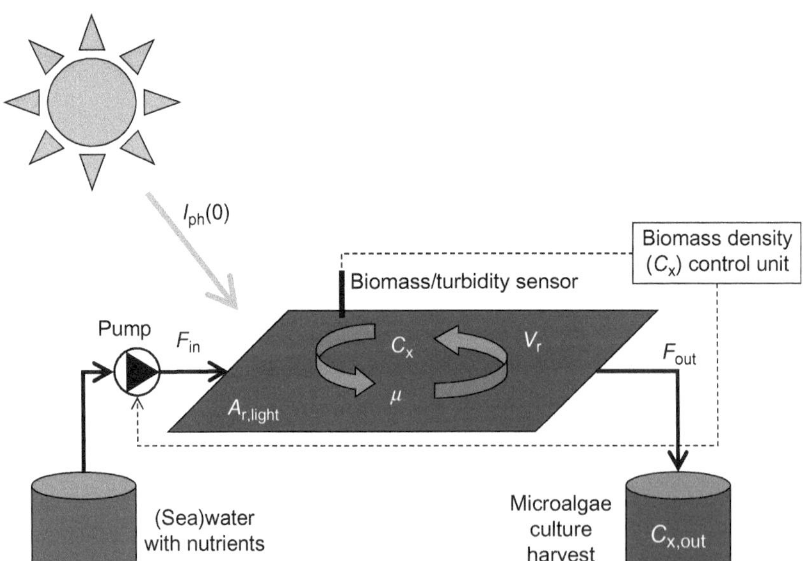

Figure 29 Schematic of turbidostat mode of operation of a microalgae cultivation system.

equal to the specific growth rate. The advantage of a turbidostat control is that the biomass concentration can be set, and maintained, at a preferred value. The actual setting for the biomass concentration can be selected based on the relations derived before coupling the average specific growth rate and the biomass concentration in a photobioreactor.

3.3.3 Chemostat and Turbidostat Operation Under Day/Night Cycles

Under outdoor conditions under day/night cycles biomass accumulation inside the photobioreactor cannot be neglected. Especially in case of chemostat operation considerable changes in biomass concentration will occur. Even in a turbidostat in the night biomass concentrations will decrease because of respiration. The biomass balance then must be written as follows:

$$\frac{\mathrm{d}C_{\mathrm{x}}}{\mathrm{d}t} = \overline{\mu} \cdot C_{\mathrm{x}} - D \cdot C_{\mathrm{x}}$$

This differential equation can be solved numerically to analyze the performance of chemostat or turbidostat cultures under day/night cycles and results are shown in Fig. 30. The average specific growth rate is calculated based on the integration of local photosynthetic rates employing the Jassby and Platt model for photosynthesis and the Lambert–Beer model for light penetration including the spectral distribution of sunlight and light absorption. It must be emphasized that this modeling approach is not optimized yet for microalgal growth under day/night cycles as it does not include the storage of carbohydrates (starch) during daytime and their use during the night. In the night microalgal growth can continue based on the use of these carbohydrate reserves. More specifically, microalgae synchronize their cell cycle with the day/night cycle and cell division takes place at the end of the day or the beginning of the night (Vítová et al, 2015). In the simplified modeling approach introduced in this chapter it is assumed that all sugar produced in photosynthesis is immediately used in the light, thus in daytime. In the night only maintenance-based sugar consumption is included. In reality part of the growth-related sugar consumption is taking place in the night as these sugars are stored as starch during the day. The simulations shown in Fig. 30 therefore overestimate biomass growth during the day period to some extent, but biomass loss in the night is underestimated. Both effects are assumed to cancel out over the whole day/night period.

For both the chemostat and turbidostat simulations the photobioreactor was not diluted during the night. In the chemostat and turbidostat culture

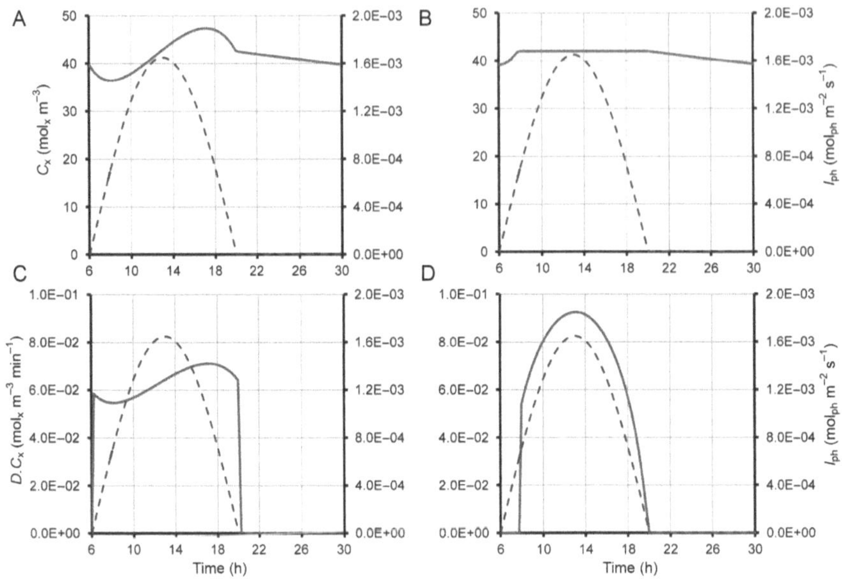

Figure 30 Simulations of chemostat (A and C) and turbidostat (B and D) operation under day/night cycles for the production of *Chlorella sorokiniana*. In *upper* graphs (A and B) the biomass concentration C_x is shown (*solid blue* (*dark gray* in the print version) *lines*), and in the *lower* graphs (C and D) the reactor productivity is shown (*solid blue* (*dark gray* in the print version) *lines*) expressed as $D{\cdot}C_x$. In all graphs the photon flux density is shown as a reference (*dashed red* (*gray* in the print version) *lines*). Light during the day was simulated with a sine curve representing a clear-sky summer day: $I_{ph,max} = 1.65 \times 10^{-3}$ mol$_{ph}$ m^{-2} s^{-1}; $t_{day} = 14$ h. The biomass balance was solved based on a numerical routine taking fixed time steps of 1 min. Simulation was based on the Jassby and Platt model including spectral resolution. The optical path in Lambert–Beer equation was corrected with a factor 1.5 to simulate the effect of nonperpendicular light entry along the day, and the effect of light scattering. Photobioreactor: $d = 0.03$ m. Setpoint chemostat: $D = 0.09$ h^{-1} (day period only). Setpoint turbidostat: $C_x = 42$ mol$_x$ m^{-3} (day period only). Parameter values: $a_x = 3.33$ m^2 mol$_x^{-1}$; $E_{n,\lambda}$ for sunlight; $Y^c_{s/ph,m} = 0.10$ mol$_s$ mol$_{ph}^{-1}$; $q^c_{s,m} = 1.25 \times 10^{-4}$ mol$_s$ mol$_x^{-1}$ s^{-1}; $Y_{x/s} = 0.625$ mol$_x$ mol$_s^{-1}$; $m_s = 3.0 \times 10^{-6}$ mol$_s$ mol$_x^{-1}$ s^{-1}.

the biomass concentration decreases in the night due to maintenance losses. During the day in the chemostat culture the biomass concentration first decreases to a minimal level of 36.4 mol$_x$ m^{-3} because the rate of dilution is higher than the specific growth rate at these low light levels. When the photon flux density increases growth catches up and the biomass concentration increases to a maximum of 47.3 mol$_x$ m^{-3}. Only at the end of the afternoon the biomass concentration decreases again due to decreasing sunlight levels. In the turbidostat culture the biomass concentration remains constant throughout the day period. The biomass harvested from the chemostat ($D \cdot C_x$, unit $mol_x\ m^{-3}\ s^{-1}$) varies together with the biomass concentration. In the turbidostat the harvest is even more variable since the harvest directly follows the dilution rate which depends on the specific growth rate which is dictated again by the sunlight intensity.

Photoacclimation and a change in microalgal optical cross section were not included in these simulations. In these examples the biomass concentrations were relatively low ($42\ mol_x\ m^{-3} = 1.0\ g\ L^{-1}$) for a photobioreactor with a 3-cm optical path. It was therefore assumed that the microalgae were high light acclimated. The overall daily productivity simulated under these conditions was $39\ g\ m^{-2}\ day^{-1}$ for both operational strategies corresponding to a biomass yield on light ($Y_{x/ph}$) of $0.74\ g\ mol^{-1}$. These values correspond well to data obtained for *C. sorokiniana* under similar conditions in laboratory-scale experiments (Cuaresma et al, 2011; Zijffers et al, 2010).

Despite the fact that productivity of chemostat operation is similar to turbidostat operation the turbidostat is preferred especially at locations with more day-to-day variation in irradiance. Under chemostat operation biomass concentration will slowly decrease on consecutive days with cloud cover. If followed by a clear-sky day, the culture is vulnerable to photoinhibition because of its low biomass concentration. Turbidostat operation allows for a direct control of biomass concentration. Washout of a microalgae culture in a turbidostat is by definition not possible. In a chemostat this could occur in situations the dilution rate is not adjusted in time. Turbidostat operation is thus more robust and allows for reliable automation of outdoor microalgae production (Bosma et al, 2014).

4. IMPROVING THE ESTIMATION OF PHOTOBIOREACTOR PRODUCTIVITY

4.1 Understanding and Prediction Photoacclimation

Photoacclimation has already been discussed. It has been shown that microalgae can adjust their optical cross section as a response to a change in light regime. It is, however, complicated to predict this photoacclimation response. Excellent attempts have been made by several research groups (García-Camacho et al, 2012; Geider et al, 1998; Ross and Geider, 2009), but these result in complex models requiring the fitting of additional model parameters. These models are defined cell specific in order to take into account cell history and they will require recalculation to metrics required on a photobioreactor level. Specifically in the situation of the diurnal change in light conditions modeling, or predicting, photoacclimation is complicated because the timescale of the change in light level is in the same order as the timescale of the photoacclimation response. As an example the change in optical cross section (i.e., specific light absorption coefficient) is shown in Fig. 31 as was measured for *C. sorokiniana* upon a shift from high light condition to low light conditions. Within 8 h after the shift specific light absorption already approached that of low-light acclimated microalgae.

Although it is challenging to incorporate photoacclimation in photobioreactor production models it is important for future improvement of microalgae production strategies. The size of the optical cross section (i.e., the specific light absorption coefficient) determines specific light absorption and the specific rate of photosynthesis. As such, also the specific growth rate and the light-use efficiency strongly depend on the specific light absorption coefficient of the microalgae. This is exemplified for the green microalga *C. sorokiniana* in Fig. 32. Cells with a large specific light absorption coefficient are photosaturated at lower photon flux density and express a lower biomass yield on photons at high photon flux density when compared

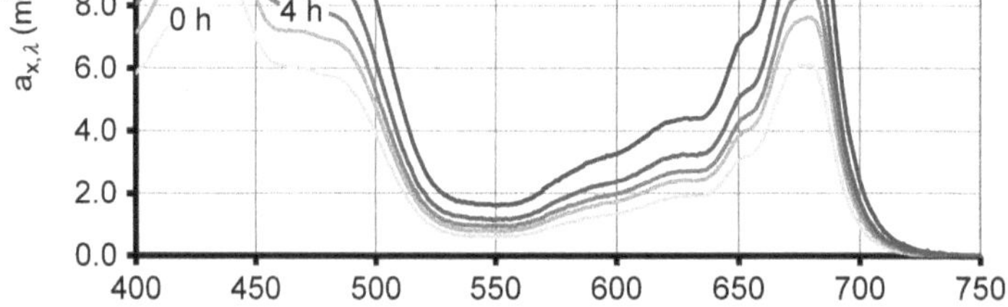

Figure 31 Dynamics of photoacclimation: change of the specific light absorption coefficient $a_{x,\lambda}$ as a response to a shift from high light to low light. The example given is based on the green microalga *Chlorella sorokiniana*. The microalgae were grown in a 14-mm plate-type photobioreactor in turbidostat mode with ingoing photon flux density of 980×10^{-6} mol_{ph} m^{-2} s^{-1} and an outgoing photon flux density of 430×10^{-6} mol_{ph} m^{-2} s^{-1}. At $t=0$ h the ingoing photon flux density was decreased stepwise to 60×10^{-6} mol_{ph} m^{-2} s^{-1} and the specific absorption coefficient was measured after 2, 4, 8, and 28 h.

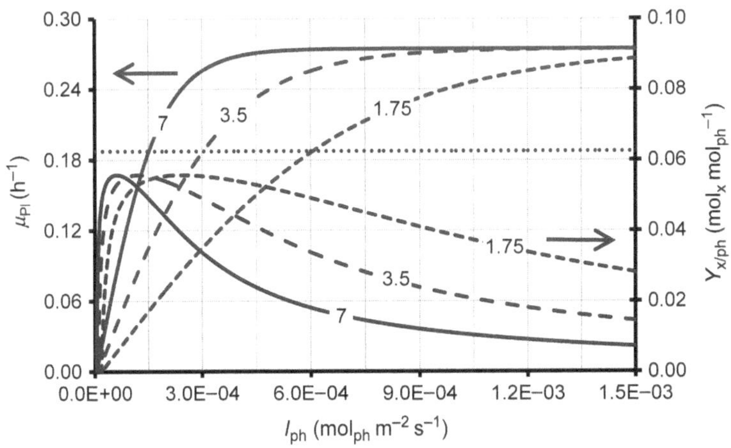

Figure 32 Instantaneous specific growth rate μ_{PI} and biomass yield on photons $Y_{x/ph}$ as a function of photon flux density I_{ph} for different values of the specific light absorption coefficient a_x. The model of Jassby and Platt was used and spectral effects were neglected, thus assuming a constant a_x. The *dotted red* (*dark gray* in the print version) *line* represents the hypothetical maximal biomass yield on photons $Y_{x/ph,m}$ ($m_s = 0$). Parameter values based on *Chlorella sorokiniana*: $Y^c_{s/ph,m} = 0.10$ mol_s mol_{ph}^{-1}; $q^c_{s,m} = 1.25 \times 10^{-4}$ mol_s mol_x^{-1} s^{-1}; $Y_{x/s} = 0.625$ mol_x mol_s^{-1}; $m_s = 3.0 \times 10^{-6}$ mol_s mol_x^{-1} s^{-1}; $M_x = 24$ g mol_x^{-1}. Values of a_x given in figure: 7.0 m^2 mol_x^{-1} (low light acclimated); 3.5 m^2 mol_x^{-1} (high light acclimated); 1.75 m^2 mol_x^{-1} (hypothetical antenna size mutant).

to cells with small absorption coefficient. In Fig. 32 trends are shown for high-light acclimated cells (7.0 m^2 mol_x^{-1}) and low-light acclimated cells (3.5 m^2 mol_x^{-1}). In addition, a model prediction is included reflecting a hypothetical mutant strain with a reduced antenna size ($a_x = 1.75$ m^2 mol_x^{-1}). Specifically the latter "strain" is interesting because it only reaches its maximal specific growth rate at 1.5×10^{-3} mol_{ph} m^{-2} s^{-1} and still expresses a good biomass yield on photons (≈50% of maximal yield).

The area in Fig. 32 enclosed by the curve of $Y_{x/ph}$ vs I_{ph} and the x-axis ($Y_{x/ph} = 0$) represents the areal productivity of a photobioreactor P_x:

$$P_x = \int_{I_{ph}(0)}^{I_{ph}(d)} Y_{x/ph} \cdot dI_{ph} \left(mol_x m^{-2} s^{-1}\right)$$

This is illustrated in more detail in Fig. 33. The point on the x-axis where the $Y_{x/ph}$ trend crosses ($Y_{x/ph} = 0$) represents the compensation point of photosynthesis $I_{ph,c}$ and this would be the optimal photon flux density at the backside of a photobioreactor with a depth d, $I_{ph}(d)$. On the right hand side the area is not enclosed. In this specific example the x-axis runs until

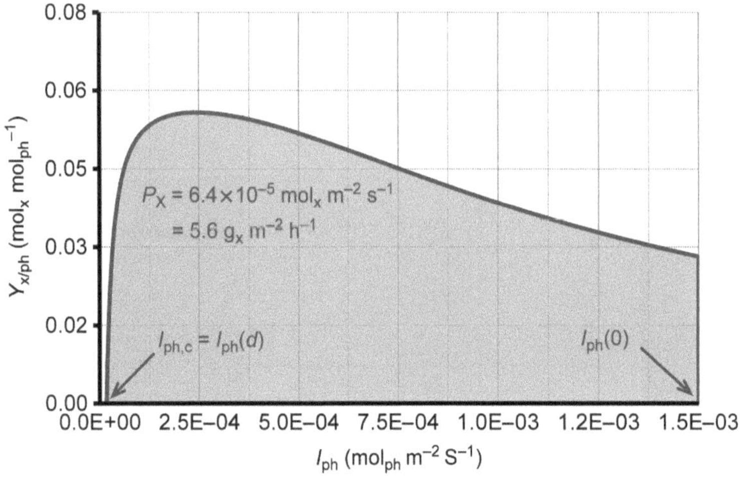

Figure 33 Illustration of the graphical estimation of photobioreactor productivity expressed per m^2 of illuminated surface. Graph is based on hypothetical antenna size mutant with $a_x = 1.75$ m^2 mol$_x^{-1}$. See Fig. 32 for more details.

1500×10^{-6} mol$_{ph}$ m^{-2} s^{-1}. Taking the area up to this photon flux density would mean assuming the photobioreactor to be exposed to an ingoing photon flux density of 1.5×10^{-3} mol$_{ph}$ m^{-2} s^{-1}, $I_{ph}(0)$. This analysis only holds for flat systems and neglects spectral effects with respect to light absorption. Nevertheless comparing these surfaces underneath the $Y_{x/ph}$ curves for different values of the specific light absorption coefficient (Fig. 32) shows the potential effect of reduced antenna size on photobioreactor productivity. Productivity could be more than doubled when the antenna size of the photosystems is reduced with a factor 4 in comparison to the low-light acclimated microalgae.

Antenna size reduction unfortunately is not trivial and current antenna size mutants of *Chlamydomonas reinhardtii*, for example, were not able to outperform the wild type (De Mooij et al, 2014). It is important to stress that only light absorption should be reduced and that the capacity to process photons and perform photosynthesis should remain unaltered. In other words, the number of photosystems must remain maximal and only the size of the antenna complex of the photosystems should be reduced (Formighieri et al, 2012; Wobbe and Remacle, 2014).

4.2 Mixing-Induced Light/Dark Cycles in Photobioreactors

In the simplified model approach to predict photobioreactor productivity adopted in this chapter it was assumed that microalgae immediately respond to the local photon flux density. It was assumed that the specific rate of photosynthesis at any point in time and space is only dependent on the actual photon flux density at that point in time, and at that position within the photobioreactor. Of course this is a simplification of reality and its implications will be qualitatively discussed here.

Mixing of the microalgal suspension in photobioreactors results in a movement of individual microalgal cells over the light gradient resulting in light/dark cycles. The timescale of these events depends on reactor design and reactor operation. An evaluation of studies based on CFD modeling shows that for typical photobioreactors shapes (tubes, plates, and columns) the average cycle time is in the order of seconds (Gómez-Pérez et al, 2015; Moberg et al, 2011; Olivieri et al, 2015). Only when including static mixers, or very short optical paths in combination with high liquid velocities, light/dark cycles between 0.1 and 1 s can be obtained (Moberg et al, 2011; Perner-Nochta and Posten, 2007; Zhang et al, 2012). This type of mixing over the light gradient of a microalgal culture in a photobioreactor could affect photosynthesis.

4.2.1 The Flashing Light Effect—Light Dilution

On a very short timescale it is hypothesized that microalgae can store the reducing power liberated during photosynthesis within the electron carriers associated to the photosynthetic electron transport chain (Vejrazka et al, 2011): plastoquinone, cytochrome, plastocyanin, ferredoxin, and NADP. In addition, power can be stored by means of the electrochemical proton gradient generated over the thylakoid membranes. The capacity of this battery will only suffice to "store" high light flashes in the order of 1.0×10^{-3} $mol_{ph}\ m^{-2}\ s^{-1}$ for about a millisecond (Vejrazka et al, 2015). In a following dark period, which is in the order of 10 ms, the reducing power then can be used. The consequence of this storage effect is that high light flashes still can be used at high photosynthetic efficiency and this is also referred to as the "flashing light" effect (Phillips and Myers, 1954). In theory this effect could be exploited by rapid mixing of microalgal suspensions in photobioreactors such that microalgae are only exposed shortly to high light at the photobioreactor surface and have sufficient time in the darker zones to process the power stored. In other words, oversaturating light at the reactor surface can be "diluted" over the whole reactor volume. However, considering the short timescales of the flashing light effect (<10 ms), the flashing light effect cannot be exploited in large-scale photobioreactors which have mixing times in the order of seconds.

Nevertheless, there are reports that increased and more systematic mixing enhances photobioreactor productivity (Degen et al, 2001; Kong et al, 2013; Laws et al, 1983, 1986; Qiang and Richmond, 1996). These are all intensively mixed systems with an estimated mixing time along the light gradient in the order of a second. It is attractive to attribute this to a flashing light effect, but it is more likely that other effects play a role. Downstream of the photosynthetic electron transport chain other metabolites might form another storage effect. It could be hypothesized that the capacity of the photosynthetic electron transport chain is somewhat higher than the capacity of the Calvin–Benson–Bassham cycle for carbon dioxide reduction to sugar (Stitt, 1986). Such an effect, however, must not be overestimated. Consensus is growing for the concept that the capacity of photosynthetic electron transport closely matches the rate of carbon dioxide reduction by means of tight control of the actual rates and the maximal capacities of both pathways (Foyer et al, 2012).

4.2.2 Photoinhibition and Light/Dark Cycling

A better explanation of the observed increase in photobioreactor productivity as a result of increased and more systematic mixing is the reduction of the effect of photoinhibition. Under oversaturating light photosystems are continuously inactivated by surplus light absorbed. At the same time inactivated photosystems are continuously repaired by a dedicated pathway: the PSII repair cycle (Nixon et al, 2010). The actual fraction of photosystems inactivated depends on the photon flux density, duration of light exposure, and the rate of the repair process (Campbell and Tyystjärvi, 2012). Increased mixing along the light gradient will decrease the light dose received during light exposure and will thus result in less photoinactivation. At the same time more frequent visits to the darker zones of the photobioreactor will allow the repair process to reactivate the inactivated photosystems again. This effect has already been integrated in a model description of photosynthesis by Eilers and Peeters (1988), the so-called three-state model. This model has been used to predict the effect of rapid light/dark fluctuations on microalgal growth (García-Camacho et al, 2012; Olivieri et al, 2015; Rubio Camacho et al, 2003). Specifically Olivieri concluded from their modeling efforts that the phenomenon of second scale light/dark cycling in combination with the PSII repair cycle could explain frequent observations that high-density microalgal culture in photobioreactors do not seem to be inhibited at high photon flux density (Olivieri et al, 2015). Light/dark fluctuations in the order of seconds could thus reduce photoinhibition and this effect should be clearly distinguished from the "flashing light effect" which is based on a millisecond potential to store reducing power within the photosynthetic electron carriers. This flashing light effect cannot be reasonably exploited in large-scale photobioreactors.

4.2.3 Light Dissipation, Photoacclimation, and Light/Dark Cycling

In case the timescale of mixing along the light gradient (light/dark cycling) is 10 s or more negative effects of light/dark cycles have been reported. Specifically long dark periods have been reported to reduce the efficiency of light utilization, or photobioreactor productivity (Barbosa et al, 2003; Janssen et al, 2000a; Takache et al, 2015). These observations might be related to the process of photoacclimation, to the activation and deactivation of dissipative pathways for excess light absorption, or to conformational changes of the photosystem antennae (i.e., state transitions).

The biomass density in combination with the optical path of the photobioreactor determines the size of dark volume in a photobioreactor. An increase in the relative size of the dark volume will induce photoacclimation and an increase of the optical cross section (Janssen et al, 2000a, 2000b). This will result in more oversaturation and light wasting in the light volume of the photobioreactor. One could argue that in the ideal situation there is no real dark zone and that the photon flux density at the backside should be around the compensation point photobioreactors. In practice, however, photon flux density changes within the day and it will also change from day to day. For this reason it is impossible to maintain optimal conditions for each point in time. Consequently, during many hours of the day biomass density is above the optimal value resulting in a dark zone stimulating photoacclimation and an increase of the optical cross section.

Besides adapting the optical cross section which is a relatively long process, microalgae have other tools at hand to quickly adapt to changing light conditions. Within the antenna complexes of the photosystems conformational changes can be activated to allow safe dissipation of absorbed light energy as heat in the situation that the photosystems are not able to process the energy by means of photosynthesis. These dissipative processes are usually referred to as "nonphotochemical quenching" (NPQ) (Müller et al, 2001; Niyogi and Truong, 2013). NPQ can be activated and deactivated on a timescale of multiple seconds to minutes. In the situation NPQ is not sufficient to dissipate excess light; absorbed photosystems will be photoinactivated as discussed before.

Besides safe dissipation of light in the antennae complexes the antennae complexes itself can detach from one type of photosystem (PSII) and possibly attach again to the other type of photosystems (PSI), or vice versa (Finazzi et al, 2002; Mullineaux and Emlyn-Jones, 2005; Ünlü et al, 2014). This process is called "state transitions" and allows photosynthesis to adjust the ratio of ATP production vs NADPH production by adjusting the rates of linear electron transport through PSII and PSI, yielding NADPH and ATP, and cyclic electron transport around PSI, yielding ATP only. Also state transitions can occur on a timescale of multiple seconds to minutes. State transitions and NPQ will play a role in microalgae cultivation in photobioreactors with mixing times along the light gradient of several seconds or more. These biological regulatory processes are subject of in-depth studies and it is a huge challenge to summarize these effects in simplified models which can be used to improve model predictions of photobioreactor productivity.

Based on this evaluation of the effect of mixing along the long gradient of microalgal cultures in photobioreactors the model description of photobioreactor productivity adopted in this chapter can be evaluated. In the photobioreactor production model it was assumed that the specific rate of photosynthesis is only dependent on the actual photon flux density at that point in time, and at that position within the photobioreactor. A flashing light effect based on a limited storage capacity of electrons and diluting light over time is thus excluded, and it is fully acknowledged within the model description that light/dark cycling in the order of 10 ms is not possible in large-scale outdoor photobioreactors. In addition, photoinhibition is not included in the model description meaning that it is implicitly assumed that biomass density and mixing along the light gradient are sufficiently fast to prevent significant photodamage under oversaturating photon flux density.

4.3 Diurnal Variations in Light Intensity

On longer timescales of minutes to hours other light variations can be identified: the daily increase and decrease of light intensity, and increasing or decreasing cloud cover. These changes will activate or deactivate the same regulatory mechanisms associated to the safe dissipation of excess light energy absorbed as discussed earlier for light/dark cycles in the ranges of seconds to minutes. In addition, regulatory mechanisms can be activated allowing for an adjustment of the ATP:NADPH ratio required by the microalgae. On a timescale of hours photoacclimation will take place which also has been discussed already. A good summary of the biological potential of phototrophic organisms to respond to changes in light level, and light quality, is provided by Foyer et al (2012). Preferably the most dominant among all these mechanisms are included in a model description for photobioreactor productivity. Partly such effects are already included in cell-based models such as those of Garcia-Camacho and Ross (García-Camacho et al, 2012; Ross and Geider, 2009).

The diurnal change of irradiance has been included in the model description of photobioreactor productivity under day/night cycles as visualized in Fig. 30. It was assumed that all sugar produced in photosynthesis is immediately used in the light, thus in daytime. In the night only maintenance-based sugar consumption is included. In reality part of the growth-related sugar consumption is taking place in the night as these sugars are stored as starch during the day. Including these dynamics will introduce additional complexity to the model description of photobioreactor productivity, but again it will improve the accuracy of model predictions. Moreover it will be important to be able to predict respiratory oxygen consumption in the night as well in order to optimize the gas exchange systems of photobioreactors.

4.4 Light Direction and Light Scattering and Photobioreactor Productivity

The model description of microalgal growth in photobioreactors was based on a simplified description of the light field within microalgal cultures. The Lambert–Beer law was employed while assuming a flat photobioreactor surface (flat panels, raceway ponds, sloped cascade systems). In addition it was assumed that light entrance was always perpendicular to the light-exposed surface. These assumptions allowed for a relatively simple mathematical description of photobioreactor productivity. In this manner important aspects of microalgal production processes could be highlighted. An improved description of the light field, however, will make the model predictions more reliable and is a step which must be taken when describing productivity in real outdoor microalgal production systems. The two most important aspects which must be taken into account when modeling the light field are the angle of incidence at the photobioreactor surface and the effect of light scattering by the microalgae. In relation to that also reflection and refraction events at photobioreactor surfaces must be included, as well as photobioreactor shape and orientation. These aspects will be covered in detail in other chapters of this book. Below the impact of the angle of incidence and light scattering will be discussed.

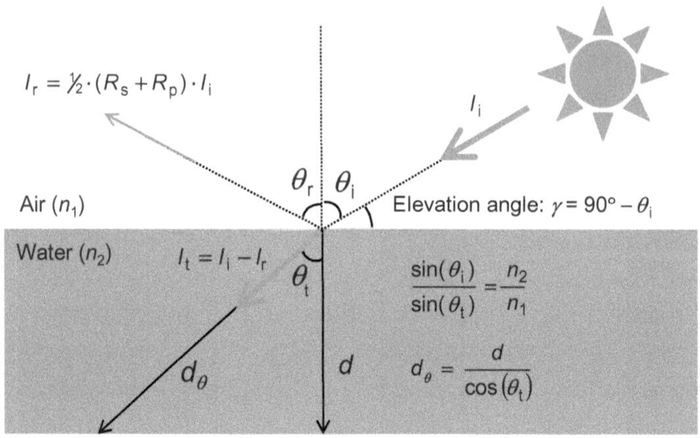

Figure 34 Illustration of the relation between angle of incidence θ_i on a photobioreactor and the light path through a microalgal culture d_θ. In this example the situation in a microalgal pond system is illustrated. In addition to light path also the effect of light refraction and reflection at the culture surface is illustrated including Snell's law. Factors R_s and R_p must be calculated with the Fresnel equations.

The angle of incidence from direct sunlight depends on the geographical location and varies over the day and seasons. The illustration in Fig. 34 exemplifies that this results in a light path (d_θ in Fig. 34) which is substantially longer than the depth of the system (d in Fig. 34). The actual light path determines the optimal biomass concentration as discussed before in this chapter. In an outdoor raceway pond, for example, the optimal biomass concentration is much lower in early morning than around solar noon. Consequently, it is almost impossible to maintain optimal conditions throughout a day. In general the biomass concentrations will be above optimal in early morning and late afternoon when the sun is low on the horizon leading to a substantial dark zone in the microalgal culture.

The daily change in light path clearly has implications for the light field within the culture and thus the local rates of photosynthesis, and this effect must be taken into account when aiming for a reliable prediction of outdoor photobioreactor productivity. For accurate predictions the photobioreactor production models must thus be resolved for every point in time during the day. In the simulations of outdoor systems presented in Fig. 30 the optical path was corrected with a factor 1.5 to partly accommodate for this effect to make the predictions more in line with the outcome of real systems. A factor of 1.5 is equivalent to an angle 48° between the light ray and the normal of the surface. The real angle of incidence of sunlight is higher due to refraction at the photobioreactor surface, which is also illustrated in Fig. 34. It is interesting to note that refraction leads to spreading of collimated sunlight and thus light dilution, which will lead to less oversaturation of individual microalgal cells at the photobioreactor surface. This effect has been included in a future outlook on large-scale microalgae production by Breuer et al (2015). The factor 1.5 adopted for the simulation presented in Fig. 30 was an arbitrary choice but it can be placed in perspective by calculating the average light path traveled by perfectly diffused (i.e., isotropic) light through a flat system. This calculation is illustrated in Fig. 35 and it results in an average light path being exactly twice as large as the shortest path through the culture along the normal of the surface. This is equivalent to an average angle between light beam and surface normal of 60°.

In Fig. 35 collimated sunlight is diffused by clouds. Clouds will reduce sunlight irradiance on ground surface and will indeed lead to less direct irradiance and a more diffuse irradiance. However, within the microalgal culture light scattering by the microalgal cells will also lead to a more isotropic character of the light field. Light scattering has been discussed before and was already illustrated in Fig. 9. Also light scattering will lead to an increase of the light path in relation to culture depth *d*. On the one hand, the increase of light path can even be higher than the factor 2 derived for isotropic light (Fig. 35). This is related to the fact that a light flux will develop within the culture in the opposite direction of the entry of light. On the other hand, it was observed in practice that a light path correction factor of 1.2–1.6 was

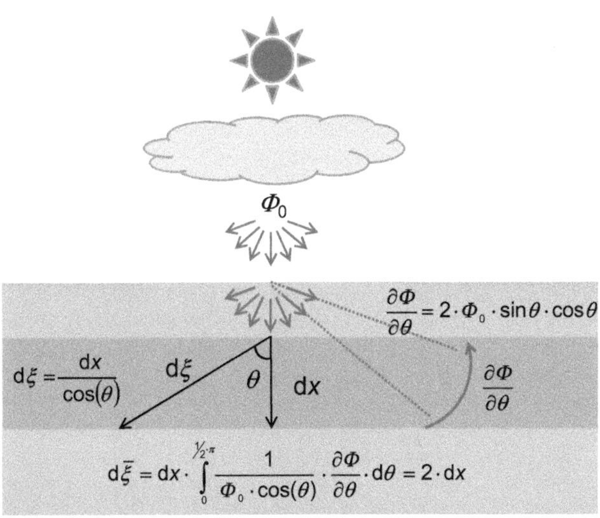

Figure 35 Diffused light and average light path through a microalgal culture. The parameter Φ represents the total radiant flux or photon flux (unit W or $mol_{ph}\ s^{-1}$).

sufficient to accommodate for this effect (unpublished data, Bioprocess Engineering, Wageningen University) when employing Algaemist laboratory-scale photobioreactors (De Mooij et al, 2014) illuminated with quasi-collimated light. Including scattering within model predictions of the light field in microalgal cultures in photobioreactors will be discussed in other chapters of this book. It is a big challenge and requires complex mathematical and numerical routines. Such an approach must also include the effect of reactor shape and orientation. An accurate description of the light field though is necessary for a more reliable description and prediction of photobioreactor productivity.

REFERENCES

Allen JF: Cyclic, pseudocyclic and noncyclic photophosphorylation: new links in the chain, *Trends Plant Sci* 8(1):15–19, 2003.

Barbera E, Sforza E, Bertucco A: Maximizing the production of *Scenedesmus obliquus* in photobioreactors under different irradiation regimes: experiments and modeling, *Bioprocess Biosyst Eng* 38(11):2177–2188, 2015. http://dx.doi.org/10.1007/s00449-015-1457-9. published online 20 August.

Barbosa MJ, Janssen M, Ham N, et al: Microalgae cultivation in air-lift reactors: modeling biomass yield and growth rate as a function of mixing frequency, *Biotechnol Bioeng* 82(2):170–179, 2003.

Björkman O, Demmig B: Photon yield of O_2 evolution and chlorophyll fluorescence characteristics at 77 K among vascular plants of diverse origins, *Planta* 170:489–504, 1987.

Blackman FF: Optima and limiting factors, *Ann Bot* 19:281–296, 1905.

Blanken W, Cuaresma M, Wijffels RH, et al: Cultivation of microalgae on artificial light comes at a cost, *Algal Res* 2(4):333–340, 2013.

Bosma R, de Vree JH, Slegers PM, et al: Design and construction of the microalgal pilot facility AlgaePARC, *Algal Res* 6:160–169, 2014.

Breuer G, Lamers PP, Janssen M, et al: Opportunities to improve the areal oil productivity of microalgae, *Bioresour Technol* 186:294–302, 2015.

Campbell DA, Tyystjärvi E: Parameterization of photosystem II photoinactivation and repair, *Biochim Biophys Acta Bioenerg* 1817(1):258–265, 2012.

Cape JL, Bowman MK, Kramer DM: Understanding the cytochrome *bc* complexes by what they don't do. The Q-cycle at 30, *Trends Plant Sci* 11(1):46–55, 2006.

Chalker BE: Modelling light saturation curves for photosynthesis: an exponential function, *J Theor Biol* 84:205–213, 1980.

Chen F, Johns MR: Effect of C/N ratio and aeration on the fatty acid composition of heterotrophic *Chlorella sorokiniana*, *J Appl Phycol* 3:203–209, 1991.

Cordier J-L, Butsch BM, Birou B, et al: The relationship between elemental composition and heat of combustion of microbial biomass, *Appl Microbiol Biotechnol* 25(4):305–312, 1987.

Cuaresma M, Janssen M, Vílchez C, et al: Horizontal or vertical photobioreactors? How to improve microalgae photosynthetic efficiency, *Bioresour Technol* 102(8):5129–5137, 2011.

Davies-Colley RJ, Pridmore RD, Hewitt JE: Optical properties of some freshwater phytoplanktonic algae, *Hydrobiologia* 133(2):165–178, 1986.

De Mooij T, Janssen M, Cerezo-Chinarro O, et al: Antenna size reduction as a strategy to increase biomass productivity: a great potential not yet realized, *J Appl Phycol* 27(3):1063–1077, 2014.

Degen J, Uebele A, Retze A, et al: A novel airlift photobioreactor with baffles for improved light utilization through the flashing light effect, *J Biotechnol* 92:89–94, 2001.

Dubinsky Z, Stambler N: Photoacclimation processes in phytoplankton: mechanisms, consequences, and applications, *Aquat Microb Ecol* 56(September):163–176, 2009.

Dubinsky Z, Falkowski PG, Wyman K: Light harvesting and utilization by phytoplankton, *Plant Cell Physiol* 27(7):1335–1349, 1986.

Eilers PHC, Peeters JCH: A model for the relationship between light intensity and the rate of photosynthesis in phytoplankton, *Ecol Model* 42:199–215, 1988.

Emerson R, Lewis CM: The dependence of the quantum yield of *Chlorella* photosynthesis on wavelength of Light, *Am J Bot* 30(3):165–178, 1943.

Evans J: The dependence of quantum yield on wavelength and growth irradiance, *Aust J Plant Physiol* 14:69–79, 1987.

Finazzi G, Rappaport F, Furia A, et al: Involvement of state transitions in the switch between linear and cyclic electron flow in *Chlamydomonas reinhardtii*, *EMBO Rep* 3(3):280–285, 2002.

Formighieri C, Franck F, Bassi R: Regulation of the pigment optical density of an algal cell: filling the gap between photosynthetic productivity in the laboratory and in mass culture, *J Biotechnol* 162(1):115–123, 2012.

Foyer CH, Neukermans J, Queval G, et al: Photosynthetic control of electron transport and the regulation of gene expression, *J Exp Bot* 63(4):1637–1661, 2012.

García-Camacho F, Sánchez-Mirón A, Molina-Grima E, et al: A mechanistic model of photosynthesis in microalgae including photoacclimation dynamics, *J Theor Biol* 304:1–15, 2012.

Geider RJ, Osborne BA: Respiration and microalgal growth: a review of the quantitative relationship between dark respiration and growth, *New Phytol* 112(3):327–341, 1989.

Geider RJ, MacIntyre HL, Kana TM: A dynamic regulatory model of phytoplanktonic acclimation to light, nutrients, and temperature, *Limnol Oceanogr* 43(4):679–694, 1998.

Gómez-Pérez CA, Espinosa J, Montenegro Ruiz LC, et al: CFD simulation for reduced energy costs in tubular photobioreactors using wall turbulence promoters, *Algal Res* 12:1–9, 2015.

Heijnen SJ: Thermodynamics of microbial growth and its implications for process design, *Trends Biotechnol* 12(12):483–492, 1994.

Heijnen JJ, Van Dijken JP: In search of a thermodynamic description of biomass yields for the chemotropic growth of microorganisms, *Biotechnol Bioeng* 42(9):833–858, 1992.

Hogewoning SW, Wientjes E, Douwstra P, et al: Photosynthetic quantum yield dynamics: from photosystems to leaves, *Plant Cell* 24(5):1921–1935, 2012.

Janssen M, Janssen M, de Winter M, et al: Efficiency of light utilization of *Chlamydomonas reinhardtii* under medium-duration light/dark cycles, *J Biotechnol* 78(2):123–137, 2000a.

Janssen M, de Bresser L, Baijens T, et al: Scale-up aspects of photobioreactors: effects of mixing-induced light/dark cycles, *J Appl Phycol* 12:225–237, 2000b.

Jassby AD, Platt T: Mathematical formulation of the relationship between photosynthesis and light for phytoplankton, *Limnol Oceanogr* 21(4):540–547, 1976.

Kirk JTO: *Light and photosynthesis in aquatic ecosystems,* ed 2, Cambridge, 1994, Cambridge University Press.

Kliphuis AMJ, de Winter L, Vejrazka C, et al: Photosynthetic efficiency of *Chlorella sorokiniana* in a turbulently mixed short light-path photobioreactor, *Biotechnol Prog* 26(3):687–696, 2010.

Kliphuis AMJ, Janssen M, van den End EJ, et al: Light respiration in *Chlorella sorokiniana*, *J Appl Phycol* 23(6):935–947, 2011a.

Kliphuis AMJ, Martens DE, Janssen M, et al: Effect of O_2:CO_2 ratio on the primary metabolism of *Chlamydomonas reinhardtii*, *Biotechnol Bioeng* 108(10):2390–2402, 2011b.

Kliphuis AMJ, Klok AJ, Martens DE, et al: Metabolic modeling of *Chlamydomonas reinhardtii*: energy requirements for photoautotrophic growth and maintenance, *J Appl Phycol* 24(2):253–266, 2012.

Kong B, Shanks JV, Vigil RD: Enhanced algal growth rate in a Taylor vortex reactor, *Biotechnol Bioeng* 110(8):2140–2149, 2013.

Kramer DM, Evans JR: The importance of energy balance in improving photosynthetic productivity, *Plant Physiol* 155(1):70–78, 2011.

Laws EA, Terry KL, Wickman J, et al: A simple algal production system designed to utilize the flashing light effect, *Biotechnol Bioeng* 25:2319–2335, 1983.

Laws EA, Taguchi S, Hirata J, et al: High algal production rates achieved in a shallow outdoor flume, *Biotechnol Bioeng* 28(698):191–197, 1986.

Lee Y-K, Ding S-Y, Hoe C-H, et al: Mixotrophic growth of *Chlorella sorokiniana* in outdoor enclosed photobioreactor, *J Appl Phycol* 8:163–169, 1996.

Ley AC, Mauzerall DC: Absolute absorption cross-sections for photosystem II and the minimum quantum requirement for photosynthesis in *Chlorella vulgaris*, *Biochim Biophys Acta Bioenerg* 680(1):95–106, 1982.

Li T, Zheng Y, Yu L, et al: Mixotrophic cultivation of a *Chlorella sorokiniana* strain for enhanced biomass and lipid production, *Biomass Bioenergy* 66:204–213, 2014.

Malkin S, Fork DC: Bill Arnold and calorimetric measurements of the quantum requirement of photosynthesis—once again ahead of his time, *Photosynth Res* 48(1–2):41–46, 1996.

McCree KJ: The action spectrum, absorptance and quantum yield of photosynthesis in crop plants, *Agric Meteorol* 9(3):191–216, 1971.

Merzlyak MN, Razi Naqvi K: On recording the true absorption spectrum and the scattering spectrum of a turbid sample: application to cell suspensions of the cyanobacterium *Anabaena variabilis*, *J Photochem Photobiol B* 58(2–3):123–129, 2000.

Meyer zu Tittingdorf JMW, Rexroth S, Schäfer E, et al: The stoichiometry of the chloroplast ATP synthase oligomer III in *Chlamydomonas reinhardtii* is not affected by the metabolic state, *Biochim Biophys Acta Bioenerg* 1659(1):92–99, 2004.

Moberg AK, Ellem GK, Jameson GJ, et al: Simulated cell trajectories in a stratified gas–liquid flow tubular photobioreactor, *J Appl Phycol* 24(3):357–363, 2011.

Müller P, Li XP, Niyogi KK: Non-photochemical quenching. A response to excess light energy,, *Plant Physiol* 125:1558–1566, 2001.

Mullineaux CW: Co-existence of photosynthetic and respiratory activities in cyanobacterial thylakoid membranes, *Biochim Biophys Acta Bioenerg* 1837(4):503–511, 2014.

Mullineaux CW, Emlyn-Jones D: State transitions: an example of acclimation to low-light stress, *J Exp Bot* 56(411):389–393, 2005.

Nixon PJ, Michoux F, Yu J, et al: Recent advances in understanding the assembly and repair of photosystem II, *Ann Bot* 106(1):1–16, 2010.

Niyogi KK, Truong TB: Evolution of flexible non-photochemical quenching mechanisms that regulate light harvesting in oxygenic photosynthesis, *Curr Opin Plant Biol* 16(3):307–314, 2013.

Nogales J, Gudmundsson S, Knight EM, et al: Detailing the optimality of photosynthesis in cyanobacteria through systems biology analysis, *Proc Natl Acad Sci USA* 109(7): 2678–2683, 2012.

Oldenhof H, Zachleder V, Van Den Ende H: Blue- and red-light regulation of the cell cycle in *Chlamydomonas reinhardtii* (Chlorophyta), *Eur J Phycol* 41(3):313–320, 2006.

Olivieri G, Gargiulo L, Lettieri P, et al: Photobioreactors for microalgal cultures: a Lagrangian model coupling hydrodynamics and kinetics, *Biotechnol Prog* 31(5):1259–1272, 2015. http://dx.doi.org/10.1002/btpr.2138, published online 27 July.

Perner-Nochta I, Posten C: Simulations of light intensity variation in photobioreactors, *J Biotechnol* 131(3):276–285, 2007.

Phillips JN, Myers J: Growth rate of *Chlorella* in flashing light, *Plant Physiol* 29(8):152–161, 1954.
Pirt SJ: The maintenance energy of bacteria in growing cultures, *Proc R Soc B Biol Sci* 163(991):224–231, 1965.
Pogoryelov D, Reichen C, Klyszejko AL, et al: The oligomeric state of c rings from cyanobacterial F-ATP synthases varies from 13 to 15, *J Bacteriol* 189(16):5895–5902, 2007.
Qiang H, Richmond A: Productivity and photosynthetic efficiency of *Spirulina platensis* as affected by light intensity, algal density and rate of mixing in a flat plate photobioreactor, *J Appl Phycol* 8(2):139–145, 1996.
Ross ON, Geider RJ: New cell-based model of photosynthesis and photo-acclimation: accumulation and mobilisation of energy reserves in phytoplankton, *Mar Ecol Prog Ser* 383:53–71, 2009.
Rubio Camacho F, García Camacho F, Fernández Sevilla JM, et al: A mechanistic model of photosynthesis in microalgae, *Biotechnol Bioeng* 81(4):459–473, 2003.
Seelert H, Poetsch A, Dencher NA, et al: Proton-powered turbine of a plant motor, *Nature* 405(May):418–419, 2000.
Shi XM, Chen F, Yuan JP, et al: Heterotrophic production of lutein by selected *Chlorella* strains, *J Appl Phycol* 9(5):445–450, 1997.
Stitt M: Limitation of photosynthesis by carbon metabolism: I. Evidence for excess electron transport capacity in leaves carrying out photosynthesis in saturating light and CO_2, *Plant Physiol* 81(4):1115–1122, 1986.
Sukenik A, Falkowski PG, Bennett J: Potential enhancement of photosynthetic energy conversion in algal mass culture, *Biotechnol Bioeng* 30:970–977, 1987.
Takache H, Pruvost J, Cornet JF: Kinetic modeling of the photosynthetic growth of *Chlamydomonas reinhardtii* in a photobioreactor, *Biotechnol Prog* 28(3):681–692, 2012.
Takache H, Pruvost J, Marec H: Investigation of light/dark cycles effects on the photosynthetic growth of *Chlamydomonas reinhardtii* in conditions representative of photobioreactor cultivation, *Algal Res* 8:192–204, 2015.
Tamburic B, Szabó M, Tran N-AT, et al: Action spectra of oxygen production and chlorophyll a fluorescence in the green microalga *Nannochloropsis oculata*, *Bioresour Technol* 169:320–327, 2014.
Tanada T: The photosynthetic efficiency of carotenoid pigments in *Navicula minima*, *Am J Bot* 38(4):276–283, 1951.
Ünlü C, Drop B, Croce R, et al: State transitions in *Chlamydomonas reinhardtii* strongly modulate the functional size of photosystem II but not of photosystem I, *Proc Natl Acad Sci USA* 111(9):3460–3465, 2014.
Vejrazka C, Janssen M, Streefland M, et al: Photosynthetic efficiency of *Chlamydomonas reinhardtii* in flashing light, *Biotechnol Bioeng* 108(12):2905–2913, 2011.
Vejrazka C, Janssen M, Benvenuti G, et al: Photosynthetic efficiency and oxygen evolution of *Chlamydomonas reinhardtii* under continuous and flashing light, *Appl Microbiol Biotechnol* 97(4):1523–1532, 2013.
Vejrazka C, Streefland M, Wijffels RH, et al: The role of an electron pool in algal photosynthesis during sub-second light–dark cycling, *Algal Res* 12:43–51, 2015.
Vermaas WFJ: Photosynthesis and respiration in Cyanobacteria, *Encycl Life Sci* 161:1–7, 2001.
Vítová M, Bišová K, Kawano S, et al: Accumulation of energy reserves in algae: from cell cycles to biotechnological applications, *Biotechnol Adv* 33(6):1204–1218, 2015. http://dx.doi.org/10.1016/j.biotechadv.2015.04.012. published online 16 May.
Vollmar M, Schlieper D, Winn M, et al: Structure of the c14 rotor ring of the proton translocating chloroplast ATP synthase, *J Biol Chem* 284(27):18228–18235, 2009.

Von Stockar U, Liu J-S: Does microbial life always feed on negative entropy? Thermodynamic analysis of microbial growth, *Biochim Biophys Acta Bioenerg* 1412(3):191–211, 1999.

Wobbe L, Remacle C: Improving the sunlight-to-biomass conversion efficiency in microalgal biofactories, *J Biotechnol* 201:28–42, 2014.

Zhang QH, Wu X, Xue SZ, et al: Hydrodynamic characteristics and microalgae cultivation in a novel flat-plate photobioreactor, *Biotechnol Prog* 29(1):127–134, 2012.

Zijffers J-WF, Schippers KJ, Zheng K, et al: Maximum photosynthetic yield of green microalgae in photobioreactors, *Marine Biotechnol* 12(6):708–718, 2010.

CHAPTER FIVE

Industrial Photobioreactors and Scale-Up Concepts

Jeremy Pruvost*,[1], Francois Le Borgne[†], Arnaud Artu*,[†], Jean-François Cornet[‡], Jack Legrand*
*GEPEA, Université de Nantes, CNRS, UMR6144, Bd de l'Université, Saint-Nazaire Cedex, France
[†]AlgoSource Technologies, Bd de l'Université, Saint-Nazaire Cedex, France
[‡]Université Clermont Auvergne, ENSCCF, Clermont-Ferrand, France and CNRS, Institut Pascal, Aubiere, France
[1]Corresponding author: e-mail address: jeremy.pruvost@univ-nantes.fr

Contents

Abstract

Unlike other more classical bioprocesses for heterotrophic growth (typically yeasts and bacteria) where mixing tanks have standard geometries, microalgal culture has no single standard geometry. The main reason is the need for a light supply, which (1) has

Advances in Chemical Engineering, Volume 48
ISSN 0065-2377
http://dx.doi.org/10.1016/bs.ache.2015.11.002

spurred various technologies designed to maximize light use and (2) greatly increases process complexity, as light is a complex parameter to handle. However, in-depth and long-term modeling efforts have now yielded engineering tools to design, optimize, and control photobioreactors in a predictive and rational way.

Here we discuss the parameters to consider when designing and operating microalgal cultivation systems and how appropriate engineering rules can support optimal system design and operation. Once the practical and economic constraints of the final application have been appropriately factored in, it becomes possible to set a rational design of effective technologies. This is illustrated later in this chapter in examples of successful developments, some of which are commercially available via AlgoSource Technologies. The examples chosen serve to highlight the many applications of photobioreactors from lab-scale fundamental studies to large solar industrial production, and to illustrate how a handful of engineering rules frame the various photobioreactor design options (artificial light or natural sunlight, external or internal lighting, high-cell-density culture, and more).

1. INTRODUCTION

Photosynthetic growth in standard autotrophic conditions is based on the assimilation, under illumination, of inorganic carbon and mineral nutrients dissolved in the medium. The cultivation of photosynthetic microorganisms thus requires:

- a light source (solar or artificial, with an appropriate light spectrum in the photosynthetically active radiation (PAR) range, typically 0.4–0.7 μm),
- an inorganic carbon source (such as dissolved CO_2),
- mineral nutrients (major nutrients such as N, S, P sources; micronutrients like Mg, Ca, Mn, Cu, or Fe; etc.),
- set culture conditions (pH, temperature).

Ideally, the culture system has to enable optimal control of growth conditions, but it also has to meet the many and varied practical and economic tied to different microalgae applications, from small-scale lab production to mass-scale solar culture.

Generally speaking, microalgae cultivation shares many features with bioreactors in general, such as thermal regulation, nutrient feeding procedures, pH regulation, and mixing to enhance heat and mass transfers. However, the fact that photosynthetic growth needs a light supply has repercussions all the way from culture system design to effective operation (as detailed later in this chapter). An immediate observation is that, unlike other more classical bioprocesses where mixing tanks essentially have standard geometries, microalgal cultivation is characterized by a broad diversity of systems, ranging from open ponds (open systems) to photobioreactor (PBR) technologies (closed systems).

Detailed descriptions of existing geometries can be found in the literature (Carvalho et al, 2006; Lehr and Posten, 2009; Richmond, 2004b; Ugwu et al, 2008). The aim here is not to exhaustively review the different culture systems but to describe how system design and optimal operation can be encompassed in a robust and rational engineering approach. This will be illustrated by a handful of examples illustrating how factoring in the practical and economic constraints of the final application during the engineering phase ultimately results in very different technology designs from the same rational engineering tools. The focus will be on PBR technology as it offers the greatest potential in terms of optimization.

2. PBR ENGINEERING AND SCALING RULES

2.1 Main Parameters Affecting PBR Biomass Productivity

2.1.1 Engineering Parameters

Bioprocess design starts with identifying engineering parameters affecting process efficiency. This was the purpose of a research effort aiming to establish models able to represent microalgal biomass productivity in various PBR designs. The effort focused on addressing how to represent the influence of light supply on its use for photosynthetic growth in bulk culture (see later for a detailed example of a modeling approach that proved valid in several conditions). The work of Cornet and Dussap (2009) laid the foundations, as they developed an in-depth modeling approach for setting simple engineering rules able to predict maximal biomass productivities in cultivation systems. As maximal productivities are achieved when light only limits growth, engineering parameters related to light use were clarified. This was first published for constant artificial illumination conditions and then extended to the case of solar use by introducing specific features such as effect of the incident angle θ and the diffuse-direct distribution of solar radiation on resulting conversion in the cultivation system (Pruvost and Cornet, 2012). These relations give maximal surface ($S_{X\ \max}$) and volumetric ($P_{X\ \max}$) productivities:

$$\overline{S_{X\ \max}} = (1-f_{\mathrm{d}})\rho_{\mathrm{M}}\overline{\varphi}_{x}\frac{2\alpha}{1+\alpha}\left[\frac{\bar{x}_{\mathrm{d}}K}{2}\ \ln\left[1+\frac{2\bar{q}}{K}\right] + (1-\bar{x}_{\mathrm{d}})\overline{\cos\theta}K\ \ln\left[1+\frac{\bar{q}}{K\overline{\cos\theta}}\right]\right] \tag{1}$$

with

$$\overline{P_{X\max}} = \overline{S_{X\max}}\ \frac{S_{\text{light}}}{V_{\text{R}}} = \overline{S_{X\max}}\, a_{\text{light}} \tag{2}$$

where $\overline{}$ denotes a time averaging, ie, quantities averaged over a given period of exploitation. Averaging is typically applied in solar conditions due to the variation in irradiation conditions, leading to average performances on representative periods of exploitation (ie, 24 h, month, season, year, etc.).

The parameters of Eq. (1) can be split into three groups:

- Parameters related to the cultivated species: mean mass quantum yield $\overline{\varphi}_X$, half-saturation constant for photosynthesis K, and linear scattering modulus α related to the microorganism's radiative properties (see Table 1 for an example of parameters for *Chlorella vulgaris*).
- Parameters related to the operating conditions: incident angle θ, total collected photon flux density (ie, PFD) $\bar{q}$, and corresponding diffuse fraction $\bar{x}_{\text{d}}$ (here averaged over the period of exploitation).
- Parameters related to PBR geometry: specific illuminated area a_{light} given by ratio of PBR illuminated area to total culture volume, design dark volume fraction f_{d} which represents any volume fraction of the PBR not lit by incident PFD (eg, nonlit mixing tank).

Table 1 Examples of Growth Model Parameters for *Chlorella vulgaris* (Values Are Given for Growth on Ammonia as N-Source)

Parameter	Value	Unit
ρ_{M}	0.8	—
J_{NADH_2}	1.8×10^{-3}	$\text{mol}_{\text{NADH}_2}\,\text{kg}_X^{-1}\,\text{s}^{-1}$
υ_{O_2-X}	1.13	—
$\overline{\varphi}'_{\text{O}_2}$	1.1×10^{-7}	$\text{mol}_{\text{O}_2}\,\mu\text{mol}_{h\nu}^{-1}$
$\overline{\varphi}_X$	2.34×10^{-9}	$\text{kg}_X\,\mu\text{mol}_{h\nu}^{-1}$
M_X	0.024	$\text{kg}_X\,\text{C-mol}^{-1}$
$\upsilon_{\text{NADH}_2-\text{O}_2}$	2	—
K_A	30,000	$\mu\text{mol}_{h\nu}\,\text{kg}^{-1}\,\text{s}^{-1}$
K	110	$\mu\text{mol}_{h\nu}\,\text{m}^{-2}\,\text{s}^{-1}$
K_r	150	$\mu\text{mol}_{h\nu}\,\text{kg}^{-1}\,\text{s}^{-1}$
A_c	1500	$\mu\text{mol}_{h\nu}\,\text{kg}^{-1}\,\text{s}^{-1}$
Ea	270	$\text{m}^2\,\text{kg}^{-1}$
Es	2780	$\text{m}^2\,\text{kg}^{-1}$
b	0.002	—

For a given species, parameters affecting PBR productivity are design-specific illuminated area a_{light}, design dark fraction of the reactor f_{d}, and ability of the PBR to collect light (characterized by incident PFD $\bar{q}$ and related incident angle θ and diffuse fraction $\bar{x}_{\text{d}}$). All these parameters are tied to light supply. Light collected by the PBR is obviously a function of its location and weather conditions.

These engineering formulae can be simplified, especially in the case of artificial light. In artificial light, the light source is often set to provide normal incidence ($\overline{\cos\theta} = 1$), as this also corresponds to a maximization of the light provided to the culture. It is also common practice to apply quasi-collimated light ($\bar{x}_{\text{d}} = 0$) as obtained from LED panels (ie, without combination to diffuse plate). This leads to the following simplified formula:

$$\overline{S_{X\ \max}} = (1 - f_{\text{d}})\rho_{\text{M}}\overline{\varphi}_X \frac{2\alpha}{1+\alpha} K \ln\left[1 + \frac{\bar{q}}{K}\right] \tag{3}$$

with

$$\overline{P_{X\ \max}} = a_{\text{light}}(1 - f_{\text{d}})\rho_{\text{M}}\overline{\varphi}_X \frac{2\alpha}{1+\alpha} K \ln\left[1 + \frac{\bar{q}}{K}\right] \tag{4}$$

Obviously, these formulae only provide maximal performances which, as explained later, can only be achieved if other conditions are fulfilled, especially in actual culture system operating. It also assumes that light alone limits growth, assuming all other biological needs (nutrients, dissolved carbon) and operating conditions (pH, temperature) are controlled at optimal values (Cornet, 2010; Pruvost and Cornet, 2012; Takache et al, 2010). In practice, this could prove a big challenge (especially in mass-scale outdoor production, see later), but the relations that can be easily used (ie, analytic formulae) already give highly valuable information in the preliminary engineering phase.

These relations also underline the relevant engineering parameters affecting PBR productivity, ie, specific illuminated area a_{light}, engineering dark fraction f_{d}, and light collected $\bar{q}$. Note that dark fraction must not be confused with dark volume which results from light attenuation in the culture volume due to light absorption by photosynthetic cells (see later). This reflects to unlit fractions of the culture system, resulting from the design itself, and is typically obtained when adding a dark tank in the hydraulic loop for cooling or pH regulation purposes for example, or when a non-illuminated airlift vertical tube is introduced for culture mixing and circulation. To maximize PBR performance, the design dark fraction should be minimized down to negligible or null values, but this condition is not always met in practice.

Fig. 1 shows the influence of these engineering parameters on maximal productivities as predicted by these relations, here for the case of *C. vulgaris*

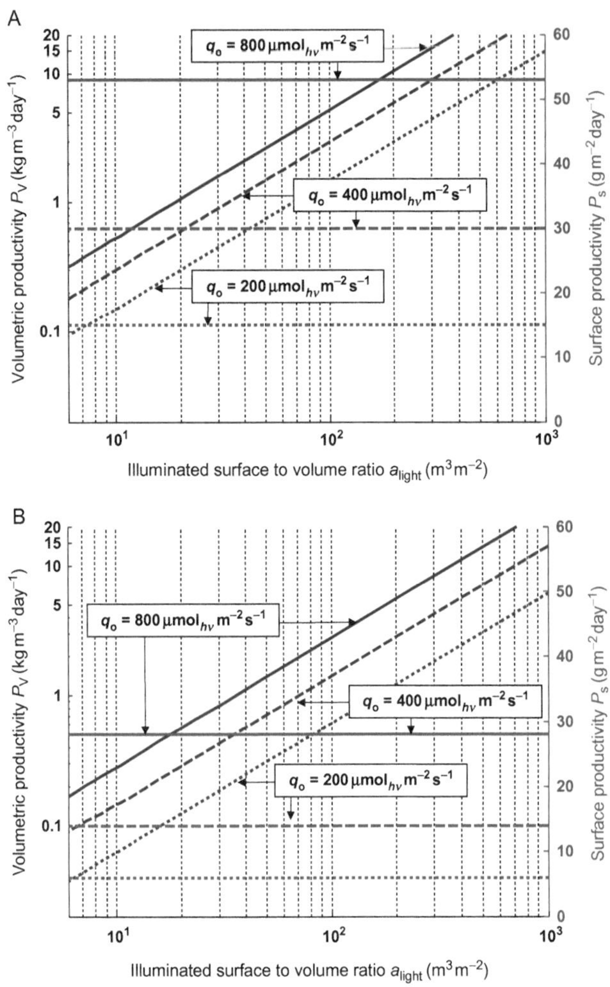

Figure 1 Influence of the illuminated surface to volume ratio (a_{light}) on PBR productivities. A direct influence on volumetric productivity is shown (two orders of magnitude of variation). Surface productivity is found independent of this engineering parameter. PFD (q_o) reveals to have a positive effect on both values. Influence of the design dark volume fraction of the PBR (f_d) is also illustrated. Panel (A) is the best design case, namely without design dark volume fraction ($f_d = 0$), and panel (B) is for a PBR design presenting 20% of its total volume in the dark ($f_d = 0.2$). Results are given for *C. vulgaris*, and all values correspond to maximal performances (ie, as obtained in continuous cultivation, light-limited conditions, luminostat "$\gamma = 1$" regime).

cultivation. In addition to the ideal condition of no dark fraction in the cultivation system ($f_d = 0$, Fig. 1A), a typical dark fraction value of 20% was also considered ($f_d = 0.2$, Fig. 1B). The figures illustrate the main guiding rules of PBR engineering:

- Specific illuminated surface a_{light} has a huge influence on volumetric productivity (two orders of magnitude are covered here) but no influence on surface productivity. Indeed, it is well known that productivity, when expressed per unit surface area and under light limitation, is independent of PBR depth as it is only dependent on the light collected in light-limited growth conditions, which is defined by the PBR collecting surface and not its volume (Cornet, 2010; Lee et al, 2014).
- PFD $\bar{q}$ is a relevant parameter as it has a positive effect on both surface and volumetric productivities. In solar conditions, the PFD will be defined by the ability of the system to collect light, which will depend on PBR geometry, geographical location, and positioning, as shown in numerous works (Acién Fernández et al, 2001; Chen et al, 2006; Chini Zittelli et al, 2000; Doucha and Livansky, 2006; Molina et al, 2001; Oswald, 1988; Pruvost, 2011; Pruvost and Cornet, 2012; Pruvost et al, 2012; Richmond and Cheng-Wu, 2001).
- The design dark volume fraction f_d has a highly negative influence on both surface and volumetric productivities. This is especially the case for microalgae presenting significant respiration activity in the dark. The dark volume fraction is not only a nonproducing volume but also contributes negatively to the overall PBR performance due to biomass catabolism in this nonilluminated volume. As a result, a dark volume fraction of 20% can decrease PBR productivities by a factor of 2 for *C. vulgaris*. Note that dark volume is usually introduced in design practice for microalgal cultivation units (ie, mixing tank in the cultivation loop of a tubular system, nonilluminated volume of an airlift PBR, etc.).

Results of Fig. 1 also show that due to the progressive saturation of photosynthetic conversion (as represented by parameter *K*, which is species dependent), an increase of PFD received on the cultivation system will increase productivity (as shown in Fig. 1) but will also decrease the thermodynamic efficiency of the process (ie, yield of conversion of light energy into biomass). This is shown in Fig. 1 by the values of surface productivities obtained for different PFDs. Increasing the PFD from 400 to 800 $\mu mol_{h\nu}\,m^{-2}\,s^{-1}$ (a 2-fold increase) leads to an increase in surface productivity from 30 to 52 $g\,m^{-2}\,day^{-1}$ (a 1.7-fold increase).

This highlights the importance of the light dilution principle, as obtained from insertion of light sources inside the culture volume (leading to what are dubbed "volumetrically lightened" systems; see Cornet, 2010), wherein the surface illuminating the culture becomes higher than the surface directly exposed to the light source (light capture surface; see Fig. 2). This results in light dilution which increases light conversion yield by photosynthesis. When expressed per unit of light capture surface, biomass productivity is higher with volumetrically lightened systems than surface-lightened systems, but these technologies carry several drawbacks, including higher technological complexity (need for optical capture devices), which can inflate costs, and the fact that efficiency hinges on proper design. The main challenges are to deliver a light flux at the required value for optimal photosynthetic conversion by the culture, and the need to engineer an optical capture device that minimizes loss of light energy when transmitting light from capture to culture. Diluting light also entails a trade-off with volumetric productivity (ie, biomass concentration) and will thus have to be compensated to a certain extent by an increase of specific illuminated surface. This leads to specific technologies such as the DiCoFluV concept, which will be detailed further (Cornet, 2010). Despite the challenges of setting these culture systems, the effort can pay off, especially in the case of solar production, where biomass productivity per unit of land area (ie, capture surface) could be a relevant factor.

2.1.2 Operating Parameters

Light, carbon and mineral nutrient supply, temperature, and pH are the main variables liable to limit photosynthetic growth and thus reduce the productivity of cultivation systems (assuming there is no predatory contamination). Except for light, these parameters can be controlled and set at optimal or near-optimal values with appropriate engineering and operating procedures. This is where PBRs, as a closed geometry, have a critical advantage, although even here the engineering of the culture system, eg, thermal regulation or carbon supply, still proves highly influential.

2.1.2.1 Thermal Regulation

Like in any biological process, temperature directly influences photosynthesis and microorganism growth. Particularly under solar illumination, closed-system PBRs tend to overheat whereas open-system PBRs can suffer evaporation of water under strong incident irradiance, explained by culture confinement and the strongly exoenergetic photosynthetic growth (Carvalho et al, 2011; Hindersin et al, 2013; Torzillo et al, 1996; Wilhelm and Selmar, 2011). In fact, thermodynamic efficiency over the

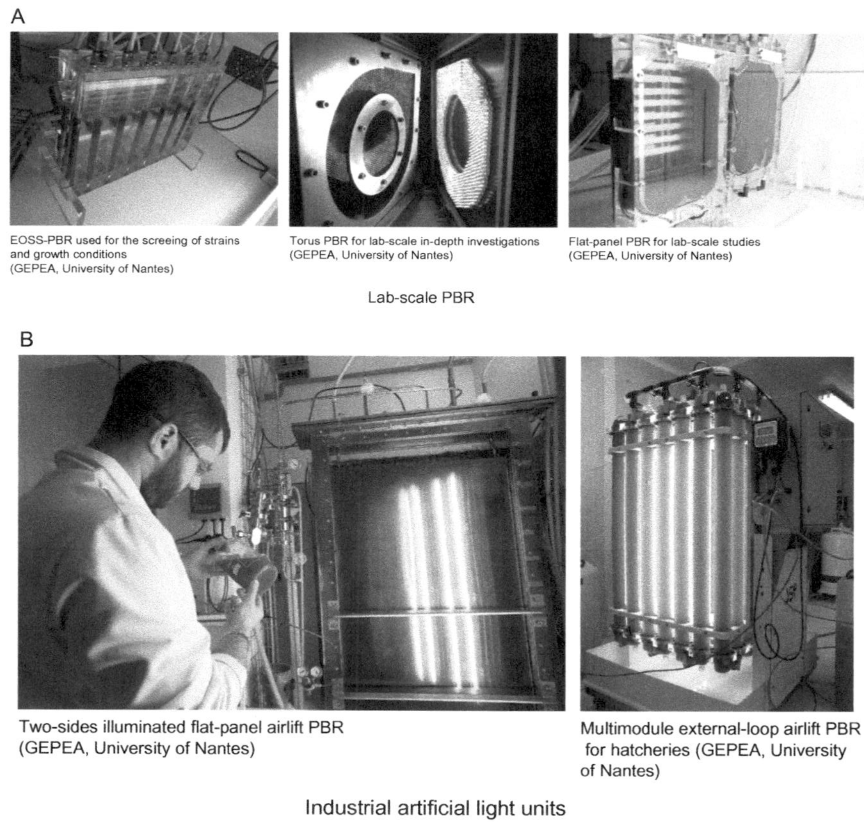

Figure 2 Examples of microalgae cultivation systems. Each system has been designed for a specific purpose and was scaled using on the engineering approach described in the text.

(Continued)

C

Flat-panel solar PBR
(GEPEA, University of Nantes)

Enclosed raceway
(AlgoSource Technologies, France)

AlgoFilm© solar PBR (ultra-thin PBR)
(GEPEA, University of Nantes)

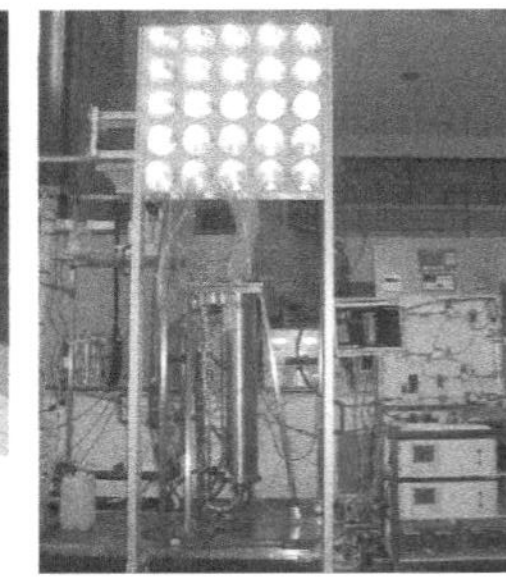

DiCoFluV© solar photobioreactor with
Fresnel lenses for sun capture and lateral
diffusing optical fibers inside the reactor
(Institut Pascal, Clermont-Fd, France)

Solar PBR

Figure 2—Cont'd

PAR region of systems working with low light typical of artificial illumination (100–300 $\mu mol_{h\nu} m^{-2} s^{-1}$) is generally below 5% (Cornet, 2010) and decreases to 2% under large solar irradiance (>500 $\mu mol_{h\nu} g^{-2} s^{-1}$). As a result, around 95% of the captured light is converted into heat by biochemical reactions and dissipation in light-collecting antennas. In fact, under outdoor conditions, around 50% of the energy in the solar radiation is contained in the near- and mid-infrared above 750 nm and directly participates in heating up the culture (Goetz et al, 2011; Hindersin, 2013; Hindersin et al, 2013, 2014).

Thermal regulation of PBRs has been widely investigated as a major issue of solar microalgal cultivation (Borowitzka, 1999; Grobbelaar, 2008; Hindersin et al, 2013, 2014). Unfortunately, without proper thermoregulation, temperatures lethal to living microorganisms can easily be reached inside the solar PBR, illustrating why PBR cooling is a usually a major engineering issue. In winter in temperate climates, excessively low temperatures can result in loss of biomass growth and productivity, so heating the culture can be beneficial (Hindersin, 2013). The appropriate temperature window is strongly dependent on species cultivated but typically in the 10–30°C range.

Various solutions are available for heating or cooling PBRs depending on PBR technology, size, and location. Water cooling and/or heating by spraying on the outside PBR surfaces or by direct PBR immersion in a pool are often used (Borowitzka, 1999). In temperate regions, cultivation systems can also be placed in greenhouses. Although efficient, these methods can increase the build and operating costs and negatively impact environmental footprint through excessive energy and water use.

Although technical solutions exist, PBR temperature control remains a challenge under solar conditions, especially if the design brief is for a cost-effective solution offering low-energy consumption and year-round operation which may need both cooling and heating. The engineering of the cultivation system is also relevant. Goetz et al (2011) experimentally and theoretically investigated the effect of various flat-panel PBR designs and found a decrease of up to one order of magnitude in PBR energy consumption depending on configuration. IR filtering, for example, was found to be especially effective at reducing culture overheat. More recently, research efforts have investigated the integration of PBR technology in building façades. This integration offers various benefits in terms of thermal management of both PBRs and buildings. Energy exchanges between the building and the PBRs can be designed so as to cool or warm each subsystem. For example, PBRs can filter sunlight in summer to reduce the thermal load on the building. In winter, excess thermal energy in the cultivation system can be used to warm the building. Note also that the added thermal mass of the building can be used to facilitate PBR thermal regulation regardless of season.

Overall, PBR thermal regulation depends on location, time of year, and strain cultivated. Cooling and/or heating requirements have to be estimated (usually in the range 50–200 W/m^2) and the associated thermal solutions should be defined and integrated well upstream in system design. For climates with large variations in outdoor temperature and solar irradiation over the course of a year, it could be beneficial to cultivate different species with optimal growths at different temperatures (Hindersin, 2013), which could significantly decrease energy needs over the period of exploitation.

2.1.2.2 Carbon and Mineral Nutrient Requirements

Assuming the growth medium of the strain is known, growth limitation by mineral nutrients can easily be avoided. The growth medium has to contain all the necessary (macro and micro) nutrients in sufficient quantities and must therefore be adjusted according to the biomass concentration planned. Stoichiometric growth equations can be used for this purpose or, more simply, concentrations can be monitored during cultivation. In specific cases, it would also be of interest to apply mineral limitation to induce specific metabolic responses, such as lipid accumulation (N source deprivation) or hydrogen production (sulfur deprivation). This is where combining mass balances on the cultivation system with stoichiometric growth equations is useful. Interested readers can refer to studies in which the method has been applied to various species (Degrenne et al, 2011b; Pruvost, 2011; Pruvost et al, 2009).

The inorganic carbon source should ideally be CO_2 dissolved in the culture medium, which makes preventing growth limitation by the carbon source more problematic. This depends on dissolved carbon concentration and thus on the gas–liquid mass transfer rate in the PBR. CO_2 dissolution also affects pH value (acidification), which in turn influences the amount and form of dissolved carbon obtained (CO_2, HCO_3^-, or CO_3^{2-}). Nutrient consumption can also interact with pH evolution (especially nitrogen source due to its significant consumption). Keeping an optimal pH value for growth while averting limitation by the carbon source may thus prove challenging. In most cases, simple CO_2 bubbling is found to suffice in the first instance for both pH regulation (acidification) and carbon feeding, but specific cases, such as when using an ammonium source (the consumption of which also leads to acidification), could present a more difficult challenge. Dissolved carbon concentration can always be monitored experimentally to forestall limitation (Degrenne et al, 2010; Le Gouic, 2013).

2.1.2.3 pH Control

Photoautotrophic microorganisms are cultivated in an aqueous solution in which the inorganic carbonaceous substrate is supplied through the dissolution of CO_2 gas in water or through the speciation of carbonates from the culture media. In most cases, the CO_2 is supplied in the form of fine bubbles. In water, the CO_2 gas forms other species such as dissolved carbon dioxide (CO_{2aq}), carbonic acid ($H_2CO_3^-$), bicarbonate (HCO_3^-), and carbonate (CO_3^{2-}), whose sum is termed TIC (total inorganic carbon). Many species of microalgae have developed mechanisms that enable both CO_{2aq} and HCO_3^- to support photosynthesis, but CO_{2aq} is still required. It is obtained by splitting the bicarbonate inside the cell ($HCO_3^- \leftrightarrow CO_{2aq} + OH^-$), a reaction that releases hydroxyl ions, causing the increase in pH. The ratio of CO_{2aq} to HCO_3^- depends closely on pH, as bicarbonate is the dominant species in solutions of pH >6.3, and the conversion of HCO_3^- to CO_{2aq} is very fast. Thus, when CO_{2aq} is removed from the medium, pH will increase. Microalgal cultivation often entails pH control by means of CO_2 gas bubbled into the reactor. This fresh supply of CO_2 will shift the equilibrium by lowering the pH. Ifrim et al (2014) proposed a global photoautotrophic growth model in which a radiative transfer model, a biological model, and a thermodynamic model are coupled. This model can accurately predict the dynamics of pH evolution.

2.1.2.4 Transfer Phenomena

Fluid dynamics in PBRs is import on several fronts. Although many studies have shown the relevance of mixing conditions in microalgal cultivation systems, there is still insufficient knowledge to provide engineering rules for their systematic optimization. Hydrodynamic conditions can have several outcomes, some of which are common to other bioprocesses (hydrodynamic shear stress, mass and heat transfer enhancement, cell sedimentation, and biofilm formation) while others are specific to microalgal cultivation systems, and especially for light–dark (L/D) cycle effects. L/D cycles result from cell displacement in the heterogeneous radiation field, such that cells experience a specific history with respect to the light they absorb, composed of variations from high irradiance level (in the vicinity of the light source) to low or quasi-nil values (deep in the culture) if biomass concentration is high.

As widely described in the literature (Janssen et al, 2000; Perner-Nochta and Posten, 2007; Pruvost et al, 2008; Richmond, 2004a; Rosello Sastre et al, 2007), this dynamic fluctuating regime can influence photosynthetic growth and thereby process efficiency. Note, however, that hydrodynamic timescales are several orders of magnitude greater than photosynthesis timescales, so the effects of L/D cycles due to hydrodynamics can in most cases be considered negligible (Pruvost et al, 2008), which is not the case for the presence of dark zones, as shown later (Section 3).

PBRs are generally considered perfect mixing systems, with homogenized nutrient concentrations and uniform biomass concentrations. An important point is to reduce the energy consumption for mixing by maintaining efficient mixing, which is contingent on the type of PBR. Numerical simulations could be one way to optimize flow configuration and mixing, including characterization of light regimes in cultivation systems by a Lagrangian simulation (Pruvost et al, 2002a, 2002b). CFD can be used to gain an in-depth understanding of the hydrodynamics/flow pattern in the PBRs and usefully inform scale-up. For bubble-flow PBRs, most published simulations have used two-phase models (air and water) and employed the Eulerian–Eulerian mixture model (Bitog et al, 2011). To increase radial mixing in flat-panel airlift systems, static mixers can be used (Subitec PBR) to direct the flow toward the light source (Bergmann et al, 2013). For stirred PBRs, the choice of impeller type is important (Pruvost et al, 2006). If species cultivated are not stress sensitive, the more efficient flow circulation and mixing impeller could be used. If not the case, a compromise must be found based on the strain's sensitivity to shear stress. Numerical simulation of the flow system can offer the ability to design a raceway before construction, saving considerable cost and time. Moreover, the impacts of various parameters, such as culture media depth, temperature, flow speeds, baffles, could be investigated to optimize operating conditions (James et al., 2013).

CO_2 mass transfer is one of the more important transfer phenomena issues in PBRs. CO_2 is the usual carbon source for photosynthetic culture of microalgae and is generally supplied by continuous or intermittent gas injection. As the carbon is consumed, oxygen is ultimately produced by photolysis of water and released into the culture medium, where it can be removed by gas stripping. Volumetric gas–liquid mass transfer, k_{La}, is related to power input per volume due to aeration (Acién et al., 2012; Chisti, 1989). The volumetric gas–liquid mass transfers for oxygen and for CO_2 are related to their diffusion coefficients in the culture media.

2.1.2.5 Residence Time and Light Attenuation Conditions

Biomass concentration has a critical influence as it directly impacts light attenuation regime in the culture volume. It can be controlled via the harvesting strategy. When operated in batch mode, the harvesting strategy consists of defining culture growth duration. For practical reasons, many mass-scale solar PBRs are operated either in batch mode with biomass harvesting at the end of the culture or in semi-continuous mode with spot harvesting of part of the culture and replacement by fresh growth medium. This means biomass concentration and thus light attenuation conditions evolve with time. In continuous mode, a steady state is achieved only if all operating parameters are maintained constant with time. This condition can be met in permanent illumination conditions (artificial light). The PBR is then operated with a constant permanent value of the residence time τ (or dilution rate $D=1/\tau$), leading to a steady state with constant biomass concentration and light attenuation conditions.

Fig. 3A shows the strong relation between harvesting strategy (here defined by the residence time value), biomass concentration, light attenuation regime (here represented by the illuminated fraction γ; see Section 3.4), and resulting biomass productivity, as illustrated in the case of continuous culture. On one hand, if biomass concentration is too low (ie, low residence time), part of the incoming photons is not absorbed and is instead transmitted through the culture. This results in a loss of biomass productivity. In addition, light received per cell is high and may lead to further decreases in productivity due to increased photosynthetic dissipation. It may also induce cell photoacclimation resulting in a decrease in algal pigment content, leading to a higher light penetration with then further increase of the light received per cell. The system consequently becomes highly unstable, usually resulting in culture washout. Such low light attenuation conditions should thus be avoided in microalgal cultivation, especially for high PFDs typically larger than 200 $\mu mol_{h\nu}\,m^{-2}\,s^{-1}$.

On the other hand, if biomass concentration is too high (ie, high residence time), a dark zone appears inside the culture. This dark zone is the direct consequence of light extinction by cells in suspension, whose effect can be positive in high-illumination conditions by reducing photoinhibition effects and thus increasing process stability (Carvalho et al, 2011; Grima et al, 1999; Richmond, 2004b). Note that for microorganisms like eukaryotic microalgae that show respiration activity under illumination, a dark zone in the culture volume promotes respiration, resulting in a loss of biomass productivity. Therefore, achieving the maximum biomass productivity in

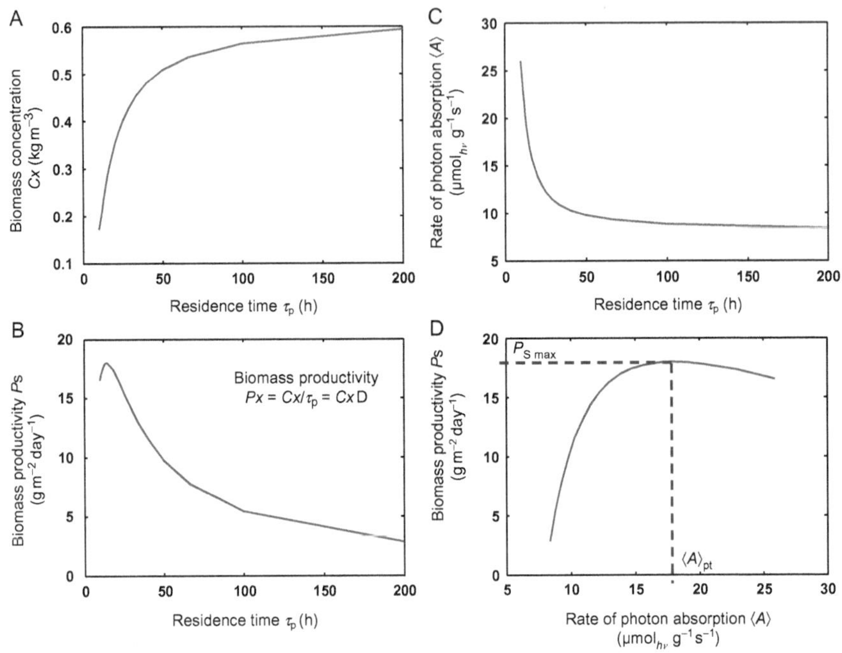

Figure 3 Evolution of biomass concentration (A), biomass productivity (B), and photon absorption rate (C) as a function of the residence time applied to the cultivation system. This illustrates the strong relation between all variables in microalgal cultivation system, as explained by the direct effect on light attenuation conditions. The example is here given for *C. vulgaris*. (D) The relation between biomass productivity and photon absorption rate.

this case requires the exact condition of complete absorption of the incident light (Takache et al, 2010) but without a dark zone in the culture volume. This condition is often referred to as luminostat mode. Note that it should not be confused with turbidostat mode, which refers to a turbidity-based regulation of a continuous culture. This condition has also been introduced as the "$\gamma = 1$" condition, γ denoting the ratio between illuminated volume and total cultivation system volume (see Section 3.4). For microorganisms with negligible respiration activity under illumination, such as prokaryotic cyanobacteria cells, fulfilling the condition of complete light absorption ($\gamma \leq 1$) will be enough to reach maximum biomass productivity.

2.1.2.6 Specific Rate of Photon Absorption A

Another way to represent the strong correlation between light attenuation conditions and the associated biomass productivity is to calculate the specific rate of photon absorption, noted A (here expressed per unit of biomass, ie, in $\mu mol_{h\nu} s^{-1} kg^{-1}$). Surprisingly, even though this value has been found beneficial in numerous studies devoted to photoreaction, it is rarely used in microalgal culture (Cassano et al, 1995). A is obtained by integrating the product of spectral values of local irradiance G_λ (see Section 3.4) and microalgae mass absorption coefficient Ea_λ (see Section 3.5) on the PAR region (Aiba, 1982; Cassano et al, 1995; Kandilian et al, 2013):

$$A = \int_{PAR} Ea_\lambda G_\lambda d\lambda \tag{5}$$

This value have been demonstrated as useful in photoreactor or PBR modeling (Dauchet, 2012; Kandilian et al, 2014; Pruvost and Cornet, 2012), to relate light absorption conditions to (biological) reactions. The rate of photon absorption represents the light effectively absorbed by the cells, which is a combination of light received (irradiance G) and the ability of the cells to absorb light (absorption coefficient Ea). As light absorption by cells depends of their pigment content, which is highly variable, rate of photon absorption was found more representative (both for kinetic modeling and cells regulation mechanisms) than considering the irradiance value alone.

Kandilian et al (2013) have shown the direct relation of the specific rate of photon absorption with lipid accumulation in the condition of nitrogen starvation, which is known to trigger lipid reserve (ie, TAG) accumulation but also to strongly decrease pigment content, thus altering light absorption by cells. The authors found that a given value of specific rate of photon absorption A was necessary to trigger TAG overaccumulation, and also that TAG synthesis rate was strongly related to A.

More recently, Soulies et al (accepted) investigated the influence of specific lighting conditions such as a change in light spectrum or incident angle. Introducing specific rate of photon absorption A was again found useful to relate these conditions to growth kinetics and thus make it possible to capture the respective influences of absorption rates and growth of red and white lights and non-normal incident angles. A key finding here was that white light decreases the negative effect of dark volumes. In contrast to red light, whose wavelengths were almost uniformly and rapidly absorbed in the culture volume, a part of the white light spectrum (ie, green light) was found to penetrate deeper in the culture volume meaning that at similar biomass concentration, white light showed a higher rate of absorption in the culture depth than red light. The net result was that this tended to decrease the expected positive effect of red light on biomass productivity. Those authors also reported marked pigment acclimation in the studied strain (ie, *C. vulgaris*) which tended to compensate the fast decrease in available light with culture depth (in the case of red light) but also non-normal incident angle. The rate of photon absorption was found to help efficiently represent all effects, and was then proven as a value of interest in microalgal culture optimization.

Generally speaking, introducing specific rate of photon absorption A could find applications for any case where light absorption rates are potentially relevant. This could be in the optimization of light attenuation conditions for achieving maximal biomass productivity, but also in solar operations where light conditions tend to be oversaturating, leading to possible photoinhibition. These features are introduced in a typical example given in Fig. 3B. Increasing biomass concentration in the cultivation system will decrease the rate of photon absorption due to stronger light attenuation, thus resulting in smaller irradiance G. As a result, peak biomass productivity will be obtained at an optimal photon absorption rate value. For the case simulated in Fig. 3, this optimal value is typically situated around 15–20 $\mu mol_{h\nu}\, g^{-1}\, s^{-1}$ (Fig. 3B). Note that this representation is consistent with the condition of luminostat regime ($\gamma = 1$), and both approaches can be used to maximize the biomass productivity of any cultivation system.

3. MODELING PBRs

3.1 Introduction

The previous engineering rules (Eq. 1) make it possible to calculate the maximal performances of a given culture system as a function of design, light received, and cultivated strain. Such information is highly valuable for scaling the system as a function of operational constraints, ie, objective of biomass production, algae farming resources available (land area, irradiation conditions, etc.). In many cases, this information is considered sufficient for the engineer to estimate, for example, the number/size of production units and the allied capital and operational costs (ie, CAPEX and OPEX). Bear in mind that these relations give theoretical maximal productivities whereas, in practice, productivities will be lower for many reasons: nonideal culture conditions (temperature or pH, dissolved carbon or medium, contamination), the strong influence of daytime culture conditions variation on growth kinetics (ie, weather conditions, day–night cycles), partial shading by other units or surrounding buildings or trees, nonoptimized harvesting strategies, and poor control of the irradiation field leading, for example, to photoinhibition phenomena.

The following section provides a knowledge model able to predict what is called "light-limited growth" (Pruvost and Cornet, 2012; Takache et al, 2012) where biomass production rate is only a function of light received (no mineral limitations, optimal pH, and temperature values). As discussed previously, appropriate engineering and operation of the cultivation system could make it possible to attain light-limited growth, but as culture systems can be limited by several other parameters, then quantitative information like biomass productivity will obviously be overpredicted. In some cases, this will be acceptable, as modeling is generally used to give a first estimation of process operation. In other cases, the model will have to be consolidated by adding equations related to effects of relevant parameters. However, as light will always influence growth (even in the case of severe limitation, like for nitrogen deprivation; see Kandilian et al, 2014), the model described in the following section will be able to serve as a basis for further model development work.

By definition, a light-limited growth model is able to couple light attenuation conditions to photosynthetic growth rate. This can prove invaluable when looking to further optimize the cultivation system, as it allows an in-depth understanding of this coupling which governs the culture response. More practically, it also serves to determine information of primary relevance like time course of biomass concentration (or biomass productivity) as a function of microalgal cultivation systems design and operating parameters. The interested reader is invited to refer to Pruvost et al (2011a) for a fuller description of the solar PBR model and to further work by Pruvost et al (Pruvost and Cornet, 2012; Pruvost et al, 2011a, 2011b, 2012) for more detailed investigations. This model is the culmination of years of development and has proved efficient in several settings including artificial and sunlight conditions (Cornet and Dussap, 2009; Pruvost et al, 2011a, 2012, 2015; Takache et al, 2012) to the scaling and optimization of PBRs of various shapes (Cornet, 2010; Loubiere et al, 2011), biomass optimization of different microalga and cyanobacteria strains (Cornet and Albiol, 2000; Cornet et al, 1992b, 1998, 2003; Farges et al, 2009; Pruvost et al, 2011b; Takache et al, 2010).

3.2 Overview of Light-Limited Growth Modeling in a PBR

Takache et al (2012) introduced a generic model for light-limited growth in PBRs. This model was recently slightly revised to take into account the specific rate of photon absorption A in place of irradiance G which was found more relevant for coupling light absorption influence to photosynthetic growth response. Specific rate of photon absorption A (Eq. 5) represents the light effectively absorbed by cells, which is the combination of light received (irradiance G) and ability of the cells to absorb light (absorption coefficient Ea_λ).

The light-limited growth model is based on the coupling between a kinetic photosynthetic growth model and a radiative transfer model to represent light attenuation in a PBR volume as a function of parameters affecting light transfer, ie, biomass concentration, microalgae radiative properties, and light emission characteristics (spectrum, PFD, incident angle). The coupling between radiative and kinetic growth models makes it possible to calculate the resulting mean volumetric biomass production rate $\langle r_X \rangle$ and then biomass concentration and productivity. An overview of the model is given in Fig. 4. The following section gives details for each subpart of the model.

3.3 Kinetic Growth Model

In light-limited conditions, the kinetic growth model is able to relate the heterogeneous light radiation field in the PBR to local photosynthetic growth rate. Photosynthetic growth can be expressed first from the local specific rate of oxygen production or consumption J_{O_2}, considered here at the scale of intracellular organelles, close to the primary photosynthetic and respiration events. When considering oxygen evolution/consumption, it is useful to introduce the compensation point of photosynthesis A_C (Cornet and Dussap, 2009; Cornet et al, 1992a; Takache et al, 2010). By definition, values of specific rate of photon absorption higher than A_C are necessary for a net positive photosynthetic growth (strictly, a net oxygen evolution rate). Values below the compensation point of photosynthesis have different effects depending on whether the cells are eukaryotic (microalgae) or prokaryotic (cyanobacteria). As cyanobacteria have their respiration inhibited by light, then for short-residence-time exposure to dark (Gonzalez de la Vara and Gomez-Lojero, 1986; Myers and Kratz, 1955), we can assume a nil oxygen evolution rate for irradiances below the A_C value. For eukaryotic microalgae, photosynthesis and respiration operate separately in chloroplasts and mitochondria. Hence microalgae, unlike cyanobacteria, present respiration both in the dark and in light. Oxygen consumption rates will thus be obtained for values below the compensation point of photosynthesis.

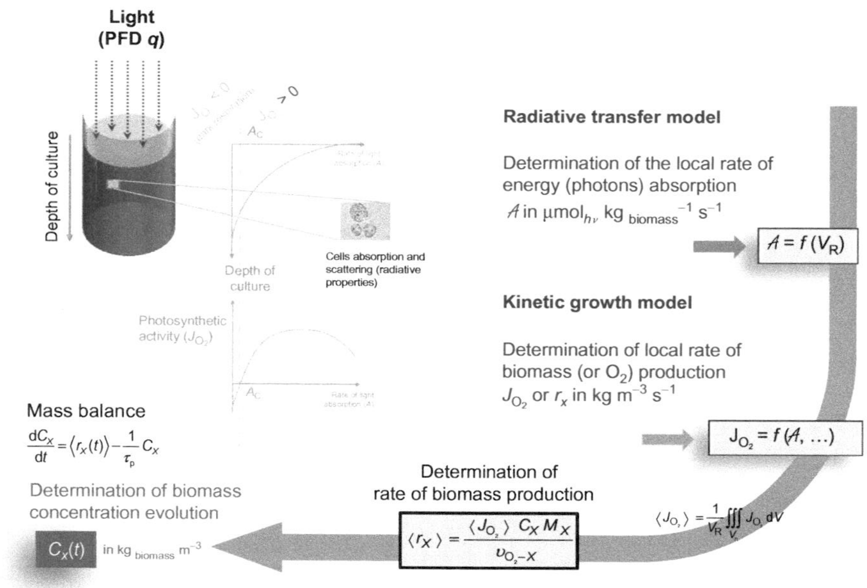

Figure 4 Overview of the general modeling approach used to simulate PBR.

The kinetic response needs to be related to the heterogeneous light distribution in cultivation systems, represented here by the specific rate of photon absorption A ($\mu mol_{h\nu} s^{-1} kg^{-1}$). As previously explained, Eq. (6) on the inhibition of respiration by light was proposed for cyanobacteria by Cornet and Dussap (2009):

$$J_{O_2} = \rho \overline{\varphi}'_{O_2} A \, \mathcal{H}(A - A_C) = \rho_M \frac{K}{K + G} \overline{\varphi}'_{O_2} A \, \mathcal{H}(A - A_C) \tag{6}$$

where $\mathcal{H}(A - A_C)$ is the Heaviside function ($\mathcal{H}(A - A_C) = 0$ if $A < A_C$ and $\mathcal{H}(A - A_C) = 1$ if $A > A_C$), $\rho = \rho_M \frac{K}{K + A}$ is the energetic yield for photon conversion of maximum value ρ_M (demonstrated as roughly equal to 0.8; Table 1), $\overline{\varphi}'_{O_2} = \upsilon_{O_2-X} \overline{\varphi}'_X$ is the molar quantum yield for the Z-scheme of photosynthesis as deduced from the structured stoichiometric equations (see Cornet et al, 1998; Pruvost and Cornet, 2012), and K is the half-saturation constant for photosynthesis depending on the microorganism considered.

Takache et al (2012) completed this formulation for the specific case of microalgae with an additional term (right-hand term in Eq. 6) to consider respiration activity in light (Takache et al, 2012), which was to be found especially necessary if a dark zone appears in the culture volume due to the significant contribution of respiration to resulting growth in the whole PBR. By introducing the specific rate of photon absorption $\mathcal{A}$ in place of irradiance G, as explained earlier, Eq. (7) can thus be used for microalgae:

$$\begin{aligned} J_{O_2} &= \left[\rho \overline{\varphi'}_{O_2} \mathcal{A} - \frac{J_{NADH_2}}{\upsilon_{NADH_2-O_2}} \times \frac{K_r}{K_r + \mathcal{A}} \right] \\ &= \left[\rho_M \frac{K_A}{K_A + \mathcal{A}} \overline{\varphi'}_{O_2} \mathcal{A} - \frac{J_{NADH_2}}{\upsilon_{NADH_2-O_2}} \times \frac{K_r}{K_r + \mathcal{A}} \right] \end{aligned} \tag{7}$$

where J_{NADH_2} is specific rate of cofactor regeneration on the respiratory chain, here linked to oxygen consumption by the stoichiometric coefficient $\upsilon_{NADH_2-O_2}$ (the stoichiometric coefficient of cofactor regeneration on the respiratory chain). Note that the effect, well known to physiologists, of the radiation field on the respiratory activity term was taken into account as an adaptive process of cell energetics (Cogne et al, 2011; Cournac et al, 2002; Peltier and Thibault, 1985). The decrease in respiration activity with respect to light was modeled here by an irradiance-dependent relation in a preliminary approach by simply introducing an inhibition term with a constant K_r describing the decreased respiration in light. We stress that this parameter is entirely determined by the knowledge of the compensation point of photosynthesis A_C ($J_{O_2}(A_C) = 0$) when the specific respiration rate J_{NADH_2} is known (roughly equal to $14 \times 10^{-3} \text{mol}_{NADH_2} \text{kg}_X^{-1} \text{s}^{-1}$, with A_C in the range 1500–3000 $\mu\text{mol}_{h\nu} \text{kg}^{-1} \text{s}^{-1}$ for eukaryotic cells and 200–500 $\mu\text{mol}_{h\nu} \text{kg}^{-1} \text{s}^{-1}$ for prokaryotic cells).

As a direct result of the light distribution inside the culture, the kinetic relation (Eq. 6 for cyanobacteria or Eq. 7 for microalgae) is of the local type. This implies calculating the corresponding mean value by averaging over the total culture volume V_R:

$$\langle J_{O_2} \rangle = \frac{1}{V_R} \iiint_{V_R} J_{O_2} \mathrm{d}V \quad (8)$$

For a cultivation system with Cartesian one-dimensional light attenuation (such as flat-panel PBRs), this consists of a simple integration along the depth of culture z:

$$\langle J_{O_2} \rangle = \frac{1}{L} \int_{z=0}^{z=L} J_{O_2} \mathrm{d}z, \quad (9)$$

where L is reactor depth. Once $\langle J_{O_2} \rangle$ is known, the mean volumetric biomass growth rate $\langle r_X \rangle$ can be deduced directly as:

$$\langle r_X \rangle = \frac{\langle J_{O_2} \rangle C_X M_X}{\upsilon_{O_2-X}} \quad (10)$$

where M_X is C-molar mass of the biomass and υ_{O_2-X} is the stoichiometric coefficient of oxygen production (see Table 1 for an example of parameters set). Hence the mass balance equations (Eqs. 5 or 6) can be solved for any light-limited growth operating conditions.

Finally, once the mean volumetric growth rate is known, the resolution of the mass balance equation for biomass can serve to calculate biomass concentration and productivity as a function of operating parameter (lighting conditions and dilution rate D—or residence time $\tau_p = 1/D$—resulting from the liquid flow rate of the feed):

$$\frac{\mathrm{d}C_X}{\mathrm{d}t} = \langle r_X \rangle - \frac{C_X}{\tau_p} \quad (11)$$

3.4 Modeling of Radiative Transfer

Solving Eq. (6) or (7) entails determining the field of the specific rate of photon absorption A, which is obtained from radiative transfer modeling. This modeling is highly dependent on cultivation system geometry and can range from simple one-dimensional (Cornet, 2010; Pottier et al, 2005) to complex three-dimensional PBR geometries (Dauchet et al, 2013; Lee et al, 2014). Luckily, most cultivation systems present light attenuation along only one main direction (ie, the depth of culture z), which makes it possible to apply a hypothesis of one light attenuation direction and thus apply a simplified model like the two-flux model that has already proved efficient in several studies (Cornet et al, 1995, 1998; Lee et al, 2014; Takache et al, 2012).

A full description can be found in Pottier et al (2005) and Pruvost et al (2011a) for the more general case of solar irradiation (direct and diffuse radiation, non-normal incidence angle). A typical solution is given below as a function of the incident angle θ to take into account the general case of oblique irradiation with any incident light spectrum ($\cos\theta = 1$ in the usual case of normal incidence):

$$\frac{G_\lambda}{q_\lambda} = \frac{2}{\cos\theta} \frac{(1+\alpha_\lambda)\exp[-\delta_\lambda(z-L)] - (1-\alpha_\lambda)\exp[\delta_\lambda(z-L)]}{(1+\alpha_\lambda)^2\exp[\delta_\lambda L] - (1-\alpha_\lambda)^2\exp[-\delta_\lambda L]} \quad (12)$$

where q_λ is PFD measured perpendicular to the illuminated surface, $\alpha_\lambda = \sqrt{\frac{Ea_\lambda}{(Ea_\lambda + 2b_\lambda Es_\lambda)}}$ is the spectral linear scattering modulus (also see Table 1 for PAR-averaged values), and $\delta_\lambda = \frac{\alpha C_X}{\cos\theta}(Ea_\lambda + 2b_\lambda Es_\lambda)$ is the spectral two-flux extinction coefficient, where Ea_λ and Es_λ are spectral values of mass absorption and mass scattering coefficients, respectively, for the cultured photosynthetic microorganism, and b_λ is the back-scattered fraction. As described elsewhere (Pottier et al, 2005; Soulies et al, submitted), Eq. (12) can be simplified for a spectrally averaged resolution to reduce the calculation effort.

Once the light attenuation profile is known, the illuminated fraction γ can be obtained (Cornet et al, 1994; Takache et al, 2010). The illuminated fraction γ is given by the depth of culture z_c where the value of specific rate of photon absorption for compensation $A(z_c) = A_C$ is obtained, with A_C being the minimum value required to obtain net positive photosynthetic growth ($J_{O_2} = 0$ for $A = A_C$ in Eqs. 6 & 7). In the case of cultivation systems with one-dimensional light attenuation, we obtain:

$$\gamma = \frac{V_{\text{light}}}{V_r} = \frac{z_c}{L} \quad (13)$$

A γ value below 1 indicates that all light available for a net photosynthetic growth is absorbed by the culture. Conversely, when the illuminated fraction is greater than 1 (a hypothetical representation because at maximum $V_{light} = V_r$), some of the light is transmitted.

3.5 Determination of Radiative Properties

The radiative transfer calculation implies determining spectral radiative properties of microalgal suspension (Ea_λ, Es_λ, b_λ). Despite obvious relevance to PBR modeling (and engineering), these properties remain tricky to determine as there are still only a handful of studies proposing robust determinations. Determining these properties is effectively far from a trivial task. Future developments should solve this problem by providing engineers with approaches that are easier to apply in practice.

As things stand, there are two approaches. Radiative properties can be determined experimentally from measurements with aspectrophotometer equipped with an integrated sphere. This method was proposed by Pilon et al and has already been applied on several species (Berberoglu et al, 2008; Kandilian et al, 2013; Pilon et al, 2011). Alternatively, radiative properties can be determined from theory (see Dauchet et al, 2015 for a revised description). However, the input parameters are the pigment contents and the size distributions of the cells, which have to be determined experimentally. For example, it was shown that *C. vulgaris* and *Chlamydomonas reinhardtii* could be assimilated to spherical equivalent particles with an average radius of 2 μm. Size distribution was found to be log-normal with a standard deviation of 1.218 μm for *C. vulgaris* and 1.17 μm for *C. reinhardtii* (Pottier et al, 2005).

3.6 Solar PBR Modeling

Modeling is especially useful to solar culture as it can relate the complex phenomena involved in these conditions, such as time variations in sunlight in terms of intensity, beam-diffuse radiation partitioning, or collimated angle onto the PBR surface, and their effects on radiative transfer in the culture volume and the resulting photosynthetic conversion and biomass growth. Several recent studies have modeled solar PBR operation in attempts to optimize productivities as a function of PBR design, location, and/or cultivated species (Pruvost et al, 2011a, 2012; Quinn et al, 2011; Slegers et al, 2011, 2013a, 2013b). In general, current models mainly aim to relate sunlight conditions obtained from meteorological databases to growth kinetics in order to predict PBR performances (Pruvost et al, 2011a; Quinn et al, 2011; Slegers et al, 2011). These models can provide valuable predictions of productivity for PBRs operated during an entire year. They can also assess the influence of various parameters, such as PBR location, harvesting strategy, strains cultivated, and the effects of night and day cycles. However, they may be regarded as oversimplified given the complexity and numbers of different parameters affecting PBR operation and productivity in outdoor conditions. As discussed earlier, numerous features can impair bioprocess production, from mineral or carbon limitation to nonideal temperature or pH control, nonoptimized harvesting strategies, contamination, and more. There is a clear need to pursue with efforts to develop a set of robust tools for solar cultivation optimization to achieve better accuracy and extend their applicability to other solar PBR-related challenges. For example, Slegers et al (2013a) integrated a thermal model able to predict the time-course evolution of culture temperature under solar conditions and assess its influence on growth. Temperature was found to strongly influence growth rate and the resulting biomass productivity.

For the so-called light-limited regime where only light limits growth, the model can be adapted to solar case. The main modifications compared to artificial light reside in the consideration of sunlight characteristics (non-normal incidence, direct and diffuse light) in radiative transfer calculation and the need to introduce a time resolution due to the time-changing irradiation conditions, with day and night periods requiring special consideration. This model is already described elsewhere (Pruvost and Cornet, 2012; Pruvost et al, 2011a, 2012, 2015), so only the main features are reported here. Note that all these features were proved relevant in the predictions of solar PBR performances.

The example in this section applies to cultivation systems presenting a flat illuminated surface (ponds, rectangular PBR, etc.). The one-dimensional and azimuth-independence assumptions can then be used to describe the irradiance field in the culture bulk, making it possible to apply the two-flux radiative model with its corresponding analytical solutions (Pottier et al, 2005). Application to the solar case implies factoring in non-normal incidence (thus introducing the incident angle θ) with a separate treatment of the direct and diffuse components of the radiation due to their difference in angular distribution over the PBR surface (Pruvost et al, 2011a). Total hemispherical incident light flux density (or PFD, see next section) q is divided into direct $q_{//}$ and diffuse $q_{\cap}$ components $\left(q = q_{//} + q_{\cap}\right)$. Total irradiance (representing the amount of light received in the culture bulk) is given by summing the resulting contribution of collimated and diffuse radiation:

$$G(z) = G_{\mathrm{col}}(z) + G_{\mathrm{dif}}(z) \tag{14}$$

where G_{col} is the irradiance field for collimated radiation, as given by:

$$\frac{G_{\text{col}}(z)}{q_{//}} = \frac{2}{\cos\theta}\frac{(1+\alpha)\exp[-\delta_{\text{col}}(z-L)] - (1-\alpha)\exp[\delta_{\text{col}}(z-L)]}{(1+\alpha)^2\exp[\delta_{\text{col}}L] - (1-\alpha)^2\exp[-\delta_{\text{col}}L]} \tag{15}$$

and G_{dif} the irradiance field for diffuse radiation:

$$\frac{G_{\text{dif}}(z)}{q_{\cap}} = 4\frac{(1+\alpha)\exp[-\delta_{\text{dif}}(z-L)] - (1-\alpha)\exp[\delta_{\text{dif}}(z-L)]}{(1+\alpha)^2\exp[\delta_{\text{dif}}L] - (1-\alpha)^2\exp[-\delta_{\text{dif}}L]} \tag{16}$$

In these equations, $\delta_{\text{col}} = \frac{\alpha C_X}{\cos\theta}(Ea + 2bEs)$ and $\delta_{\text{dif}} = 2\alpha C_x(Ea + 2bEs)$ are the two-flux collimated and diffuse extinction coefficients, respectively.

Determining the irradiance field makes it possible to determine the corresponding local photosynthetic growth rate in the culture volume. The same kinetic relations (Eq. 6 or 7) can be applied here, making it possible to calculate mass volumetric biomass growth rate $\langle r_X \rangle$ (Eq. 11). The only restriction is that Eqs. (6) and (7) are valid insofar as the culture is illuminated (ie, during daytime). At night, long dark periods of several hours trigger a switch to respiratory metabolism which results in biomass catabolism (Le Borgne and Pruvost, 2013; Ogbonna and Tanaka, 1996). This biomass catabolism is species dependent and differs strongly between eukaryotic (microalgae) and prokaryotic (cyanobacteria) cells. For *Arthrospira platensis* and *C. reinhardtii*, values of $\langle r_X \rangle / C_X = \mu = 0.001$ and $0.004\ \text{h}^{-1}$, respectively, were recorded at their optimal growth temperature, ie, 308K for *A. platensis* and 293K for *C. reinhardtii* (Cornet, 1992; Le Borgne, 2011).

Finally, the determination of the mean growth rate allows the mass balance equation, here for biomass, to be solved (Eq. 11). The variable PFD in sunlight conditions means that the irradiance field inside the culture bulk and the resulting local and mean volumetric growth rates vary continuously, and hence steady state cannot be assumed in Eq. (11). This implies solving the transient form of the mass balance equation. Once the time course of biomass concentration has been determined, the corresponding biomass productivity can be calculated, as well as surface productivity P_S (g m^{-2} day^{-1}) which is a useful variable to extrapolate to land-area production (Eq. 2).

4. OPTIMIZATION OF PBR OPERATION

4.1 Understanding Light-Limited Growth

In practice, the control of culture conditions such as pH and temperature can prove challenging, especially in outdoor conditions (Borowitzka, 1999;

Grobbelaar, 2008; Richmond, 2004a; Torzillo et al, 1996). These challenges can, however, be overcome with adequate engineering and control of the cultivation system. As technical solutions are highly dependent on culture system technology, these aspects will not be discussed in detail here. The main point is that if all cultivation conditions are kept at optimal values and nutrients are provided in adequate amounts, then light-limited conditions should eventually occur, which is crucial given that the light-limited regime has several major features.

The first consequence of light-limited conditions achievement is that, by definition, the culture is not subject to any further limitation other than light use. Thus, maximum biomass productivity can be achieved and is determined by the amount of light provided and its use by the culture (Pruvost, 2011; Pruvost and Cornet, 2012; Pruvost et al, 2011b, 2012; Takache et al, 2010). Any limitation other than light limitation would result in further decreases of biomass productivity, whereas maximizing the PFD received on the culture system increases its productivity. Note that this remains valid in the case of high PFD leading to photoinhibition of the photosynthetic apparatus (PFD grossly greater than 400 $\mu mol_{h\nu}\, m^{-2}\, s^{-1}$). Special attention should be paid to light attenuation conditions to avoid or at least greatly reduce photoinhibition phenomena by operating the PBR to achieve complete light extinction in the culture, as described in detail in the next section.

A second important consequence is that in the light-limited regime, controlling the incident light and its effect on the process equates to controlling aggregate cultivation system performance. This is the so-called physical limitation in chemical engineering, where the process is limited by one parameter which, if controlled, enables control of the entire process. This feature is essential to the efficient design and operation of photobiological cultivation systems. The role of light in the rational design of microalgal cultivation systems has been touched on earlier and will be explored in greater depth later in this chapter by actual examples of technologies. Implications in terms of operation are discussed later.

4.2 Optimizing Light Attenuation Conditions for Maximal Biomass Productivities in PBRs

Although a necessary condition, the light-limited regime alone is not sufficient to obtain maximal biomass productivities, which also hinge on controlling radiative transfer conditions inside the culture (Cornet and Dussap, 2009; Pruvost, 2011; Takache et al, 2010). As already discussed, if biomass concentration is too low, some of the light gets transmitted through the culture, and if biomass concentration is too high, a dark zone appears deep in the culture. For eukaryotic cells like microalgae that demonstrate respiration in light, a dark zone in the culture volume where respiration is predominant will result in a loss of productivity due to respiratory activity. Maximal productivity will then require the specific condition of full absorption of all light received but without a dark zone in the culture volume—in other words the *luminostat* regime (Pruvost and Cornet, 2012; Takache et al, 2010). As a result, unlike processes based solely on surface conversion (eg, photovoltaic panels), optimizing the amount of light collected on the microalgal cultivation system surface is still not sufficient. As light conversion by photosynthetic microorganisms occurs within the culture bulk, transfer of the collected light flux inside the bulk has to be taken into account.

Light attenuation conditions can be controlled by adjusting biomass concentration in the cultivation system (see Section 2.1.2.5), which can be done in continuous mode by modifying the residence time τ_p applied to the system (or dilution rate $D=1/\tau_p$). In practice, maintaining optimal light attenuation conditions is no easy task, especially in the case of solar production which adds a degree of complexity to the optimization and control of the cultivation system compared to artificial illumination. The process is fully dynamic and driven by an uncontrolled input, ie, solar incident flux. Under sunlight, biomass growth rate is insufficient to compensate for the rapid changes in sunlight intensity. Consequently, light attenuation conditions that are fixed by biomass concentration are never optimal. A compromise has to be found on the conditions thus applied, for example, by defining a residence value that will maximize biomass productivity over the period of operation by acting on biomass concentration time course and the related light attenuation conditions.

4.2.1 The Role of Light Attenuation Conditions in Culture Stability

Although a dark volume has an impact on respiration activity (see next), high light attenuation conditions are also well known to have a positive effect on culture stability (Carvalho et al, 2011; Grima et al, 1999; Hindersin et al, 2013; Richmond, 2004b; Torzillo et al, 1996). Light transmission also corresponds to a high light received per cell (ie, high specific absorption rates, see Section 2.1.2.5), which could induce culture drift by oversaturation of the photosynthetic chain (Grima et al, 1996; Hindersin et al, 2013; Wu and Merchuk, 2001). Note also that this generally also results in photoacclimation and a decrease of pigment content (Zonneveld, 1998), which in turn increases light penetration in the culture depth, and thus the light received per cell, thereby increasing culture drift. In practice, the culture will become highly unstable when transmission occurs, especially if PFD is higher than 200 $\mu mol_{h\nu}\,m^{-2}\,s^{-1}$. A typical result is given in Fig. 5, which depicts a *C. vulgaris* culture in a lab-scale PBR. The same PFD is applied in all experiments, and only light attenuation conditions are modified through changing biomass concentration by adjusting residence time. In the light transmission conditions, pigment content decreased as a result of a higher specific photon energy absorption rate. This decrease was especially visible for chlorophylls, where it results in a higher carotenoids-to-chlorophylls ratio (as shown by the yellow (light gray in the print version) color of the culture). In practice, it also generally marks the appearance of biofilm, despite the lower biomass concentration obtained, leading to a progressive culture drift up to potential washout.

For the operator, a general rule will be to promote full-light attenuation conditions. Note that this condition will be difficult to fulfill in some cases, such as in species presenting low pigment content (such as strains with small

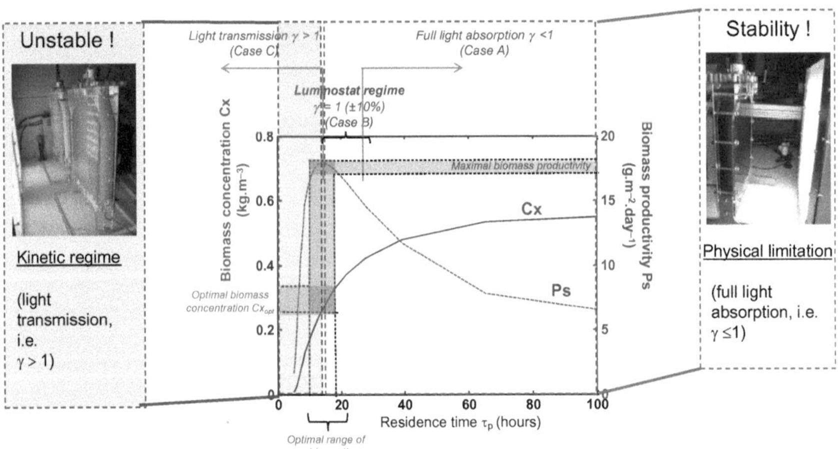

Figure 5 Effects of light attenuation conditions on culture stability and biomass productivity. For low residence time, low biomass concentration results in light transmission and high rate of photon absorption (ie, high light received per cell) inducing possible culture drift. For high residence time, the dark volume then generated can have a negative effect on biomass productivity due to the promotion of respiration activity, but also results in more stable culture because of a lower rates of photon absorption (ie, lower light received per cell).

antennas; Berberoglu et al, 2008) due to their lower absorption, or in solar conditions due to the time-course changes in light attenuation conditions, as we will see later.

4.2.2 Microalgae vs Cyanobacteria

If the biomass is too high, a dark zone appears in the culture. Here is an important distinction to make between eukaryotic (microalgae) and prokaryotic (cyanobacteria) cells. In cyanobacteria cultivation, as the cells have common electron carrier chains and no short-time respiration in the dark (Gonzalez de la Vara and Gomez-Lojero, 1986), a dark zone will be sufficient ($\gamma \leq 1$) to guarantee maximal productivity (Cornet, 2010; Cornet and Dussap, 2009). For eukaryotic cells presenting respiration in light (microalgae), a dark zone in the culture volume where respiration is predominant will result in a loss of productivity due to biomass catabolism. Achieving maximal productivity will thus be contingent on the γ fraction meeting the exact condition $\gamma = 1$ (the "luminostat" regime), corresponding to full absorption of all light received but without a dark zone in the culture volume (Takache et al, 2010).

In practice, maintaining an optimal value of the γ parameter is not easy, especially with microalgae (where the condition $\gamma = 1$ has to be met). Some illustrations are given below for both batch and continuous production modes. Because the regime does not allow full absorption of the light captured, light transmission always leads to a loss of efficiency, in addition to possible culture drift due to an excess of light received per cell, as discussed earlier ($\gamma > 1$). This regime is, however, usually encountered at the beginning of a batch production run (Fig. 6A). Biomass growth means that the light attenuation conditions will continuously evolve and the γ value will progressively decrease down to a value below 1. For prokaryotic cells, as soon as full absorption is obtained, the maximal value of the mean volumetric growth rate will be achieved and then remain constant (until a large dark zone is formed, inducing another possible shift in cell metabolism). For eukaryotic cells, the condition $\gamma = 1$, and thus the maximal value of the mean volumetric growth rate $\langle r_X \rangle$ will only be transitorily satisfied (mean volumetric growth rate being represented by the slope of $C_X(t)$, see Eq. 11 with $1/\tau_p = 0$). The increase in the dark volume will then progressively decrease the mean volumetric growth rate.

In continuous mode, light attenuation conditions can be controlled by modifying the dilution rate to adjust the in-system biomass concentration. For cyanobacteria (Fig. 6B), there will be an optimal range of biomass

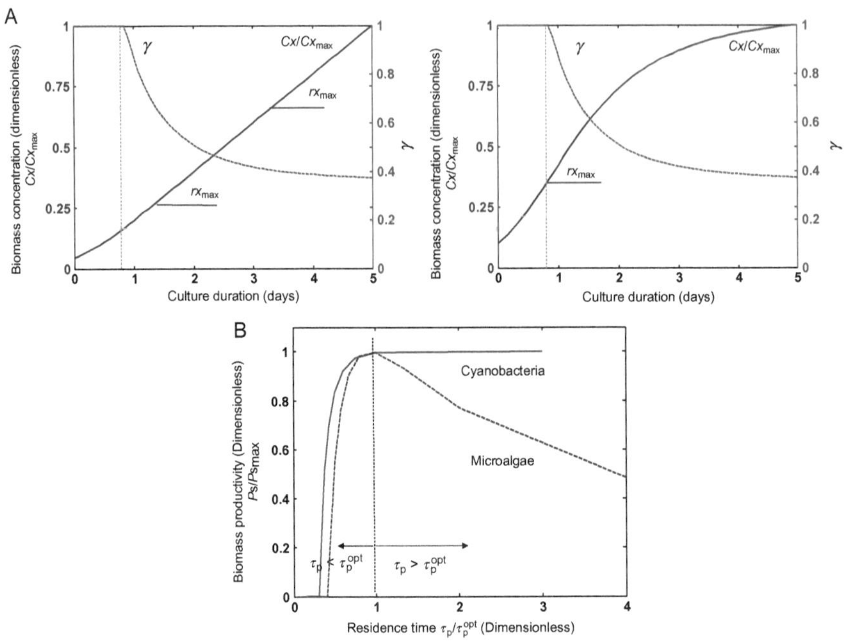

Figure 6 Typical evolution of biomass concentration during a batch cultivation of cyanobacteria and microalgae in light-limited conditions (A). Biomass productivity and concentration in continuous mode are given in (B).

concentrations to meet the condition $\gamma \leq 1$. For microalgae, the $\gamma = 1$ condition will require an optimal biomass concentration (C_X^{opt}) corresponding precisely to the condition of full-light absorption but no dark zone (as shown in Takache et al, 2010, a deviation of the γ value in the range $\gamma = 1 \pm 15\%$ is tolerable in practice).

Whichever production mode (continuous or batch) is used, the control of light attenuation conditions, represented here by the illuminated fraction (with $\gamma \leq 1$ for cyanobacteria and $\gamma = 1 \pm 15\%$ for microalgae), makes it possible to obtain the maximum biomass productivity of the cultivation system in light-limited conditions (volume and surface). If radiative transfer conditions are known (using a radiative transfer model, as already described), then the optimal biomass concentration can be determined theoretically, or else experimentally simply by varying the residence time and measuring corresponding biomass concentration and productivity (Takache et al, 2010).

4.3 Optimizing Light Attenuation in Solar Cultivation

Outdoor conditions and the use of sunlight as primary energy source pose several challenges to the engineering design and control of outdoor cultivation systems. Sunlight is characterized by a wide, rapid, and uncontrolled variation in irradiation conditions. On a single day, the PFDs received onto a cultivation system surface can range from null (night) to potentially damaging levels for the photosynthetic chain of growing cells (high PFDs typically larger than 1000 μmol m^{-2} s^{-1}, which are commonly encountered in most locations on Earth in summer). Strong light attenuation in the PBR is in this case known to have a positive effect as it decreases the amount of light energy received per cell along the depth of the PBR (Carvalho et al, 2011; Hindersin et al, 2013; Torzillo et al, 1996).

The amount of direct and diffuse solar incident irradiance as well as the strongly time-dependent incident PFD and the associated incident angle have also been found to significantly dictate process efficiencies (Pruvost et al, 2011a, 2012). Consequently, although the luminostat regime is the ideal case leading to maximum biomass productivity, it cannot be maintained under solar conditions due to how much faster light varies with time than biomass concentration (Hindersin et al, 2013; Pruvost et al, 2011a, 2012). The net result is that there is a design and operation compromise to be found.

In continuous or semi-continuous PBRs, this can be achieved by defining, for example, a residence time that maximizes yearly biomass productivity through control over the temporal evolution of the biomass concentration and light attenuation in the PBR. Modeling can prove invaluable here by simulating PBR operation over a whole-year period as a function of various key parameters such as (1) PBR location, design, inclination, and orientation; (2) PBR operating parameters (harvesting strategy for instance); and (3) species cultivated.

Fig. 7 gives examples of yearly biomass productivities as a function of residence time applied on the cultivation system (see Pruvost et al, 2015 for details). As was the case for continuous light, an optimal value exists, but it corresponds to the value that gives the maximal productivity over a given time period. Simply optimizing the residence time in the cultivation system is not enough to maintain the ideal luminostat regime condition ($\gamma = 1$) because the illumination conditions vary so much faster than the kinetics of photosynthetic growth. The optimal residence time can only be regarded as the best compromise to maximize productivity on a given cultivation period (a full year period here). The immediate consequence is that it will result in large variation of light attenuation conditions with time in the cultivation system.

Obviously, the residence time value can be optimized all along the year. In winter, for example, increasing the residence time can prove beneficial for

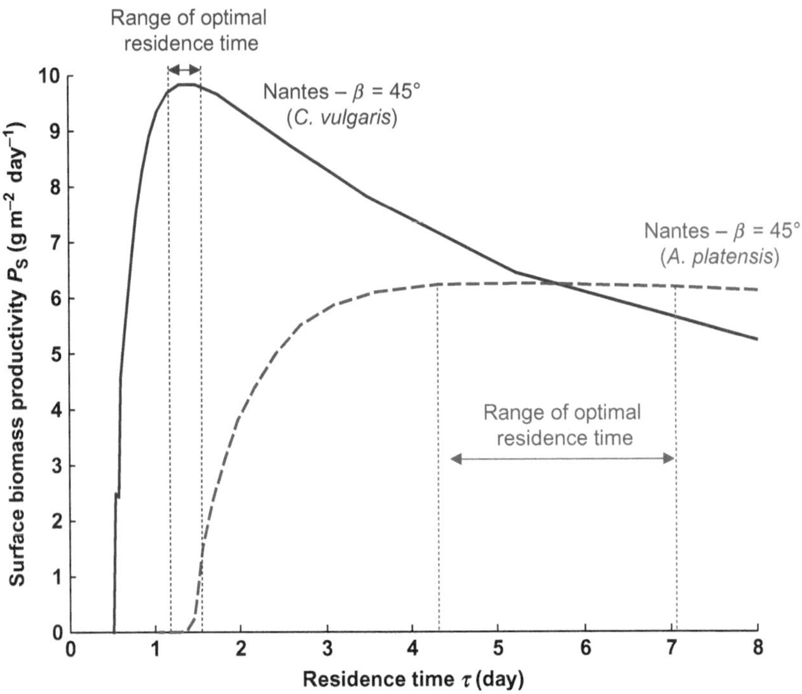

Figure 7 Yearly average areal productivity of an inclined flat-panel PBR (45 degree) as a function of the residence time applied on the cultivation system operated in continuous mode (Nantes locations, France). Values are given for the microalga *C. vulgaris* and for the cyanobacteria *A. platensis*, illustrating the narrower range of residence time to maximize productivity for eukaryotic cells as explained by their sensitivity to dark volumes induced by too high values of residence time values.

microalgae due to their lower growth, which means longer residence times for this specific period can have positive impacts on net biomass productivity. Modeling is again valuable here, as it can be used to calculate biomass productivity for any residence time value and to define an optimal year-long residence time course. Looking at the *C. vulgaris* growth presented in Fig. 7, ideally, higher values should have been applied in winter (up to $\tau_p = 2.3$ days), and lower values applied in summer (down to $\tau_p = 0.8$ day).

Fig. 7 also compares biomass productivities between microalgae (ie, *C. vulgaris*) and cyanobacteria (ie, *A. platensis*). The same type of evolution is achieved for both species at low residence times (rapid decrease of surface productivity toward culture washout for low residence time values, ie, high dilution rate), but *C. vulgaris* showed significantly different productivity at high residence time values whereas *A. platensis* showed little impact.

As a consequence, maximum values of surface productivities for *C. vulgaris* were only found for a narrow range of residence times. This important difference between the two microorganisms is explained by the negative influence of dark volume on microalgae growth kinetics. High residence times result in higher biomass concentrations and light attenuation conditions. As already observed in continuous light conditions, the impact is negligible for cyanobacteria but not for microalgae due to their respiration activity in the dark. This result has important practical implications: a harvesting strategy that maximizes biomass productivity is fairly easy to find for cyanobacteria ($\tau_p \leq \tau_p^{opt}$) but very difficult to find for the microalga *C. vulgaris*.

Another important issue resides in the light regimes obtained in the culture volume when operated in solar conditions. Once the residence time is defined, the year-long time course of biomass concentration can be calculated and thus the corresponding time evolutions of light attenuation conditions. Variations in incident irradiation mean that a wide range of light attenuation conditions can be encountered inside the culture volume over the course of a day, which can affect process stability, as described in Pruvost et al (2015, in press). As shown in Section 2.1.2.5, harvesting strategy (ie, residence time) will directly affect these light regimes. For example, promoting a higher residence time will increase biomass concentration and light attenuation (ie, decreasing photon absorption rates). This could reveal beneficial for periods where oversaturating light is encountered, such as at noon in summer. However, as increasing light attenuation conditions could also result in a decrease in biomass productivity, particularly with species that show significant respiration activity under illumination, then it will almost certainly be necessary to find a trade-off between process productivity, stability, and robustness. Here again, models can help. Modeling combined with in-depth investigations of the effect of oversaturating light on culture stability could serve as a foundation to advanced control strategies able to maintain the appropriate trade-off between biomass productivity maximization and robust culture operation, which is currently a big challenge for optimal solar culture system operation.

5. DEVELOPMENT OF COMMERCIAL TECHNOLOGIES BASED ON PBR ENGINEERING RULES

5.1 Introduction

There is a wide variety of PBR technologies available, including tubular, cylindrical, and flat-panel systems (some examples are given in Fig. 2). This diversity of PBR designs is the result of various attempts to optimize light capture while satisfying other practical constraints related to (1) engineering design, including system integration, scale of production, materials selection, and project costs; and (2) system operation factors such as CO_2 bubbling, oxygen removals, temperature and pH regulation, nutrient delivery. The literature counts an array of reports and publications on the various PBR technologies available (Borowitzka, 1999; Carvalho et al, 2006; Grima et al, 1999; Morweiser et al, 2010; Pruvost, 2011; Pulz, 2001; Ugwu et al, 2008), all of which have advantages and limitations in terms of control of culture conditions, culture confinement, hydrodynamic conditions, scalability, construction cost, biomass productivity, and energy efficiency. Regardless of the PBR concept employed, the goal is to provide sufficient control of the culture conditions to make the process only limited by the amount of light supplied and the photosynthetic process in the culture (ie, the "light-limited" regime presented in Section 3.1). PFD incident onto the PBR surface and PFD locally available inside the culture are major parameters. Although maximizing light intercepted is an obvious consideration of any microalgal cultivation system (as with any light-driven process), there are other constraints to also consider. For example, using the airlift method for mixing will preclude horizontal geometries. Shading needs to be accounted for when arranging vertical or tilted solar systems on a given land area. As a result, what characterizes microalgae culture more than any other bioprocess is the wide range of constraints involved, from light use optimization to cost of production, which ultimately makes an optimal culture technology impossible to define. These features have fueled the idea that the development of microalgal culture systems is more or less empirical. Nevertheless, as seen earlier in this chapter, there are several engineering tools now available, making it possible to propose rational and robust methods for the design of optimal geometries taking into account the application constraints. This is illustrated in the following sections by specific examples of PBR developments. The examples come from a community of groups that now share the same engineering tools (ie, as tools described here). Some of these examples are still promising lab-scale prototypes, while others are industrial units commercialized by French company AlgoSource Technology (www.algosource.com).

5.2 Artificial Light Culture Systems

5.2.1 Lab-Scale Technologies

Lab-scale technologies are very useful for fundamental studies as well as in investigations for parameter-setting process models or operating protocols for cultivation system scaling and optimization. The main constraint in the lab-scale design is to achieve well-controlled conditions, with the appropriate scale of production to allow sufficient sampling during the experiment (usually leading to around 1 L of culture volume). Note that the setting of experiments in well-controlled conditions is a general constraint of any kind of biological study. This is usually achievable in simpler systems, such as flasks (ie, studies on bacteria and yeasts for example), but in studies on photosynthetic microorganisms, the strong influence of light supply and especially light attenuation conditions makes it difficult to use flask-type systems for well-defined lab-scale investigations. At best, PFD received on the flask surface can be defined as an operating parameter. However, as seen many times above, growth rate is not only a function of PFD but also of light attenuation conditions, which are nigh impossible to determine in flask systems (3D geometries). Ideally, for investigations on photosynthetic microorganisms, engineers should opt for geometries enabling to control (and ideally, determine) light attenuation in the culture volume. Lab-scale systems should then be required to fulfill this condition for accurate and representative results. Note that this consideration applies not only to bioprocess investigations (ie, PBR optimization) but also to fundamental biology investigations where photosynthesis and thus light absorbed is a relevant factor (ie, most studies devoted to photosynthetic microorganisms).

5.2.1.1 The Torus-Shaped PBR

The torus-shaped PBR is a typical example of a microalgal culture system designed for lab-scale experiments requiring firm control of culture conditions (Fig. 2). The main characteristic is its torus shape, as the culture is circulated by the rotation of a marine impeller. The combination of the impeller and the loop configuration of the torus geometry allows good mixing without dead volume while keeping shear stress in a reasonable range (Pruvost et al, 2006). The light-supplying device (LED panel) is placed in front of the PBR. The plane front surface and the square-sectioned torus channel mean that the PBR presents no curved surfaces perpendicular to the light source. This prevents optical distortion along the light emission direction, which enables easy calculation of light transfer, as light attenuation occurs along only one main direction, and leads to the so-called one-dimensional hypothesis. This configuration makes it possible to accurately determine light attenuation conditions (using Eq. 12 for example) and then local rate of photon absorption (Eq. 5), which can be related to photosynthetic growth (Eq. 7) or averaged over the culture volume to investigate its role as a process parameter (Kandilian et al, 2014).

The torus-shaped PBR has been used for several studies in recent years, both to model and optimize microalgal biomass productivity (Takache et al, 2010, 2012) and to investigate the coupling between hydrodynamics and photosynthetic conversion (the "light/dark cycles effect"; Takache et al, 2015). As it employs mechanical mixing, the torus-shaped PBR was also found valuable for studies requiring accurate gas analysis. When combined with online gas analysis, the obtained setup was able to yield kinetics information on culture evolution as a function of culture conditions, which was impossible or at least less accurate with air injection-mixed technologies (ie, airlift PBR) due to gas dilution with the gas-flow carrier. A typical example is the investigations on H_2 production from *C. reinhardtii* (Fouchard et al, 2008), where online monitoring of gas released and consumed (O_2, CO_2, H_2) combined with biotic-phase (total biomass and biomass composition in sugars, proteins, lipids, pigments) and abiotic-phase (carbon and mineral compound consumption) measurements made it possible to determine the effects of sulfur deprivation to induce anoxic conditions and starch accumulation and subsequently H_2 production by microalgae. This work enabled a kinetic model to be set that, when combined with the highly controllable conditions of the torus-shaped PBR, led to the development of an optimized H_2 production protocol (Degrenne et al, 2008, 2010, 2011a, 2011b; Fouchard et al, 2005, 2009). This setup was recently extended to the investigation of CO_2 fixation by microalgae (Le Gouic, 2013). Another example of the use of the torus-shaped PBR can be found in Martzolff et al (2012). The good mixing performance, and especially the plug flow behavior encountered in the torus loop, was used for isotopic nonstationary 13C-metabolic flux analysis. This enabled the characterization of the kinetics of 13C-labeling incorporation, which helped to define the biochemical reaction network of *C. reinhardtii* (Cogne et al, 2011).

All those examples illustrate the large interest of using lab-scale PBR presenting a high control of culture conditions for fundamental studies, which encircles in-depth investigations of microalgae metabolism and physiology, and the setting and optimization of culture protocol for applications of interest.

5.2.1.2 Efficient Overproducing Screening System–Photobioreactor

A cultivation system specially adapted to screening microalgae in reliable conditions was recently developed (Fig. 2). This system, named efficient overproducing screening system–photobioreactor (EOSS-PBR), was used to evaluate cell growth and productivity to compare strain performances or characterize the effects of specific culture conditions on a given strain culture. It consists of six small-scale PBRs (bubble columns) operated in parallel. Each tube has a volume $V_r = 30$ mL and an illuminated area $S_L = 0.008$ m^2. The EOSS-PBR was fully automated in terms of medium injection and biomass harvesting to allow semi-continuous cultivation with the ability to set different feeding or harvesting sequences in each tube.

EOSS-PBR enables both batch and semi-continuous cultivation. Semi-continuous cultivation was found relevant for allowing strains to progressively adapt to the growth conditions of PBR cultivation. For example, after receiving strains from collections, it was necessary to wait for several weeks before reaching a stable biomass production, indicating progressive adaptation of the cultivated strain to the conditions applied. The same behavior was observed when comparing various growth media on a given strain. As a result, this simple and easy-to-use system was found useful for adapting strains to PBR growth conditions to compare algae performances in reliable conditions. Examples of its use can be found in Taleb et al (2015).

5.3 Industrial Technologies

5.3.1 Introduction

The state of art in industrial-scale technologies (Carvalho et al, 2006; Janssen et al, 2003) covers a broad gamut of geometrical configurations, led by pneumatically agitated vertical column reactors, tubular reactors, and flat-panel reactors. The main reason for this diversity is that there is still no ideal starting geometry for microalgal culture systems, as illustrated here with two examples of artificial light PBRs designed by researchers and marketed by AlgoSource for different practical applications.

5.3.2 Multimodule External-Loop Airlift PBR for Hatcheries

The first culture system was designed for continuous microalgal production in mollusk hatcheries. The context of mollusk hatcheries poses two main constraints on PBR design. First, the microalgae commonly cultivated for mollusks are usually stress-sensitive species, such as *Isochrysis affinis galbana*, *Chaetoceros calcitrans*, *Thalassiosira pseudonana*, *Skeletonema marinoï* (Borowitzka, 1997). Second, microalgal production in hatcheries is mainly performed by batch cultures in systems consisting of vertical aerated column reactors. Their major drawbacks are only partial control of biomass quality and quantity, low productivity, contamination, manpower intensiveness (frequent handling and cleaning operations), and biofouling. To improve this situation, the PBR project brief was to enable continuous production adjustable to hatchery requirements, in a closed and artificially lit system, with the simplest possible design to minimize manufacturing cost, price, and floorspace; to be robust (marine atmosphere); and easy to clean. To respond to these constraints, the PBR consisted of a succession of elementary modules, each composed of two transparent vertical interconnected columns. Liquid-phase circulation was performed pneumatically, ie, by gas injectors placed at the column bottom and uniformly dispatched across the whole PBR. Each elementary module can thus be seen as an external-loop airlift PBR, except that the outlet of the downcomer is connected to the adjacent module. In addition, as already used in annular PBRs (Muller-Feuga et al, 2003a, 2003b; Pruvost et al, 2002a) and torus PBRs (Pottier et al, 2005; Takache et al, 2010), a swirling motion was generated (tangential inlets) in order to minimize biofilm formation on the walls while keeping shear stress within a reasonable range. As a result, a PBR technology adapted to the specific case of microalgal production in mollusk hatcheries was proposed. The major utility of this design was not only to enable easy continuous microalgal culture in a closed, artificially illuminated system but also to offer volume modularity without scale-up/down calculations. Indeed, the illumination device was designed so as to conserve the same illuminated specific surface (ratio of illuminated surface to culture volume) whatever the number of modules, implying that, for a given incident flux and a fixed dilution rate, volumetric productivity remains identical for either a monomodule or multimodule PBR. Consequently, the number of elementary modules is only dictated by the microalgal production required. A complete description of the design procedure can be found in Loubiere et al (2009).

5.3.3 Two-Side Illuminated Flat-Panel Airlift PBR

The culture system described earlier is based on cylindrical tubes, which makes it difficult to calculate radiative transfer in the culture volume, which has to be solved numerically (Lee et al, 2014). As already described, the "one-dimensional hypothesis" where light attenuation occurs along only one main direction serves to obtain analytical relations to represent the light attenuation field (as with the two-flux model, Eq. 12). This enables accurate and easy determination of light attenuation conditions for any operating conditions and thus greater system control. Based on this statement, researchers designed a specific PBR. Like the multimodule external-loop airlift PBR, this system is of industrial size (130 L), but the unit is a flat panel with front illumination so as to respond to the one-dimensional hypothesis. It is also illuminated on both sides to increase specific illuminated area

($a_{light}=18.2\ m^{-1}$). The culture is pneumatically agitated. To allow steam sterilization, the PBR is made of 316L-grade stainless steel and runs in continuous culture mode with full regulation of growth parameters such as pH and temperature. This technology has proven especially suitable for cases where biomass production necessitated well-controlled conditions. An example of its use for optimized lipid production can be found in Pruvost et al (2011b). This technology is currently used in the AlgoSolis R&D facility to continuously produce microalgal biomass of constant quality (axenic conditions) so as to inoculate large outdoor culture systems (such as enclosed raceways, see next).

5.4 Solar Technologies

5.4.1 Surface and Volumetrically Lighted Systems

Light can be supplied in two general ways: by directly lighting the culture system or by distributing light sources inside the culture volume. Next, there are either surface-illuminated systems or volumetrically illuminated systems. Most cultivation systems fall in the simpler surface-illuminated category (Carvalho et al, 2006; Morweiser et al, 2010; Richmond, 2004a; Ugwu et al, 2008). As with any other solar process, various positioning options have been considered, including systems positioned horizontally (Acién Fernández et al, 2001; Molina et al, 2001; Oswald, 1988), vertically (Chini Zittelli et al, 2000, 2006; Pulz, 2001), and even tilted (Doucha and Livansky, 2006; Lee and Low, 1991; Richmond and Cheng-Wu, 2001). However, maximizing the incident solar radiation flux is no easy task. For a start, it depends on longitude and latitude of system location and on the day of the year. For example, horizontal systems are best suited for locations close to the equator (latitude 0 degree). For higher latitudes, it is necessary to tilt the exposed system surface to maximize the amount of light collected. Roughly speaking, the optimum inclination angle with respect to the Earth's surface to maximize light capture over the year on a fixed PBR corresponds to the latitude of the PBR location (Duffie and Beckman, 2006; Hu et al, 1996; Pruvost et al, 2012; Richmond and Cheng-Wu, 2001). Inclination angle can also be adjusted as a function of time to optimize light capture. Flat panels equipped with sun-tracking systems were tested by Hindersin et al (2013), and the method not only maximized light capture during the day but also prevented excessive incident irradiation on the systems around noon by temporarily setting the illuminated surface of the PBR perpendicular to the sun's collimated irradiation.

Volumetrically illuminated systems require more complex technologies than surface-illuminated systems, but they do enable optimization of light delivery and use in the culture. First, inserting light sources in the volume of the culture guarantees maximal use of the collected or emitted photons. Second, and more interestingly, internal lighting allows light to be "diluted." As discussed earlier, increasing PFD leads to higher volumetric productivity but also a progressive decrease in conversion yield due to photosynthesis saturation. By diluting the light incident on the system's surface into the volume of the culture, a larger yield can be maintained. This is of particular interest in outdoor PBRs exposed to sunlight. In this case, the solar radiation incident on a given surface is collected using a parabolic solar collector for example, and is then delivered to the culture in a controlled manner, using optical fibers (Cornet, 2010; Csogör et al, 2001) or light guides (Pilon et al, 2011) for example. The characteristically high PFD of solar conditions means surface productivity can be increased. Note that the optical connection between the light collection device and the light delivery system needs to be carefully designed as it can be a source of major optical losses. Furthermore, light dilution can be combined with a solar tracking system, offering an added possibility of optimization by maximizing light intercepted as the sun travels in the sky (Hindersin et al, 2013). A full description of such a principle has been described by Cornet (2010) with a volumetrically lightened PBR based on the "DiCoFluV" concept (more details are given in next sections). Volumetrically illuminated PBRs hold great promise, but there are still rare few examples in the literature (Cornet, 2010; Csogör et al, 2001; Hsieh and Wu, 2009; Ogbonna et al, 1996; Zijffers et al, 2008), mainly due to the technological complexity involved and the difficulty scaling up PBR systems to large surface areas.

5.4.2 Examples of Surface-Lightened PBRs

5.4.2.1 Covered Raceway

The raceway is a horizontal planar cultivation system that has gained currency as a mature cultivation system (Becker, 1994; Oswald, 1962, 1988). Raceway technologies are effectively easy to scale-up and cheap to build. Different materials can be used, from clay to PVC (Ben-Amotz, 2008). Over the last few years, optimizations have been proposed, such as using computational fluid dynamics to optimize mixing with a paddlewheel design (Chiaramonti et al, 2013; Hadiyanto et al, 2013; Hreiz et al, 2014; Liffman et al, 2013), to maximize biomass productivity and light distribution in the culture volume, and to reduce power consumption (Lundquist et al, 2010).

Like for any open system, raceway technologies are well adapted to producing extremophile microalgae like the cyanobacteria *A. platensis* which is able to grow under high pH and high-temperature conditions (reducing the risk of bacterial contaminations). On the other hand, if the goal of the system is to produce sensitive microorganisms, then a closed system is needed. AlgoSource recently developed an enclosed raceway (Fig. 2). Covering a raceway system allows to control internal growth conditions for better productivity. It also enables control of the gas phase, such as reducing CO_2 desorption, making this technology suitable for flue gas treatment applications. An covered raceway technology was recently implemented at AlgoSolis R&D facility for this purpose.

5.4.2.2 AlgoFilm Technologies

Fig. 1 illustrates the utility of increasing both specific illuminated surface (or decreasing the culture depth, ie, $a_{light} = S_{light}/V_R = 1/L$ for a flat panel) and PFD to increase volumetric productivity (or biomass concentration, the two being linked). This introduces the basic concepts of PBR intensification. More specifically, the interest of working in a thin film ($a_{light} > 100\ m^{-1}$, $L < 0.01$ m) is clearly demonstrated here: compared to usual geometries (a_{light} around 20 m^{-1} for a PBR of depth 0.05 m, 0.3 m^{-1} for raceway of depth 0.3 m), two orders of magnitude can be gained on volumetric productivity, making it possible to work in high-cell-density culture ($C_X > 10$ kg m^{-3}) leading to the advent of high volumetric productivity PBR (HVP-PBR). Note also that increasing the PFD will lead to a further increase in biomass productivity (but with a decrease in thermodynamic yield of photosynthetic conversion, as previously discussed). As surface productivity is independent of specific illuminated surface (Eq. 2), a specific feature of PBR technology is the possibility of drastically increasing volumetric productivity while maintaining surface productivity. This was the basic statement behind the design of AlgoFilm© technology (Fig. 2) which aims to propose very high volumetric productivity (HVP-PBR) at the current performance ceiling while keeping the maximal conversion of incoming light permitted by the direct illumination principle (surface-lightened system, without light dilution).

The direct advantage of intensifying volumetric productivity is that it reduces the system size needed to achieve a given production requirement. In the general framework of a global industrial exploitation, energy consumption in several processes is directly linked to culture volume (pumping, mixing, temperature control, harvesting, etc.). Increasing volumetric productivity can thus drastically reduce energy needs for a given operation. This holds primary relevance in biofuel production for example, where both surface and volumetric productivities can be increased by appropriately engineering PBRs. For example, the prototype presented in Fig. 2 has a depth of around 0.002 m which corresponds to $a_{light}=500\ m^{-1}$ (or 2 L of culture per m^3). For a given PFD of 200 $\mu mol_{h\nu}\ m^{-2}\ s^{-1}$ (corresponding roughly to the yearly averaged value of irradiation in Paris, France), a volumetric productivity P_V around 3.3 kg m^{-3} day^{-1} could be obtained with a surface productivity of 14 g m^{-2} day^{-1}. A PBR of higher depth could provide the same maximal surface productivity but with substantial volumetric productivity (0.04 kg m^{-3} day^{-1} for a PBR of 0.15 m depth, ie, $a_{light}=0.07\ m^{-1}$ or 150 L of culture per m^3). Note that the AlgoSolis R&D facility recently integrated the AlgoFilm technology as a platform for research projects in microalgal liquid biofuel production (biodiesel, biokerosene).

5.4.3 Example of a Volume-Lightened PBR

5.4.3.1 Introduction

Internal illumination is often discussed as a way to improve the productivity of PBRs. However, although the concept appears simple, the design of an efficient system is far from trivial. An inappropriate design can lead to lower productivities than easier-to-build surface-lightened systems. The optical device used for light collection and its transmission to the culture is a critical factor. Even with an optimized optical device, the arrangement and sizing of the light guides used to diffuse light to the culture have to respect some keys parameters, which were identified and fixed by Cornet (2010). An example is given here with cylindrical light guides.

The main parameters in this case are distance between light guides d_i, light-guide diameter d_s, and volume fraction of the light guides inside the total reactor volume ε. For example, Cornet (2010) defined an optimal spatial distribution $\varepsilon_{opt}=0.2267$ for cylindrical structures (see later). In addition, as ideal transmission cannot be obtained from optical devices, two energetic transmission yields have to be considered. The first concerns the transmission efficiency from the light-collecting surface to all inlet sections of the light guides η_0, while the second depends on how efficiently the light guides transport and deliver light from the inlet section to the lateral surface (η_1). In these conditions, the light flux delivered to the culture (q_2) is defined by Eq. (17):

$$q_2 = \eta_0 \eta_1 q \frac{S_0}{\sum S_2} \quad (17)$$

where S_0 is the light-collecting surface and $\sum S_2$ is the outlet surface of light guides.

Once the light flux received by the culture is known, theoretical maximal performances of volume-lightened PBRs can be determined, using Eq. (2) presented earlier (using the light flux q_2 effectively received by the culture instead of the collected light flux, q). An example given here to emphasize the difficulty of designing an efficient technology is for *A. platensis* (Cornet and Dussap, 2009) with irradiation conditions obtained in Paris, France (year-averaged PFD). Some favorable assumptions were also retained:

- Ideal transmission efficiencies ($\eta_0 = \eta_1 = 1$, which means light transmission from light source to the culture is equal to 1). In practice, any optical device will introduce a loss of transmitted light, and so the collecting surface S_o has to be increased accordingly.
- Both direct and diffuse components of the solar radiation are transmitted into the culture volume ($q_{sunlight} = q_{//} + q_{\cap}$).
- Optimal spatial distribution ε_{opt} of light guides in the culture volume.
- No biomass loss during the night.

To illustrate the impact of the light-guide characteristics, an example is given here for a system having a light-collecting surface S_0 equal to the total footprint surface of the cultivation system. Three values were fixed for the diameter and height of the light guides, ie, 0.5, 1, 2 m and 3, 7, 10 m, respectively. In these conditions, surface productivities obtained from Eq. (2) ranged from 28 to 48 t/(ha year), while volumetric productivities ranged from 0.5 to 1.8 kg/(m^3 year). Note that the maximal volumetric and surface productivities are not obtained in the same geometric configurations, which illustrates the difficulty of co-optimizing both the surface and volumetric productivities of volume-lightened PBR. This is roughly explained by the positive effect of the light dilution principle on surface productivity, which conversely decreases volumetric productivity (Fig. 1). The maximal surface productivity reached in volume-lightened PBRs is almost twofold higher than in surface-lightened PBRs (enclosed raceways and AlgoFilm© technologies). Around 25 t/(ha year) would be achieved in surface-lightened PBRs using the same simulation conditions. However, the maximal volumetric productivity is one order of magnitude higher in enclosed raceways (depth of 15 cm, $P_V = 17$ kg/(m^3 year)) and almost three orders of magnitude higher using the AlgoFilm technology (mean depth of 2 mm, $P_V = 1200$ kg/(m^3 year)).

Various designs can be easily simulated based on engineering equations (Eq. 2). Solar concentration devices can be excluded to keep the technology simple. In practice, this can be achieved by simply immersing optical devices for light dilution in the culture volume, in which case the collecting surface will then be equal to the total footprint surface multiplied by ε_{opt}. The surface productivities of volume-lightened PBRs then range from 9 to 12 t/(ha year), while volumetric productivities range from 0.15 to 0.50 kg/(m^3 year). Light dilution and the absence of solar concentration then mean that the PFD (q_2) received by photosynthetic microorganisms is very low, close to the compensation point of microalgae (A_C in the range of 1–3 $\mu mol_{h\nu} g^{-1} s^{-1}$), which leads to very low biomass concentration and volumetric productivity. This clearly demonstrates that volume-lightened PBRs must integrate solar collectors to make the technology viable and efficient in practice. This is the concept of DiCoFluV, which is presented in next section.

5.4.3.2 The DiCoFluV PBR

The DiCoFluV concept (Cornet, 2010) is based on internal volumetric illumination of the culture medium with the optimized light dilution principle. To compensate for the decrease in volumetric productivity due to light dilution, light guides are arranged to provide a very high value of specific illuminated surface ($a_{light} > 300$ m^{-1}) obtained from the use of thin optical fibers with lateral diffusion of light (diameter typically of few millimeters). The high internal illuminating surface then obtained makes it necessary to introduce a preliminary stage of solar concentration to keep sufficient light entering the culture system. By applying engineering rules for optimal light dilution, this principle enables engineers to work with classical volume bioreactor technologies and to operate very close to the thermodynamic optimum for the solar-to-biomass conversion process, using low incident light fluxes by dilution of the actual full outdoor sunlight.

The development of the corresponding technology requires several stages. First, the conception of the layout for the optical fibers with lateral diffusion of light used inside the culture volume has to be optimized (providing light and diluting the incident solar flux captured outdoor with a high illuminated specific area). This can be achieved by using the constructal approach (leading to the ε_{opt} value given in the previous section; Bejan,

2000; Bejan and Lorente, 2012) or, in the future, by analyzing the geometric sensitivities provided by an integral Monte Carlo formulation of the kinetic coupling with radiative transfer (Dauchet et al, 2013). The concept also imposes working with a low PFD at the surface of the fibers to achieve high thermodynamic efficiency (around 15% in the PAR). This requires models of light transfer for simple one-dimensional (Cornet, 2010) or complex three-dimensional PBR geometries (Dauchet et al, 2013; Lee et al, 2014). Second, the optimum solar capture area needs to be determined. As explained earlier, this makes it necessary to consider the transmission efficiencies of optical devices used for solar concentration and light transport in light guides up to delivery to the culture, but also to use kinetic models coupling the local light absorption rate $\mathcal{A}$ with biomass growth rates to predict the productivities achieved by the PBR as a function of irradiation conditions encountered over a period of exploitation.

This approach was recently adopted to build a DiCoFluV PBR with a total volume of 30 L and a capture surface using 25 Fresnel lenses (Fig. 2). The optimal light dilution factor of the incident PFD (full sunlight) was found to be relatively constant for any location on Earth. Nevertheless, the concept was clearly demonstrated as more interesting in locations with strong direct illumination. Relatively good volumetric biomass productivities are made possible by the large illuminated surface a_{light} of roughly 350 $m^2\ m^{-3}$ compensating for the low incident diluted PFD, ensuring high thermodynamic efficiency of solar energy conversion, ie, a lower footprint for this technology. Note that this technology is mainly conceived as an optimal surface biomass productivity concept capable of a fivefold increase in surface productivity (by unit footprint) in solar conditions compared to conventional direct illumination systems (considering losses in the light transmission chain). This corresponds to the maximum thermodynamic efficiency of photosynthesis. Actual system performance depends on the optical efficiency of the capture/concentration/filtration/distribution of light inside the culture vessel. On the demonstrator represented in Fig. 2, transmission efficiency reaches 30% and can probably be further increased to 50%. Another important advantage of this technology is that the complete spectrum of the sun can be used postconcentration by splitting visible and infrared radiation and converting the infrared to provide the necessary mechanical work to the PBR (pumps, mixing, and so on). This is a crucial point that is generally omitted in most PBR efficiency calculations. With this kind of technology, it could be possible to provide high-value biomass at a thermodynamic efficiency reaching 15% (defined on the whole incident solar spectrum), ie, with the same efficiency as current industrial photovoltaic devices producing only electricity.

6. CONCLUSION

This chapter discussed the parameters to consider when designing and operating microalgal cultivation systems and how a robust and rational engineering approach can support optimal system design and operation. In-depth and long-term modeling efforts have produced engineering rules and formulae to design, optimize, and control PBRs in a predictive and rational way. This was illustrated here by giving examples of recent published PBR developments for both artificial light sources and sunlight and for various purposes from lab-scale fundamental research to industrial exploitation. It was shown that factoring practical and economic constraints of the final application into the engineering phase culminates in very different technologies despite sharing the same rational engineering tools at the outset. This emphasizes how microalgal cultivation systems, unlike more classical bioprocesses for heterotrophic growth (ie, yeast, bacteria, etc.) that can work with stand-geometry mixing tanks, have no standard geometry to work to, mainly because light supply has such a big influence on process performances that various technologies have emerged in a battle to maximize light use. However, with appropriate consideration of all the constraints, as illustrated here, it is possible to set a rational design of effective technologies, which is obviously of primary interest for microalgae-based industries.

ACKNOWLEDGMENTS

This work was supported by several projects, and especially by the French National Research Agency within the framework of the DIESALG (ANR-12-BIME-0001-02) and BIOSOLIS projects. This work is also connected to R&D activities led at the AlgoSolis R&D facility (www.algosolis.com).

REFERENCES

Acién Fernández FG, Sevilla JM, Sánchez Pérez JA, Molina Grima E, Chisti Y: Airlift-driven external-loop tubular photobioreactors for outdoor production of microalgae: assessment of design and performance, *Chem Eng Sci* 56(8):2721–2732, 2001.

Acién FG, Fernández JM, Magan JJ, Molina E: Production cost of a real microalgae production plant and strategies to reduce it, *Biotechnol Adv* 30(6):1344–1353, 2012.

Aiba S: Growth kinetics of photosynthetic microorganisms, *Adv Biochem Eng Biotechnol* 23:85–156, 1982.

Becker EW: *Microalgae: biotechnology and microbiology*, vol 10, Cambridge, 1994, Cambridge University Press.

Bejan A: *Shape and structure, from engineering to nature*, Cambridge, 2000, Cambridge University Press.

Bejan A, Lorente S: The physics of spreading ideas, *Int J Heat Mass Transf* 55(4):802–807, 2012.

Ben-Amotz A: *Large-scale open algae ponds,* NREL-AFOSR joint workshop on algal oil for jet fuel production, 2008.

Berberoglu H, Pilon L, Melis A: Radiation characteristics of Chlamydomonas reinhardtii CC125 and its truncated chlorophyll antenna transformants tla1, tlaX and tla1-CW+, *Int J Hydrogen Energy* 33(22):6467–6483, 2008.

Bergmann P, Ripplinger P, Beyer L, Trösch W: Disposable flat panel airlift photobioreactors, *Chem Ing Tech* 85(1–2):202–205, 2013.

Bitog JP, Lee IB, Lee CG, et al: Application of computational fluid dynamics for modeling and designing photobioreactors for microalgae production: a review, *Comput Electron Agr* 76(2):131–147, 2011.

Borowitzka MA: Microalgae for aquaculture: opportunities and constraints, *J Appl Phycol* 9:393–401, 1997.

Borowitzka MA: Commercial production of microalgae: ponds, tanks, and fermenters, *Prog Ind Microbiol* 35:313–321, 1999.

Carvalho AP, Meireles LA, Malcata FX: Microalgal reactors: a review of enclosed system designs and performances, *Biotechnol Prog* 22:1490–1506, 2006.

Carvalho AP, Silva SO, Baptista JM, Malcata FX: Light requirements in microalgal photobioreactors: an overview of biophotonic aspects, *Appl Microbiol Biotechnol* 89(5):1275–1288, 2011.

Cassano AE, Martin CA, Brandi RJ, Alfano OM: Photoreactor analysis and design: fundamentals and applications, *Ind Eng Chem Res* 34(7):2155–2201, 1995.

Chen C-Y, Lee C-M, Chang J-S: Feasibility study on bioreactor strategies for enhanced photohydrogen production from Rhodopseudomonas palustris WP3-5 using optical-fiber-assisted illumination systems, *Int J Hydrogen Energy* 31:2345–2355, 2006.

Chiaramonti D, Prussi M, Casini D, et al: Review of energy balance in raceway ponds for microalgae cultivation: re-thinking a traditional system is possible, *Appl Energy* 102:101–111, 2013.

Chini Zittelli GC, Pastorelli R, Tredici MR: A modular flat panel photobioreactor (MFPP) for indoor mass cultivation of Nannochloropsis sp. under artificial illumination, *J Appl Phycol* 12(3–5):521–526, 2000.

Chini Zittelli G, Rodolfi L, Biondi N, Tredici MR: Productivity and photosynthetic efficiency of outdoor cultures of Tetraselmis suecica in annular columns, *Aquaculture* 261(3):932–943, 2006.

Chisti MY: *Airlift bioreactors,* ed 1, London, England and New York, USA, 1989, Elsevier Applied Science.

Cogne G, Rugen M, Bockmayr A, et al: A model-based method for investigating bioenergetic processes in autotrophically growing eukaryotic microalgae: application to the green alga Chlamydomonas reinhardtii, *Biotechnol Prog* 27(3):631–640, 2011.

Cornet JF: *Etude cinétique et énergétique d'un photobioréacteur. Etablissement d'un modèle structuré. Applications à un écosystème clos artificiel,* Orsay, 1992, Université Paris XI.

Cornet J-F: Calculation of optimal design and ideal productivities of volumetrically lightened photobioreactors using the constructal approach, *Chem Eng Sci* 65(2): 985–998, 2010.

Cornet JF, Albiol J: Modeling photoheterotrophic growth kinetics of Rhodospirillum rubrum in rectangular photobioreactors, *Biotechnol Prog* 16:199–207, 2000.

Cornet JF, Dussap CG: A simple and reliable formula for assessment of maximum volumetric productivities in photobioreactors, *Biotechnol Prog* 25:424–435, 2009.

Cornet JF, Dussap CG, Cluzel P, Dubertret G: A structured model for simulation of cultures of the cyanobacterium Spirulina platensis in photobioreactors. 1. Coupling between light transfer and growth kinetics, *Biotechnol Bioeng* 40(7):817–825, 1992a.

Cornet JF, Dussap CG, Cluzel P, Dubertret G: A structured model for simulation of cultures of the cyanobacterium Spirulina platensis in photobioreactors. 2. Identification of kinetic parameters under light and mineral limitations, *Biotechnol Bioeng* 40(7):826–834, 1992b.

Cornet JF, Dussap CG, Gros JB: Conversion of radiant light energy in photobioreactors, *AIChE J* 40(6):1055–1066, 1994.

Cornet JF, Dussap CG, Gros JB: A simplified monodimensional approach for modeling coupling between radiant light transfer and growth kinetics in photobioreactors, *Chem Eng Sci* 50(9):1489–1500, 1995.

Cornet JF, Dussap CG, Gros JB: Kinetics and energetics of photosynthetic micro-organisms in photobioreactors: application to Spirulina growth, *Adv Biochem Eng Biotechnol* 59:155–224, 1998.

Cornet JF, Favier L, Dussap CG: Modeling stability of photoheterotrophic continuous cultures in photobioreactors, *Biotechnol Prog* 19(4):1216–1227, 2003.

Cournac L, Musa F, Bernard L, Guedeney G, Vignais P, Peltier G: Limiting steps of hydrogen production in Chlamydomonas reinhardtii and Synechocystis PCC 6803 as analysed by light-induced gas exchange transients, *Int J Hydrogen Energy* 27:1229–1237, 2002.

Csogör Z, Herrenbauer M, Schmidt K, Posten C: Light distribution in a novel photobioreactor—modelling for optimization, *J Appl Phycol* 13:325–333, 2001.

Dauchet J: *Analyse radiative des photobioreacteurs,* PhD thesis, Clermont-Ferrand, France, 2012, Universite Blaise Pascal (n° ordre 2304—in French).

Dauchet J, Blanco S, Cornet JF, El Hafi M, Eymet V, Fournier R: The practice of recent radiative transfer Monte Carlo advances and its contribution to the field of microorganisms cultivation in photobioreactors, *J Quant Spectrosc Radiat Transf* 128:52–59, 2013.

Dauchet J, Blanco S, Cornet J-F, Fournier R: Calculation of the radiative properties of photosynthetic microorganisms, *J Quant Spectrosc Radiat Transf* 161:60–84, 2015.

Degrenne B, Cogne G, Pruvost J, Legrand J: Role of acetate and light attenuation in hydrogen production by Chlamydomonas reinhardtii, *J Biotechnol* 136:S559, 2008.

Degrenne B, Pruvost J, Christophe G, Cornet JF, Cogne G, Legrand J: Investigation of the combined effects of acetate and photobioreactor illuminated fraction in the induction of anoxia for hydrogen production by Chlamydomonas reinhardtii, *Int J Hydrogen Energy* 35(19):10741–10749, 2010.

Degrenne B, Pruvost J, Legrand J: Effect of prolonged hypoxia in autotrophic conditions in the hydrogen production by the green microalga Chlamydomonas reinhardtii in photobioreactor, *Bioresour Technol* 102:1035–1043, 2011a.

Degrenne B, Pruvost J, Titica M, Takache H, Legrand J: Kinetic modeling of light limitation and sulfur deprivation effects in the induction of hydrogen production with Chlamydomonas reinhardtii. Part II: definition of model-based protocols and experimental validation, *Biotechnol Bioeng* 108(10):2288–2299, 2011b.

Doucha J, Livansky K: Productivity, CO2/O2 exchange and hydraulics in outdoor open high density microalgal (Chlorella sp.) photobioreactors operated in a Middle and Southern European climate, *J Appl Phycol* 18:811–826, 2006.

Duffie JA, Beckman WA: *Solar engineering of thermal processes,* ed 3, New York, 2006, John Wiley & Sons.

Farges B, Laroche C, Cornet J-F, Dussap C-G: Spectral kinetic modeling and long-term behavior assessment of Arthrospira platensis growth in photobioreactor under red (620 nm) light illumination, *Biotechnol Prog* 25:151–162, 2009.

Fouchard S, Hemschemeier A, Caruana A, et al: Autotrophic and mixotrophic hydrogen photoproduction in sulfur-deprived Chlamydomonas cells, *Appl Environ Microbiol* 71(10):6199–6205, 2005.

Fouchard S, Pruvost J, Degrenne B, Legrand J: Investigation of H2 production using the green microalga Chlamydomonas reinhardtii in a fully controlled photobioreactor fitted with on-line gas analysis, *Int J Hydrogen Energy* 33(13):3302–3310, 2008.

Fouchard S, Pruvost J, Degrenne B, Titica M, Legrand J: Kinetic modeling of light limitation and sulphur deprivation effects in the induction of hydrogen production with Chlamydomonas reinhardtii Part I: model description and parameters determination, *Biotechnol Bioeng* 102(1):132–147, 2009.

Goetz V, Le Borgne F, Pruvost J, Plantard G, Legrand J: A generic temperature model for solar photobioreactors, *Chem Eng J* 175:443–449, 2011.

Gonzalez de la Vara L, Gomez-Lojero C: Participation of plastoquinone, cytochrome c553 and ferredoxin-NADP+ oxido reductase in both photosynthesis and respiration in Spirulina maxima, *Photosynth Res* 8:65–78, 1986.

Grima ME, Sevilla FJM, Perez JAS, Camacho FG: A study on simultaneous photolimitation and photoinhibition in dense microalgal cultures taking into account incident and averaged irradiances, *J Biotechnol* 45:59–69, 1996.

Grima ME, Fernandez AFG, Camacho GF, Chisti Y: Photobioreactors: light regime, mass transfer, and scaleup, *J Biotechnol* 70:231–247, 1999.

Grobbelaar JU: Factors governing algal growth in photobioreactors: the "open" versus "closed" debate, *J Appl Phycol* 21(5):489–492, 2008.

Hadiyanto H, Elmore S, Van Gerven T, Stankiewicz A: Hydrodynamic evaluations in high rate algae pond (HRAP) design, *Chem Eng J* 217:231–239, 2013.

Hindersin S: *Photosynthetic efficiency of microalgae and optimization of biomass production in photobioreactors,* Hamburg, 2013, Universitaät Hamburg, Dissertation.

Hindersin S, Leupold M, Kerner M, Hanelt D: Irradiance optimization of outdoor microalgal cultures using solar tracked photobioreactors, *Bioprocess Biosyst Eng* 36(3):345–355, 2013.

Hindersin S, Leupold M, Kerner M, Hanelt D: Key parameters for outdoor biomass production of Scenedesmus obliquus in solar tracked photobioreactors, *J Appl Phycol* 26:2315–2325, 2014.

Hreiz R, Sialve B, Morchain J, Escudié R, Steyer J-P, Guiraud P: Experimental and numerical investigation of hydrodynamics in raceway reactors used for algaculture, *Chem Eng J* 250:230–239, 2014.

Hsieh CH, Wu WT: A novel photobioreactor with transparent rectangular chambers for cultivation of microalgae, *Biochem Eng J* 46(3):300–305, 2009.

Hu Q, Guterman H, Richmond A: A flat inclined modular photobioreactor for outdoor mass cultivation of photoautotrophs, *Biotechnol Bioeng* 51(1):51–60, 1996.

Ifrim GA, Titica M, Cogne G, Boillereaux L, Legrand J, Caraman S: Dynamic pH model for autotrophic growth of microalgae in photobioreactor: a tool for monitoring and control purposes, *AIChE J* 60(2):585–599, 2014.

James SC, Janardhanam V, Hanson DT, Mock T: Simulating pH effects in an algal-growth hydrodynamics model, *J Phycol* 49(3):608–615, 2013.

Janssen M, Janssen MGJ, De Winter M, et al: Efficiency of light utilization of Chlamydomonas reinhardtii under medium-duration light/dark cycles, *J Biotechnol* 78:123–137, 2000.

Janssen M, Tramper J, Mur LR, Wijffels RH: Enclosed outdoor photobioreactors: light regime, photosynthetic efficiency, scale-up, and future prospects, *Biotechnol Bioeng* 81(2):193–210, 2003.

Kandilian R, Lee E, Pilon L: Radiation and optical properties of Nannochloropsis oculata grown under different irradiances and spectra, *Bioresour Technol* 137:63–73, 2013.

Kandilian R, Pruvost J, Legrand J, Pilon L: Influence of light absorption rate by Nannochloropsis oculata on triglyceride production during nitrogen starvation, *Bioresour Technol* 163:308–319, 2014.

Le Borgne F: *Développement d'un photobioréacteur solaire intensifié en vue de la production à grande échelle de biomasse microalgale,* Saint-Nazaire, 2011, Université de Nantes.

Le Borgne F, Pruvost J: Investigation and modeling of biomass decay rate in the dark and its potential influence on net productivity of solar photobioreactors for microalga

Chlamydomonas reinhardtii and cyanobacterium *Arthrospira platensis*, *Bioresour Technol* 138:271–276, 2013.

Lee Y-K, Low C-S: Effect of photobioreactor inclination on the biomass productivity of an outdoor algal culture, *Biotechnol Bioeng* 38(9):995–1000, 1991.

Lee E, Pruvost J, He X, Munipalli R, Pilon L: Design tool and guidelines for outdoor photobioreactors, *Chem Eng Sci* 106:18–29, 2014.

Le Gouic B: *Analyse et optimisation de l'apport de carbone en photobioréacteur,* PhD thesis, 2013, University of Nantes.

Lehr F, Posten C: Closed photo-bioreactors as tools for biofuel production, *Curr Opin Biotechnol* 20(3):280–285, 2009.

Liffman K, Paterson DA, Liovic P, Bandopadhayay P: Comparing the energy efficiency of different high rate algal raceway pond designs using computational fluid dynamics, *Chem Eng Res Des* 91(2):221–226, 2013.

Loubiere K, Olivo E, Bougaran G, Pruvost J, Robert R, Legrand J: A new photobioreactor for continuous microalgal production in hatcheries based on external-loop airlift and swirling flow, *Biotechnol Bioeng* 102(1):132–147, 2009.

Loubiere K, Pruvost J, Aloui F, Legrand J: Investigations in an external-loop airlift photobioreactor with annular light chambers and swirling flow, *Chem Eng Res Des* 89(2):164–171, 2011.

Lundquist TJ, Woertz IC, Quinn NWT, Benemann JR: *A realistic technology and engineering assessment of algae biofuel production,* Berkeley, California, 2010, Energy Biosciences Institute.

Martzolff A, Cahoreau E, Cogne G, et al: Photobioreactor design for isotopic non-stationary 13C-metabolic flux analysis (INST 13C-MFA) under photoautotrophic conditions, *Biotechnol Bioeng* 109(12):3030–3040, 2012.

Molina E, Fernández J, Acién FG, Chisti Y: Tubular photobioreactor design for algal cultures, *J Biotechnol* 92(2):113–131, 2001.

Morweiser M, Kruse O, Hankamer B, Posten C: Developments and perspectives of photobioreactors for biofuel production, *Appl Microbiol Biotechnol* 87(4):1291–1301, 2010.

Muller-Feuga A, Le Guédes R, Pruvost J: Benefits and limitations of modeling for optimization of Porphyridium cruentum cultures in an annular photobioreactor, *J Biotechnol* 103(2):153–163, 2003a.

Muller-Feuga A, Pruvost J, Le Guedes R, Le Dean L, Legentilhomme P, Legrand J: Swirling flow implementation in a photobioreactor for batch and continuous cultures of Porphyridium cruentum, *Biotechnol Bioeng* 84(5):544–551, 2003b.

Myers J, Kratz WA: Relation between pigment content and photosynthetic characteristics in a blue-green algae, *J Gen Physiol* 39(1):11–22, 1955.

Ogbonna JC, Tanaka H: Night biomass loss and changes in biochemical composition of cells during light/dark cyclic culture of Chlorella pyrenoidosa, *J Ferment Bioeng* 82(6):558–564, 1996.

Ogbonna JC, Yada H, Masui H, Tanaka H: A novel internally illuminated stirred tank photobioreactor for large-scale cultivation of photosynthetic cells, *J Ferment Bioeng* 82(1):61–67, 1996.

Oswald WJ: The coming industry of controlled photosynthesis, *Am J Public Health Nations Health* 52(2):235–242, 1962.

Oswald WJ: Large-scale algal culture systems (engineering aspects). In Borowitska M, editor: *Microalgal biotechnology*, Cambridge, 1988, Cambridge University Press, pp 357–394.

Peltier G, Thibault P: Uptake in the light in Chlamydomonas. Evidence for persistent mitochondrial respiration, *Plant Physiol* 79:225–230, 1985.

Perner-Nochta I, Posten C: Simulations of light intensity variation in photobioreactors, *J Biotechnol* 131(3):276–285, 2007.

Pilon L, Berberoglu H, Kandilian R: Radiation transfer in photobiological carbon dioxide fixation and fuel production by microalgae, *J Quant Spectrosc Radiat Transf* 112(17):2639–2660, 2011.

Pottier L, Pruvost J, Deremetz J, Cornet JF, Legrand J, Dussap CG: A fully predictive model for one-dimensional light attenuation by Chlamydomonas reinhardtii in a torus photobioreactor, *Biotechnol Bioeng* 91(5):569–582, 2005.

Pruvost J: Cultivation of algae in photobioreactors for biodiesel production. In Pandey A, Larroche C, Ricke SC, Dussap CG, editors: *Biofuels: alternative feedstocks and conversion processes*, USA, 2011, Academic Press, pp 439–464.

Pruvost J, Cornet JF: Knowledge models for engineering and optimization of photobioreactors. In Posten C, Walter C, editors: *Microalgal biotechnology*, Berlin/Boston, 2012, Walter De Gruyter GmbH & Co. KG, pp 181–224.

Pruvost J, Legrand J, Legentilhomme P, Muller-Feuga A: Simulation of microalgae growth in limiting light conditions—flow effect, *AIChE J* 48:1109–1120, 2002a.

Pruvost J, Legrand J, Legentilhomme P, Muller-Feuga A: Lagrangian trajectory model for turbulent swirling flow in an annular cell. Comparison with RTD measurements, *Chem Eng Sci* 57(7):1205–1215, 2002b.

Pruvost J, Pottier L, Legrand J: Numerical investigation of hydrodynamic and mixing conditions in a torus photobioreactor, *Chem Eng Sci* 61(14):4476–4489, 2006.

Pruvost J, Cornet JF, Legrand J: Hydrodynamics influence on light conversion in photobioreactors: an energetically consistent analysis, *Chem Eng Sci* 63:3679–3694, 2008.

Pruvost J, Van Vooren G, Cogne G, Legrand J: Investigation of biomass and lipids production with *Neochloris oleoabundans* in photobioreactor, *Bioresour Technol* 100:5988–5995, 2009.

Pruvost J, Cornet JF, Goetz V, Legrand J: Modeling dynamic functioning of rectangular photobioreactors in solar conditions, *AIChE J* 57(7):1947–1960, 2011a.

Pruvost J, Van Vooren G, Le Gouic B, Couzinet-Mossion A, Legrand J: Systematic investigation of biomass and lipid productivity by microalgae in photobioreactors for biodiesel application, *Bioresour Technol* 102:150–158, 2011b.

Pruvost J, Cornet JF, Goetz V, Legrand J: Theoretical investigation of biomass productivities achievable in solar rectangular photobioreactors for the cyanobacterium *Arthrospira platensis*, *Biotechnol Prog* 28(3):699–714, 2012.

Pruvost J, Cornet JF, Le Borgne F, Goetz V, Legrand J: Theoretical investigation of microalgae culture in the light changing conditions of solar photobioreactor production and comparison with cyanobacteria, *Algal Res* 10:87–99, 2015.

Pruvost J, Le Gouic B, Lepine O, Legrand J, Le Borgne F: Microalgae culture in building-integrated photobioreactors: biomass production modelling and energetic analysis, *Chem Eng J* 284:850–861, 2016.

Pulz O: Photobioreactors: production systems for phototrophic microorganisms, *Appl Microbiol Biotechnol* 57(3):287–293, 2001.

Quinn J, De Winter L, Bradley T: Microalgae bulk growth model with application to industrial scale systems, *Bioresour Technol* 102:5083–5092, 2011.

Richmond A: *Handbook of microalgal culture: biotechnology and applied phycology*, Oxford, UK, 2004a, Blackwell Sciences Ltd.

Richmond A: Principles for attaining maximal microalgal productivity in photobioreactors: an overview, *Hydrobiologia* 512:33–37, 2004b.

Richmond A, Cheng-Wu Z: Optimization of a flat plate glass reactor for mass production of Nannochloropsis sp. outdoors, *J Biotechnol* 85(3):259–269, 2001.

Rosello Sastre R, Csögör Z, Perner-Nochta I, Fleck-Schneider P, Posten C: Scale-down of microalgae cultivations in tubular photo-bioreactors—a conceptual approach, *J Biotechnol* 132(2):127–133, 2007.

Slegers PM, Wijffels RH, Van Straten G, Van Boxtel AJB: Design scenarios for flat panel photobioreactors, *J Appl Energy* 88:3342–3353, 2011.

Slegers PM, Lösing MB, Wijffels RH, van Straten G, van Boxtel AJB: Scenario evaluation of open pond microalgae production, *Algal Res* 2(4):358–368, 2013a.
Slegers PM, van Beveren PJM, Wijffels RH, van Straten G, van Boxtel AJB: Scenario analysis of large scale algae production in tubular photobioreactors, *Appl Energy* 105:395–406, 2013b.
Soulies A, Castelain C, Burghelea TI, Legrand J, Marec H, Pruvost J: Investigation and modeling of the effects of light spectrum and incident angle on the growth of Chlorella vulgaris in photobioreactors, *Biotechnol Prog*, accepted.
Takache H, Christophe G, Cornet JF, Pruvost J: Experimental and theoretical assessment of maximum productivities for the microalgae Chlamydomonas reinhardtii in two different geometries of photobioreactors, *Biotechnol Prog* 26(2):431–440, 2010.
Takache H, Pruvost J, Cornet JF: Kinetic modeling of the photosynthetic growth of Chlamydomonas reinhardtii in a photobioreactor, *Biotechnol Prog* 28(3):681–692, 2012.
Takache H, Pruvost J, Marec H: Investigation of light/dark cycles effects on the photosynthetic growth of Chlamydomonas reinhardtii in conditions representative of photobioreactor cultivation, *Algal Res* 8:192–204, 2015.
Taleb A, Pruvost J, Legrand J, et al: Development and validation of a screening procedure of microalgae for biodiesel production: application to the genus of marine microalgae Nannochloropsis, *Bioresour Technol* 177:224–232, 2015.
Torzillo G, Accolla P, Pinzani E, Masojidek J: In situ monitoring of chlorophyll fluorescence to assess the synergistic effect of low temperature and high irradiance stresses in Spirulina cultures grown outdoors in photobioreactors, *J Appl Phycol* 8(4–5):283–291, 1996.
Ugwu CU, Aoyagia H, Uchiyamaa H: Photobioreactors for mass cultivation of algae, *Bioresour Technol* 99(10):4021–4028, 2008.
Wilhelm C, Selmar D: Energy dissipation is an essential mechanism to sustain the viability of plants: the physiological limits of improved photosynthesis, *J Plant Physiol* 168(2):79–87, 2011.
Wu X, Merchuk JC: A model integrating fluid dynamics in photosynthesis and photoinhibition processes, *Chem Eng Sci* 56:3527–3538, 2001.
Zijffers JW, Janssen M, Tramper J, Wijffels RH: Design process of an area-efficient photobioreactor, *Marine Biotechnol* 10:404–415, 2008.
Zonneveld C: Photoinhibition as affected by photoacclimation in phytoplankton: a model approach, *J Theor Biol* 193(1):115–123, 1998.

INDEX

Note: Page numbers followed by "*f*" indicate figures, and "*t*" indicate tables.

O

P

R

CONTENTS OF VOLUMES IN THIS SERIAL

Venkat Ganesan and Glenn H. Fredrickson, *Fluctuation Effects in Microemulsion Reaction Media*
David B. Graves and Cameron F. Abrams, *Molecular Dynamics Simulations of Ion–Surface Interactions with Applications to Plasma Processing*
Christian M. Lastoskie and Keith E, Gubbins, *Characterization of Porous Materials Using Molecular Theory and Simulation*
Dimitrios Maroudas, *Modeling of Radical-Surface Interactions in the Plasma-Enhanced Chemical Vapor Deposition of Silicon Thin Films*
Sanat Kumar, M. Antonio Floriano, and Athanassiors Z. Panagiotopoulos, *Nanostructured Formation and Phase Separation in Surfactant Solutions*
Stanley I. Sandler, Amadeu K. Sum, and Shiang-Tai Lin, *Some Chemical Engineering Applications of Quantum Chemical Calculations*
Bernhardt L. Trout, *Car-Parrinello Methods in Chemical Engineering: Their Scope and potential*
R. A. van Santen and X. Rozanska, *Theory of Zeolite Catalysis*
Zhen-Gang Wang, *Morphology, Fluctuation, Metastability and Kinetics in Ordered Block* Copolymers

Volume 29 (2004)
Michael V. Sefton, *The New Biomaterials*
Kristi S. Anseth and Kristyn S. Masters, *Cell–Material Interactions*
Surya K. Mallapragada and Jennifer B. Recknor, *Polymeric Biomaterias for Nerve Regeneration*
Anthony M. Lowman, Thomas D. Dziubla, Petr Bures, and Nicholas A. Peppas, *Structural and Dynamic Response of Neutral and Intelligent Networks in Biomedical Environments*
F. Kurtis Kasper and Antonios G. Mikos, *Biomaterials and Gene Therapy*
Balaji Narasimhan and Matt J. Kipper, *Surface-Erodible Biomaterials for Drug Delivery*

Volume 30 (2005)
Dionisio Vlachos, *A Review of Multiscale Analysis: Examples from System Biology, Materials Engineering, and Other Fluids-Surface Interacting Systems*
Lynn F. Gladden, M.D. Mantle and A.J. Sederman, *Quantifying Physics and Chemistry at Multiple Length-Scales using Magnetic Resonance Techniques*
Juraj Kosek, Frantisek Steĕpánek, and Miloš Marek, Modelling of Transport and Transformation Processes in Porous and Multiphase Bodies
Vemuri Balakotaiah and Saikat Chakraborty, *Spatially Averaged Multiscale Models for Chemical Reactors*

Volume 31 (2006)
Yang Ge and Liang-Shih Fan, *3-D Direct Numerical Simulation of Gas–Liquid and Gas–Liquid–Solid Flow Systems Using the Level-Set and Immersed-Boundary Methods*
M.A. van der Hoef, M. Ye, M. van Sint Annaland, A.T. Andrews IV, S. Sundaresan, and J.A.M. Kuipers, *Multiscale Modeling of Gas-Fluidized Beds*
Harry E.A. Van den Akker, *The Details of Turbulent Mixing Process and their Simulation*
Rodney O. Fox, *CFD Models for Analysis and Design of Chemical Reactors*
Anthony G. Dixon, Michiel Nijemeisland, and E. Hugh Stitt, *Packed Tubular Reactor Modeling and Catalyst Design Using Computational Fluid Dynamics*

Volume 32 (2007)
William H. Green, Jr., *Predictive Kinetics: A New Approach for the 21st Century*
Mario Dente, Giulia Bozzano, Tiziano Faravelli, Alessandro Marongiu, Sauro Pierucci and Eliseo Ranzi, *Kinetic Modelling of Pyrolysis Processes in Gas and Condensed Phase*
Mikhail Sinev, Vladimir Arutyunov and Andrey Romanets, *Kinetic Models of C1–C4 Alkane Oxidation as Applied to Processing of Hydrocarbon Gases: Principles, Approaches and Developments*
Pierre Galtier, *Kinetic Methods in Petroleum Process Engineering*

Volume 48 (2016)

Kolibri 360 GmbH
Ferdinand-Jühlke-Straße 7
99095 Erfurt, Deutschland
produktsicherheit@kolibri360.de